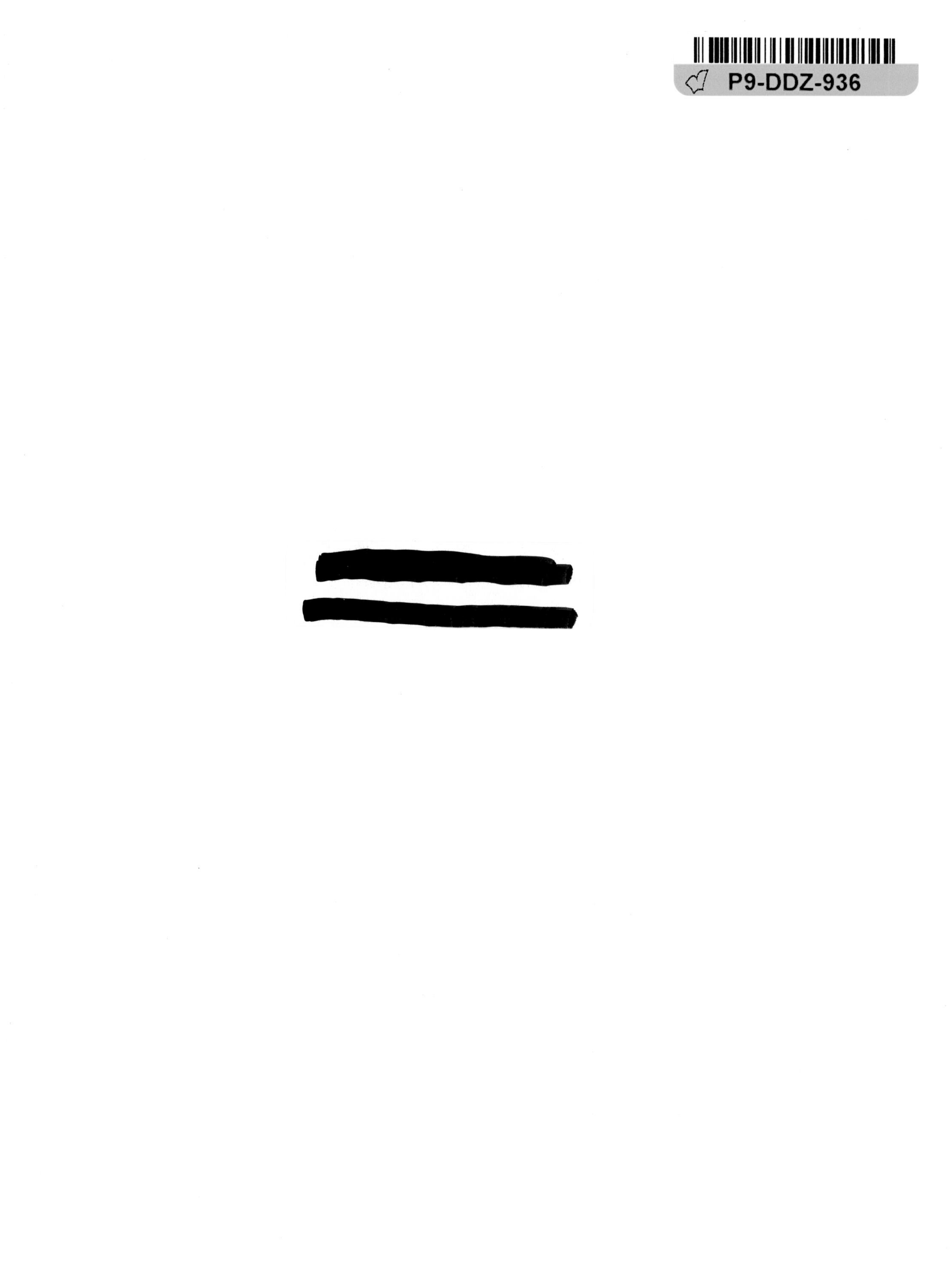

Encyclopedia of Environmental Issues

Volume I

Abbey, Edward—Environmental impact statements and assessments

Editor
Craig W. Allin
Cornell College

Project Editor
Robert McClenaghan

Salem Press, Inc.
Pasadena, California   Hackensack, New Jersey

Managing Editor: Christina J. Moose
Research Supervisor: Jeffry Jensen
Acquisitions Editor: Mark Rehn
Photograph Editor: Karrie Hyatt
Project Editor: Robert McClenaghan
Copy Editor: Doug Long
Production Editor: Janet Alice Long
Layout: William Zimmerman

∞ The paper used in these volumes conforms to the American National Standard for Permanence of Paper for Printed Library Materials, Z39.48-1992(R1997)

Library of Congress Cataloging-in-Publication Data

Encyclopedia of environmental issues / editor, Craig W. Allin; project editor, Robert McClenaghan.
p. cm.
Includes bibliographical references and index.
ISBN 0-89356-994-1 (set : alk. paper). — ISBN 0-89356-995-X (v. 1 : alk. paper) — ISBN 0-89356-996-8 (v. 2 : alk. paper) — ISBN 0-89356-997-6 (v. 3 : alk. paper)
1. Environmental sciences—Encyclopedias. 2. Pollution—Encyclopedias. I. Allin, Craig W. (Craig Willard) II. McClenaghan, Robert, 1961-
GE10 .E52 2000
363.7′003—dc21 99-046373
CIP

Second Printing

PRINTED IN THE UNITED STATES OF AMERICA

Contents

Publisher's Note

For centuries, human attitudes toward the environment were based on the assumption that the planet's resources were infinite and that the earth had an inexhaustible capacity to sustain life. The Industrial Revolution of the nineteenth century, however, inaugurated numerous technologies and practices that greatly increased the pace of environmental degradation and resource exploitation. With these technological advances came a growing number of voices that began expressing uneasiness about the rates at which wilderness was disappearing and the human population was growing, and these voices proliferated during the twentieth century. By the late 1960's, they had coalesced into the environmental movement, a collection of individuals and organizations willing to take action to maintain the health of the planet and the well-being of those species, human and otherwise, that inhabit it.

Environmentalists have concerned themselves with issues ranging from water pollution and storage of nuclear waste at the local level to problems associated with worldwide population growth and the ramifications of global warming and the greenhouse effect. Such environmental hazards have the potential to affect everyone on the planet; it has therefore become important to raise the environmental literacy of people of all nations by exposing them to information relevant to environmental issues at all points of the learning process, from kindergarten to the highest levels of postsecondary education. Such knowledge will help create new generations of informed citizens who are capable of making sound choices about how to deal with the growing list of environmental hazards. As these people move into the workforce, such knowledge will permeate government, industry and business, and the media.

Salem Press's three-volume *Encyclopedia of Environmental Issues* is a wide-ranging guide designed to meet the growing need for environmental literacy. The encyclopedia assembles information from fields of knowledge relevant to the study of environmental issues such as biology, geology, anthropology, demographics, genetics, and engineering, and it explains their interrelationships in terms that are easily understood by nonspecialists.

The *Encyclopedia of Environmental Issues* contains 475 alphabetically arranged articles that range in length from five hundred to three thousand words. All articles are signed by the academics and other experts who wrote them. They cover a wide variety of topics, including air pollution, national parks, environmental legislation, oil spills, and solar power. The articles are not limited to wilderness issues. In recognition of the fact that people spend an overwhelming proportion of their time in human-made environments, essays also cover such topics as sick-building syndrome, smog, and urban planning.

The bulk of the articles consist of overviews of issues, concepts, and terms relevant to the study of environmental matters. These articles range from broad concepts such as ecology to specific issues such as the conflict between loggers and defenders of the endangered northern spotted owl in the Pacific Northwest of the United States. The encyclopedia also includes numerous specialized articles covering biographies, events, legislation, and organizations. Among the biographical entries are discussions of John Muir, the American preservationist who founded the Sierra Club, and James Watt, U.S. president Ronald Reagan's secretary of the interior, who attempted to undo most of the environmental legislation of the previous decades. Articles detailing significant events range from disasters such as the *Exxon Valdez* oil spill and the Minimata Bay, Japan, mercury poisoning to international meetings—such as the 1992 Earth Summit in Brazil—that offer hope for global solutions to environmental problems. Also included are essays exploring laws, acts of legislation, and court cases that have affected the manner in which humans interact with the environment, including the U.S. Endangered Species Act of 1972 and *Sierra Club v. Morton*, the court case that first posed the question of whether trees have legal standing. Important organizations and movements covered include the antienvironmental Sagebrush Rebellion of the late 1970's, the Chipko Andolan movement in India, and the U.S. Fish and Wildlife Service.

Each article in the set begins with a category subhead and a summary of its relevance to environmental issues and ends with a list of cross-references to related articles in the set. Essays covering events begin with a subhead that indicates the date of occurrence. Articles that are one thousand words in length or longer also include suggestions for further reading; these are especially useful for students seeking resources for more in-depth, current information on the topic at hand. The encyclopedia includes nearly 200 photographs and 100 charts, graphs, tables, and other illustrations that illuminate the events and concepts detailed in the essays.

The *Encyclopedia of Environmental Issues* ends with several useful appendices, including a time line of environmental legislation, a directory of environmental organizations, a directory of U.S. national parks, a glossary, and an extensive bibliography listing up-to-date publications that provide new insights into ongoing issues. Also included are a comprehensive subject index and a list of entries arranged according to the following categories: agriculture and food, animals and endangered species, atmosphere and air pollution, biotechnology and genetic engineering, ecology and ecosystems, energy, forests and plants, human health and the environment, land and land use, nuclear power and radiation, philosophy and ethics, pollutants and toxins, population issues, preservation and wilderness issues, resources and resource management, the urban environment, waste and waste management, water and water pollution, and weather and climate.

Salem Press thanks Editor Craig W. Allin of Cornell College for his many contributions to the creation of these volumes. We also thank the experts in the various fields of environmental studies who wrote the articles. The *Encyclopedia of Environmental Issues* lists the full names of these scholars at the end of each article and in the list of contributors and their affiliations that appears on the following pages.

List of Contributors

Richard Adler
University of Michigan—Dearborn

Craig W. Allin
Cornell College

Emily Alward
Independent Scholar

Anita Baker-Blocker
industrialengineering.com

Ruth Bamberger
Drury College

Grace A. Banks
Chestnut Hill College

David Landis Barnhill
Guilford College

Robert B. Bechtel
University of Arizona

Alvin K. Benson
Brigham Young University

Lisa M. Benton
Colgate University

Massimo D. Bezoari
Huntingdon College

Cynthia A. Bily
Adrian College

Margaret F. Boorstein
C. W. Post College of Long Island University

Richard G. Botzler
Humboldt State University

Kristie Brewer
Huntingdon College

Kenneth H. Brown
Northwestern Oklahoma State University

Bruce G. Brunton
James Madison University

Aubyn C. Burnside
Independent Scholar

Dale F. Burnside
Lenoir-Rhyne College

Welland D. Burnside
Independent Scholar

Glenn Canyon
Independent Scholar

Roger V. Carlson
Jet Propulsion Laboratory

Robert S. Carmichael
University of Iowa

Robert E. Carver
University of Georgia, Emeritus

Thomas Clarkin
Independent Scholar

Mark Coyne
University of Kentucky

Ralph D. Cross
University of Southern Mississippi, Emeritus

Robert L. Cullers
Kansas State University

George Cvetkovich
Western Washington University

Roy Darville
East Texas Baptist University

René A. De Hon
Northeast Louisiana University

Dennis R. DeVries
Auburn University

Gordon Neal Diem
ADVANCE Education and Development Institute

John P. DiVincenzo
Middle Tennessee State University

Stephen B. Dobrow
Fairleigh Dickinson University

Gary E. Dolph
Indiana University—Kokomo

Colleen M. Driscoll
Independent Scholar

Andrew P. Duncan
New England College

John M. Dunn
Independent Scholar

Timothy C. Earle
Independent Scholar

Frank N. Egerton
University of Wisconsin—Parkside

Robert D. Engelken
Arkansas State University

Jess W. Everett
Rowan University

Jack B. Evett
University of North Carolina at Charlotte

George J. Flynn
State University of New York—Plattsburgh

Roberto Garza
San Antonio College

Soraya Ghayourmanesh
Independent Scholar

Craig S. Gilman
Coastal Carolina University

D. R. Gossett
Louisiana State University—Shreveport

Daniel G. Graetzer
University of Washington Medical Center

Hans G. Graetzer
South Dakota State University, Emeritus

Jerry E. Green
Miami University

William C. Green
Morehead State University

Phillip A. Greenberg
Independent Scholar

Wendy H. Hallows
Chestnut Hill College

Michael S. Hamilton
University of Southern Maine

Clayton D. Harris
Middle Tennessee State University

Jasper L. Harris
North Carolina Central University

Thomas E. Hemmerly
Middle Tennessee State University

Mark Henkels
Western Oregon University

Charles E. Herdendorf
Ohio State University

Jane F. Hill
Independent Scholar

Joseph W. Hinton
Independent Scholar

Laurent Hodges
Iowa State University

John R. Holmes
Franciscan University of Steubenville

Robert M. Hordon
Rutgers University

Louise D. Hose
Westminster College

Ronald K. Huch
University of Papua New Guinea

Diane White Husic
East Stroudsburg University

H. David Husic
Independent Scholar

Allan Jenkins
University of Nebraska at Kearney

Albert C. Jensen
Central Florida Community College

Jeffrey A. Joens
Florida International University

Suzanne Jones
Huntingdon College

Karen N. Kähler
Independent Scholar

Karen E. Kalumuck
The Exploratorium

Michael D. Kaplowitz
Michigan State University

Kyle L. Kayler
Kayler Geoscience, Ltd.

Carolynn Kimberly
The University of Dayton

Robert W. Kingsolver
Kentucky Wesleyan College

Ralph L. Langenheim
University of Illinois—Urbana

Eugene Larson
Los Angeles Pierce College

Thomas T. Lewis
Mount Senario College

Josué Njock Libii
Purdue University—Fort Wayne

M. A. K. Lodhi
Texas Tech University

Donald W. Lovejoy
Palm Beach Atlantic College

David C. Lukowitz
Hamline University

Larry S. Luton
Eastern Washington University

Fai Ma
University of California, Berkeley

Steven B. McBride
West Virginia University

Robert McClenaghan
Independent Scholar

David F. MacInnes, Jr.
Guilford College

Francis P. Mac Kay
Providence College

Louise Magoon
Independent Scholar

Nancy Farm Männikkö
University of Minnesota

Chogollah Maroufi
California State University, Los Angeles

Kathleen Rath Marr
Lakeland College

Richard F. Modlin
University of Alabama in Huntsville

Charles Mortensen
Ball State University

Mysore Narayanan
Miami University

Peter Neushul
California Institute of Technology

Anthony J. Nicastro
West Chester University

Martin A. Nie
University of Pittsburgh at Bradford

Oghenekome U. Onokpise
Florida A & M University

G. Padmanabhan
North Dakota State University

Beth Ann Parker
Huntingdon College

Gordon A. Parker
University of Michigan—Dearborn

John Pichtel
Ball State University

George R. Plitnik
Frostburg State University

Aaron S. Pollak
Independent Scholar

Oliver B. Pollak
University of Nebraska at Omaha

Noreen D. Poor
University of South Florida

Allison Popwell
Huntingdon College

Victoria Price
Lamar University

Syed R. Qasim
The University of Texas at Arlington

P. S. Ramsey
Independent Scholar

Ronald J. Raven
State University of New York at Buffalo

Donald F. Reaser
The University of Texas at Arlington

John Rickett
University of Arkansas at Little Rock

Raymond U. Roberts
Oklahoma Department of Environmental Quality

Gene D. Robinson
James Madison University

Jacqueline J. Robinson
Huntingdon College

James L. Robinson
University of Illinois at Urbana-Champaign

Charles W. Rogers
Southwestern Oklahoma State University

Donna L. Rogers
Arkansas Tech University

Kenneth A. Rogers
Arkansas Tech University

Keith E. Rolfe
The University of Dayton

Neil E. Salisbury
Independent Scholar

Robert M. Sanford
University of Southern Maine

Elizabeth D. Schafer
Independent Scholar

John Richard Schrock
Emporia State University

Robert B. Seaman
New England College

Rose Secrest
Independent Scholar

R. Baird Shuman
University of Illinois at Urbana-Champaign, Emeritus

Carolyn Simmons
Huntingdon College

Paul P. Sipiera
Harper College

Jane Marie Smith
Slippery Rock University

Roger Smith
Independent Scholar

Diane Stanitski-Martin
Shippensburg University

Anne Statham
University of Wisconsin—Parkside

Joan C. Stevenson
Western Washington University

Dion Stewart
Adams State College

Robert J. Stewart
California Maritime Academy

Toby Stewart
Independent Scholar

Mary W. Stoertz
Ohio University

Hubert B. Stroud
Arkansas State University

John R. Tate
Montclair State College

William R. Teska
Furman University

John M. Theilmann
Converse College

Nicholas C. Thomas
Auburn University at Montgomery

Donald J. Thompson
California University of Pennsylvania

Charles L. Vigue
University of New Haven

Joseph M. Wahome
Mississippi Valley State University

Xingwu Wang
Alfred University

John P. Watkins
Westminster College

Lynn L. Weldon
Adams State College

Robert J. Wells
Society for Technical Communication

Edwin G. Wiggins
Webb Institute

Thomas A. Wikle
Oklahoma State University

Kay R. S. Williams
Shippensburg University

Marcie Wingfield
Huntingdon College

William C. Wood
James Madison University

Lisa A. Wroble
Redford Township District Library

Jay R. Yett
Orange Coast College

Michele Zebich-Knos
Kennesaw State University

Introduction

Progress is a cultural concept—and an anthropocentric one. The term "progress" almost always means human progress: the management of global resources so as to provide greater benefits to the human species. This should come as no surprise; humans are unavoidably anthropocentric. All species strive to thrive, and some may do serious damage to their habitats in the process. Humans differ from other species in the extent of our influence on the natural environment and in our ability to think about what we are doing.

In 1800, the world population numbered fewer than one billion. Most people were engaged in subsistence agriculture and had only localized environmental impact. In 1999, fewer than two hundred years later, world population reached six billion. The increase in our numbers and the power of our technology have radically increased the impact of people on planetary resources and dramatically shifted the natural balance in our favor, at least in the short run. The numbers of people the earth can sustain has become a matter of serious debate. Thinking about the physical limits of earth resources and the increasing human impact on them has given rise to "environmental issues."

Planet Earth is the only locale in the entire universe where we know that humans can live. Although mathematics suggests there are other earthlike planets somewhere in the cosmos, science holds out little chance of our ever reaching one. We are stuck right here on the only known planet with a natural and—up to a point—self-repairing life-support system. Such perspective highlights the importance of environmental issues.

Compiling an *Encyclopedia of Environmental Issues* is an enormous challenge, and the dimensions of that challenge grow greater every year. Our understanding of environmental issues is uneven, inadequate, and ever changing. Our ability to meet the human challenges presented by environmental issues is very much in doubt. New issues emerge even as older ones remain unresolved, perhaps even unaddressed. The dynamic nature of environmental issues guarantees that this is a vastly different collection than it would have been only a few years ago. Those differences reflect fundamental changes in our understanding of environmental issues themselves.

The Scope and Growth of Environmental Issues

We have experienced an explosion in the breadth of environmental issues, from a relatively narrow focus on conservation to a far broader concept that embraces almost everything that provides context for human existence. Environmental issues have not always attracted the attention they do today; modern environmentalism is the most recent and far-reaching manifestation of a conservation movement that was born in the nineteenth century and has been growing ever since. Nineteenth century conservation was a response to population growth, urbanization, and industrialization. Late in the century, unprecedented human impact on the natural world stimulated public concern about exhausting resources. Forests were being logged faster than they could regrow, and important wildlife species that had once numbered in the millions—such as the passenger pigeon and the buffalo—were extinct or in danger of extinction. The resulting efforts at conservation emphasized the areas of greatest national concern: preservation of forests and wildlife. By 1920, governments had responded with policies creating national forests, national parks, national wildlife refuges, and national monuments.

Like the conservation movement that preceded it, modern environmentalism was a response to shortages brought about by the increasing pace of economic development. Many more resources now seemed in short supply, including clean air and water and unpolluted land. Equally important, the emerging science of ecology emphasized the interconnectedness of human life and all other life. For laypersons, the environmental era may have begun with the publication of Rachel Carson's *Silent Spring* in 1962. Carson's exploration of the consequences of pesticide use drove home the lesson that people were part of the world's ecosystems and that in destroying parts of nature, we threatened it all.

The modern concern for environmental issues still centers on the use and abuse of natural resources, but to the historic concerns of conservation—such as forests, wilderness, and wildlife—have been added clean water, clean air, energy supply, hazardous and toxic waste, environmental illnesses, nuclear safety, and a host of other quality-of-life issues. Beginning in the 1960's and 1970's, Congress responded to these new concerns with an unprecedented flood of environmental policy laws, which are covered extensively in these volumes.

SCIENCE AND ENVIRONMENTAL ISSUES

An explosion in scientific understanding has produced new environmental concerns. The discovery of previously unanticipated problems may spur a search for solutions, but there is no guarantee that science can achieve the requisite level of understanding, at least in the short run.

Global climate change serves as a kind of poster child for the complex relationship between modern science and the human environment. First, global climate change could never have become an environmental issue without sophisticated science. That emissions of greenhouse gases from the combustion of fossil fuels could result in global warming could not have been hypothesized without a degree of scientific understanding that is relatively recent. Testing this greenhouse hypothesis requires even greater scientific sophistication. We all experience weather in our daily lives, but without high-tech, globally deployed instrumentation, even the most astute scientist could not begin to measure global warming.

Second, global climate change became a political issue not because voters or citizens were concerned but because scientists were concerned, and because those concerns were publicly voiced. In the United States, global warming became a political issue on June 23, 1988, when James E. Hansen, director of the National Aeronautics and Space Administration (NASA) Goddard Institute for Space Studies, testified before the Senate Committee on Energy and Natural Resources. Hansen reported that he was 99 percent certain that global warming was under way. To many, it seemed as if NASA had spoken with all the authority of science, and suddenly a scientific controversy was a political issue as well.

Third, although science created the climate change issue, contemporary science is far from certain about its existence, causes, or consequences. The very existence of global climate change is based on an uncertain science. Although the evidence of global warming appears to be increasing, every aspect of the case is subject to scientific challenge. Conclusions necessarily rest on incomplete modern temperature data, historic temperature data that are even less reliable, and computer models that may be inaccurate in their portrayal of the relationships among variables.

If the facts of global warming are difficult to establish, the causes are even more problematic. We know that the earth's climate changes over time. Evidence of ice ages lasting thousands of years is well established, but recent studies of ice cores recovered from continental sheets in Greenland and Antarctica suggest that during human prehistory, global climate may have undergone violent short-term changes as well. Various scientists have speculated that current climate change—if it is occurring—may result from solar activity, the atmospheric debris of volcanic eruptions, or some unidentified natural cycle, as well as from the increased emissions of greenhouse gases associated with large human populations burning fossil fuels and clearing forests.

The consequences of global warming seem less uncertain. Ocean levels will rise as polar ice caps melt. Temperature zones will shift toward the poles, changing patterns of precipitation, increasing desertification and deforestation, and requiring a reorganization of world agriculture. Rapid changes will almost certainly result in huge—and only partially predictable—social, economic, and political changes for people and massive extinctions for species unable to adapt to their new environments.

THE GLOBALIZATION OF ENVIRONMENTAL ISSUES

It has become increasingly obvious that both the causes and the consequences of environmental issues are global in nature, but the international community has been slow to respond.

First, where the environmental consequences of human behavior are localized, there is no longer any guarantee that the localized effects will occur nearby. Ecologist Raymond F. Dasmann coined the term "biosphere people" to remind us that modern technology and a global marketplace have given us the capacity to do environmental damage worldwide. Until recently, the earth was populated by "ecosystem people" whose survival depended directly on their care of the ecosystems of which they were a part. Native Americans who depended upon the buffalo dared not overhunt, because the end of the buffalo would have doomed the human community as well. Ecosystem people had no choice but to live with the ecological results of their actions. Today, the world's population is interdependent in countless ways. We are "biosphere people." Resources are mobilized on a global scale—from the entire biosphere—to provide for our food, shelter, comfort, or amusement. The effects of our behavior are distant, so the link between cause and effect is obscured. Not knowing the effects of our behavior, we have no strong incentive to behave in a way that is protective of the earth's resources.

Second, many environmental consequences of human behavior are no longer localized at all. They are global. Nuclear fallout, acid precipitation, and ocean pollution move unimpeded across international boundaries. Stratospheric ozone depletion and climate change threaten the ability of the biosphere to sustain life as we have known it. Declines in global biological diversity may presage an earth environment that is eventually less hospitable to people as well.

The international community is only slowly becoming aware of its global environmental responsibilities. The history of treaties seeking to address environmental issues parallels the history of environmental issues themselves. Early efforts were aimed almost exclusively at the conservation of threatened fauna and flora. Berlin hosted an international conference on salmon fishing in the Rhine River in 1885. In 1900, an agreement was drafted in London concerning the preservation of African mammals, but it was never ratified. A 1902 conference at Paris adopted a treaty for the protection of birds useful to agriculture. The United States and Canada ratified a treaty to protect migratory ducks and geese in 1916. There were international agreements to protect the fauna and flora of Africa in 1933 and the Western Hemisphere in 1940.

By the 1970's, the focus had shifted to the broader concerns associated with environmentalism. Most of the world's industrialized democracies enacted significant environmental protection legislation during the 1970's. Given the limits of national legislation in the face of global environmental problems, it is not surprising that they should have also sought to achieve environmental objectives through international agreements.

A number of international programs have been instituted to reward individual nations for doing the right thing, an approach pioneered by the Council of Europe. In 1965, it established a European Diploma, recognizing effective protection of internationally significant areas. Six years later, the United Nations Educational, Scientific, and Cultural Organization (UNESCO) launched Man and the Biosphere, designed to improve the relationship between humans and their environment through science and education. A key feature was the Biosphere Reserve Programme, an international system of representative ecological areas established to preserve genetic diversity. Another UNESCO convention, effective in 1975, designates World Heritage Sites. These endeavors resemble the older conservation tradition, and they have had little real effect. The designations generally go to national parks and historical sites that are already well protected. The 1972 treaty to prohibit international trade in endangered species has been more effective, but poaching and illegal trafficking remain serious problems.

Since the 1970's, a series of world conferences have dealt with environmental issues. The first Conference on the Human Environment was held in Stockholm, Sweden, in 1972 and attracted representatives from 113 nations. It dramatized international environmental concerns and resulted in establishment of the United Nations Environment Programme. Two decades later, the Earth Summit in Rio de Janeiro, Brazil, drew delegates from 179 nations and emphasized the linkage between economic development in the Third World and environmental protection. The limitations of the treaty system for achieving international environmental goals was demonstrated by the refusal of the United States to accept international agreements on greenhouse gas emissions and the preservation of biological diversity. In 1997, a third international gathering at Kyoto, Japan, produced a draft treaty designed to reduce global emission of greenhouse gases, but few nations appear to be willing to undertake this difficult and costly task.

INSTITUTIONAL CAPACITY TO DEAL WITH ENVIRONMENTAL ISSUES

Dealing effectively with environmental issues may require institutions capable of global coordination over long periods of time. This institutional capacity is noticeably absent. As the recent history of international cooperation on environmental issues attests, the international arena lacks the institutions required to make and enforce any global environmental policy. Planet Earth is subdivided into nation-states, each recognizing no power superior to itself. Nation-states are reluctant to give up any portion of their sovereignty or to recognize that environmental survival may require it. Treaties and other agreements among nation-states are regarded as binding only so long as they serve each nation's interest. Weaker nations may occasionally be forced to comply by stronger ones, but this system of international might-makes-right falls far short of "government" in any meaningful sense of that term.

Domestic institutions are also seriously limited in dealing with environmental issues. With the end of the Cold War, the world became more homogeneous, both politically and economically. Politically, almost all nations claim to be democratic, though the degree to which they live up to that claim varies widely. Economically, capitalism is ascendent. Governments continue to regulate markets, but private ownership of property is increasingly the norm. In many respects, these changes are liberating, but they have not resolved our environmental issues.

Both capitalism and democracy are notoriously short-sighted. Economic theory systematically discounts the future. Corporate managers are required to maximize short-term profits to please investors. In an ideal world, this capitalistic myopia would be balanced by political leadership committed to the long-term interests of society and even to generations yet unborn. In the practical world of democracy, however, political leaders behave like corporate managers, seeking short-term benefits likely to influence the next election.

Our institutions have often been ineffective in addressing environmental issues, but they are not predestined to fail. In the final analysis, both democratic elections and free markets are driven from the bottom up by popular sentiment. Citizens in all nations have political and economic choices to make. To choose wisely, they require both information and insight. It is my hope that the *Encyclopedia of Environmental Issues* helps to fulfill those requirements.

Craig W. Allin
Cornell College

List of Articles by Category

Encyclopedia of Environmental Issues

A

Abbey, Edward

Born: January 29, 1927; Home, Pennsylvania
Died: March 14, 1989; Tucson, Arizona
Category: Preservation and wilderness issues

Edward Abbey, an environmental writer and activist, produced twenty-one volumes of fiction, essays, speeches, and letters expressing his love for the earth, his hatred for modern technological society, and his fervent belief that development was destroying the American West. The originality of his ideas and the eloquence of his rhetoric attracted the admiration of both environmentalists and those who simply appreciate good writing.

When Edward Abbey was twenty-one years old, having spent some time in the military and in college, he left his home in Pennsylvania to see the American West. He hitchhiked, rode trains, and walked over the mountains and through the desert. He claimed the desert as his spiritual home and lived in or near it for most of the rest of his life. He completed a master's degree at the University of New Mexico and wrote his Ph.D. thesis on anarchism and the morality of violence. During his ten-year college career, which included two years as a Fulbright Fellow at the University of Edinburgh in Scotland, he began a number of writing projects and published his first novel, *Jonathan Troy* (1954).

For fifteen years, during his thirties and forties, Abbey worked as a part-time ranger at various national parks in the Southwest. The two years in the late 1950's that he spent at Arches National Monument (now a national park) in Utah led to his first important book, *Desert Solitaire* (1968). This book combines beautiful descriptive passages, an unflinching look at the violence in nature, and a strong call for the preservation of desert habitats. Reminiscent of Henry David Thoreau's *Walden* (1854) in its ideas and its use of the natural year for its structure, *Desert Solitaire* brought Abbey national attention as an environmental writer.

In 1975 Abbey published *The Monkey Wrench Gang*, a novel about four rebels who set out to destroy the roads, bridges, and power lines that they believe are defacing the southwestern desert. It was loosely based on the exploits of a friend of Abbey who had committed some of the acts depicted in the novel. However, Abbey maintained that the book was primarily meant as

Edward Abbey's 1975 novel The Monkey Wrench Gang *inspired radical environmentalists to use direct action to protect wilderness areas from destruction.* (Michael Hendrickson)

humor. Nevertheless, the book helped inspire the radical environmental movement Earth First!, and Abbey did come to support that movement, praising its operations although never actually joining it.

In fact, Abbey never joined any political or environmental organizations, although he participated in political actions, especially those that expressed disapproval of the military or land development. Though most of his books are set in the wilderness of the Southwest and express his deep love for such spaces, Abbey disliked being called a "nature writer." In fact, students of his work have had difficulty attaching any label to Abbey and making it stick, so idiosyncratic are his ideas and connections.

When Abbey died in 1989, he left instructions to bury him in the desert, unembalmed, in his sleeping bag. Although this kind of burial is illegal, his friends followed his wishes.

Cynthia A. Bily

SEE ALSO: Earth First!; Monkeywrenching; Preservation.

Accounting for nature

CATEGORY: Resources and resource management

Accounting for nature means expanding economic principles and adapting financial decisions to take into consideration natural resources, ecosystem services, and values derived from human contact with the natural world.

Environmental decisions frequently set ecologists and economists on a collision course: Does society's demand for energy justify the environmental impact of mining and burning coal? Is the increased economic efficiency of large corporate farms worth the loss of a rural lifestyle? How much should the public sacrifice to protect an endangered bird or plant? These decisions are difficult because they force comparisons of "apples and oranges," pitting one set of values against another. Quantitative models (or accounting systems) that compare the costs and benefits of a course of action are frequently used to guide business and government decisions, but these have generally omitted environmental values. Accounting for nature in these models cannot produce completely objective solutions but can help coordinate ecological and economic expertise in a more balanced problem-solving approach.

Ecologist Edward O. Wilson identified three kinds of national wealth: economic, cultural, and biological. He observed that nations frequently create the illusion of a growing economy by consuming their biological or cultural "capital" to create short-term economic prosperity. For example, burning rain forests and replacing them with row crops may temporarily increase farm production in a developing nation, but tropical soils are often nutrient poor and easily degraded by exposure to the sun and rain. If the biological basis of production is ignored, the population simply transfers wealth from one category to another, ensuring ecological disaster for its children in the process.

Developed nations have also neglected biological wealth in past cost-benefit analyses. Damming wild rivers to make recreational lakes, for example, has been justified by counting the benefits of lumbering trees from the watershed but not the losses of aquatic and forest habitats. Recreational activities have been evaluated according to the money that people pay to participate in them. Thus, hikers and canoeists in a wilderness are given less consideration than waterskiers or recreational vehicle drivers in a developed area because hikers and canoeists spend less money on equipment, fuel, and supplies.

Ecologists Eugene P. Odum and Howard T. Odum addressed this issue by calculating the value of ecosystem services provided by intact biological communities. Their approach was to measure the beneficial work performed by living systems and place a value on that service based on the time and energy required to replicate the service. A living tree, they reasoned, may provide a few hundred dollars in lumber if cut; if left alive, however, the oxygen it produces, carbon dioxide it absorbs, wildlife it feeds and shelters, soil it builds, evaporative cooling it yields, and flood protection it provides are worth far more on an annual basis.

In 1972 economists William Nordhaus and James Tobin refined the concept of national wealth by developing an index of net economic welfare (NEW) to replace the more familiar measurements of economic health such as gross domestic product (GDP). Their criticism of the GDP was that it counts any expenditure as a positive contribution to national wealth, whether or not the spending improves people's lives. A toxic waste dump, for example, contributes to the GDP when the pollutants are produced, again when millions of dollars are spent to clean it, and yet again if medical costs rise because of pollution-related illness. Nordhaus and Tobin's NEW index subtracts pollution abatement and other environmental costs from the value of goods and services that actually improve living standards.

Economist Ernst Friedrich Schumacher subsequently argued that environmental costs should be "internalized," or charged to the industries that create them. This idea, also called the "polluter pays principle," not only generates funds for environmental cleanup but also encourages businesses to make environmentally sound decisions. The price of recycled paper, for example, would be more competitive if the public costs of deforestation and pollution from pulp mills were added to the price of virgin wood fiber. Proposals for internalizing environmental costs have ranged from centrally planned models, such as a carbon tax on fossil fuels, to free market trading of pollution credits. Debt-for-nature swaps, in which developing nations receive financial benefits for preserving natural ecosystems, represent environmental cost accounting on the asset side of the ledger.

A fundamental difference between economic and ecological world views is the time scale under consideration. Business strategies may look five years ahead, but ecological processes can take centuries. Thus, economic models that fail to take long-term issues into account are a frequent source of criticism by environmentalists. The U.S. decision to build nuclear fission reactors during the 1960's and 1970's is a case in point. Nuclear power appeared economically attractive over the thirty-five-year life span of a fission reactor, but the twenty-four-thousand-year half-life of radioactive plutonium 239 in spent fuel rods made skeptics wonder who would pay the costs of nuclear waste disposal for generations after the plants were closed.

Debates about growth are especially contentious. Traditional economists view the growth of populations, goods, and services as positive and necessary for economic progress and social stability. As early as 1798, however, economist Thomas Robert Malthus pointed out that on a finite earth, an exponentially expanding human population would eventually run out of vital resources. In the closing decades of the twentieth century, Paul Ehrlich and Anne Ehrlich warned that unless population growth slowed soon, each person would have to consume less space, food, fuel, and materials to avoid a global population crash. Whether they are considered economic pessimists or environmental realists, Malthus and the Ehrlichs demonstrate that taking a longer view is central to the task of accounting for nature. Sustainable development is the watchword for ecologists, economists, and political leaders attempting to create prosperity today while accounting for the welfare of future generations.

Robert W. Kingsolver

Suggested readings: *An Introduction to Ecological Economics* (1997), edited by Robert Costanza, is a well-balanced beginning text. Ernst F. Schumacher's *Small Is Beautiful: Economics as if People Mattered* (1973) endures as an environmentalist's challenge to the fundamental assumptions of economic policy. For contemporary reports on environmental economics, see the annual *State of the World* publications from the Worldwatch Institute.

See also: Debt-for-nature swaps; Environmental economics; Schumacher, Ernst Friedrich; Sustainable development.

Acid deposition and acid rain

Category: Atmosphere and air pollution

Electric utilities, industries, and automobiles emit sulfur dioxide and nitrogen oxides that are read-

pH Scale Showing Acidity of Acid Precipitation

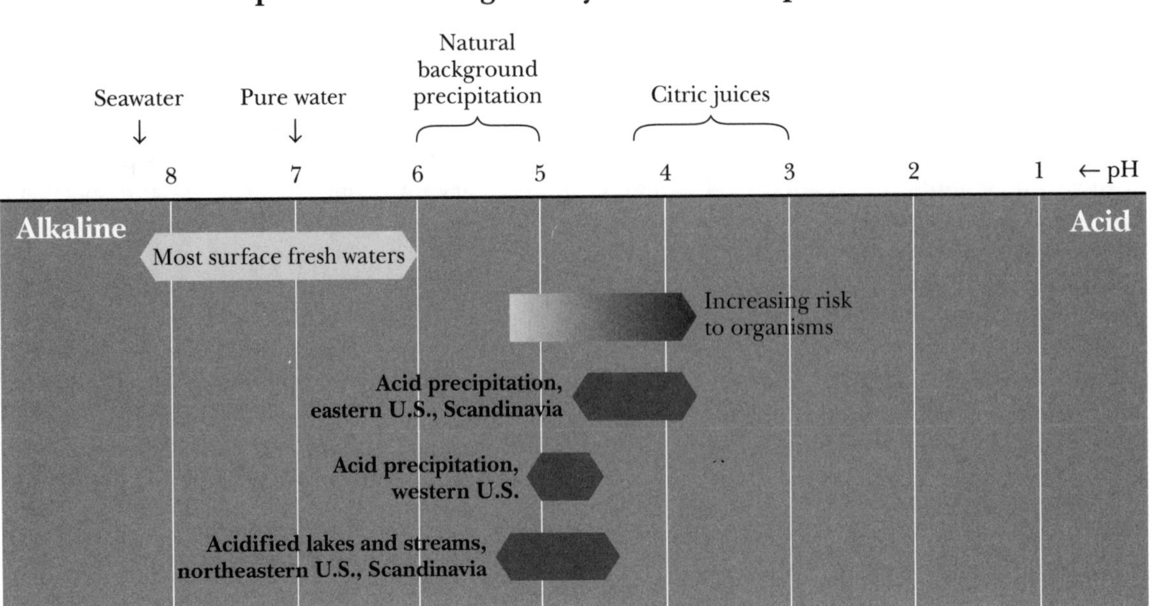

Source: Adapted from John Harte, "Acid Rain," in *The Energy-Environment Connection*, edited by Jack M. Hollander, 1992.
Note: The acid precipitation pH ranges given correspond to volume-weighted annual averages of weekly samples.

ily oxidized into sulfuric and nitric acids in the atmosphere. Long-range transport and dispersion of these air pollutants produce regional acid deposition. Acid deposition alters aquatic—and possibly forest—ecosystems and accelerates corrosion of buildings, monuments, and statuary.

In 1872 Robert Angus Smith used the term "acid rain" in his book *Air and Rain: The Beginnings of a General Climatology* to describe precipitation affected by coal-burning industries. Today, "acid rain" refers to the deposition of acidic gases, particles, and precipitation (rain, fog, dew, snow, or sleet) on the surface of the earth. The normal acidity of rain is pH 5.6, which is caused by the formation of carbonic acid from water-dissolved carbon dioxide. The acidity of precipitation collected at monitoring stations around the world varies from pH 3.8 to 6.3 (pH 3.8 is three hundred times as acidic as pH 6.3). The acidity is created when sulfur dioxide and nitrogen oxides react with water and oxidants in the atmosphere to form water-soluble sulfuric and nitric acids. Ammonia, as well as soil constituents such as calcium and magnesium that are often present in suspended dust, neutralizes atmospheric acids, which helps explain the geographical variation of precipitation acidity.

Increasing Acidity

Between the mid-nineteenth century and World War II, the Industrial Revolution led to a tremendous increase in coal burning and metal ore processing in both Europe and North America. The combustion of coal, which contains an average of 1.5 percent sulfur by weight, and the smelting of metal sulfides released opaque plumes of smoke and sulfur dioxide from short chimneys into the atmosphere.

Copper, nickel, and zinc smelters fumigated nearby landscapes with sulfur dioxide and heavy metals. One of the world's largest nickel smelters, located in Sudbury, Ontario, Canada, began operation in 1890 and by 1960 was pouring 2.6 million tons of sulfur dioxide per year into the atmosphere. By 1970 the environmental damage extended to 72,000 hectares of injured vegetation, lakes, and soils surrounding the site; within this area 17,000 hectares were barren. The land was devastated not only by acid deposition but also by the accumulation of toxic metals in the soil, the clear-cutting of forested areas for fuel, and soil erosion caused by wind, water, and frost heave.

In urban areas, high concentrations of sulfur corroded metal and accelerated the erosion of stone structures. During the winter, the added emissions from home heating and stagnant weather conditions caused severe air pollution episodes characterized by sulfuric acid fogs and thick, black soot. In 1952 a four-day air pollution episode in London, England, killed an estimated four thousand people.

After World War II, large coal-burning utilities in Western Europe and the United States built their plants with particulate control devices and tall stacks (higher than 100 meters) to improve the local air quality. Huge industrial facilities throughout Eastern Europe and the Soviet Union operated without air pollution controls for most of the twentieth century. The tall stacks increased the dispersion and transport of air pollutants from tens to hundreds of kilometers. Worldwide emissions of sulfur dioxide increased; in the United States emissions climbed from 18 million tons in 1940 to a peak of 28 million tons in 1970. Acid deposition evolved into an interstate and even an international problem.

In major cities, exhaust from automobiles combined with power plant and industrial emissions to create a choking, acrid smog of ozone, and nitric and organic acids formed by photochemical processes. The rapid deterioration of air quality in cities, with the attendant health and environmental consequences, spurred the passage of the U.S. Clean Air Act (CAA) of 1963, which was amended and expanded in 1970, 1977, and 1990. Each amendment to the CAA brought new requirements for air pollution controls.

Effects on Aquatic Ecosystems

The nature and extent of the environmental impact of acid deposition are in dispute. Landscapes or surface waters impoverished by limestone or acid-buffering soils are more sensitive to acid deposition. Regions that are both sensitive and exposed to acid deposition include the eastern United States, southeastern Canada, southern Sweden and Norway, central and Eastern Europe, the United Kingdom, southeastern China, and the northern tip of South America. Scientists hypothesize that within these regions acid rain disrupts aquatic ecosystems and contributes to forest decline.

In southern Norway, for example, fish have been virtually extinct since the late 1970's in four-fifths of the lakes and streams in an area of 2 million hectares. Records and long-term monitoring showed that the decline of fish populations began in the early twentieth century with dramatic losses in the 1950's. A strong correlation has been found between fish extinction and lake acidity. Researchers have also found that the diversity of not only fish but also phytoplankton, zooplankton, invertebrates, and amphibian species diminishes by more than 50 percent as lake water pH drops from 6.0 to 5.0. Below pH 5.6, aluminum released from lake sediments or leached from the surrounding soils interferes with gas and ion exchange in fish gills and can be toxic to aquatic life. Below pH 4.0, no fish survive.

In the United States, 11 percent of the lakes in the Adirondack Mountains in New York are too acidic to sustain fish life. Much controversy surrounds the claim that these lakes were acidified by plumes of air pollutants carried by prevailing winds from the Ohio River Valley. Like the lakes in Norway, the Adirondack lakes have low acid-neutralizing capacity. Fish declines that began in the early twentieth century and continued through the 1980's corresponded to reductions in pH. Fish kills often followed spring snowmelt, which filled the waterways with acid accumulated in winter precipitation. Historical records, field observations, and laboratory experiments contradict arguments that overfishing, disease, or water pollution killed the fish.

Effects on Forests and Cities

In areas exposed to acid rain, dead and dying trees stand as symbols of environmental change. In Germany the term *Waldsterben,* or forest death, is used to describe the rapid declines of Norway spruce, Scotch pine, and silver fir trees in the early 1980's, followed by beech and oak trees in the late 1980's, especially at high elevations in the Black and Bavarian Forests. At higher altitudes, clouds frequently shroud mountain peaks, bathing the forest canopy in a mist of heavy metals, and sulfuric and nitric acids. Under drought conditions, invisible plumes of ozone from sources hundreds of kilometers distant intercept the mountain slopes. Several forests within the United States are likewise affected, including the forests of ponderosa pine in the San Bernardino Mountains of California, balsam fir in the Smokey Mountains of North Carolina and Tennessee, and red spruce in the Green and White Mountains of New England.

After more than one decade of intensive field and laboratory investigations of forest decline in North America and Europe, the link between dead trees and acid deposition remained little more than circumstantial. Laboratory experiments often showed that acid rain had no effect, or even a fertilizing effect, on trees. Changes in foliage color, size, and shape; destruction of fine roots and associated fungi; and stunted growth are symptoms of tree stress. Many researchers attribute these symptoms and forest decline to the interactions of acid precipitation, ozone, excessive nitrogen deposition, land management practices, climate change, drought, and pestilence.

Ambient air concentrations of sulfur dioxide and nitrogen oxides are typically higher in major cities, a result of the high density of emission sources. The acids they form accelerate the weathering of exposed stone, brick, concrete, glass, metal, and paint. For example, the calcite in limestone and marble reacts with water and sulfuric acid to form gypsum (calcium sulfate). The gypsum washes off stone with rain or, if eaves protect the stone, accumulates as a soot-darkened crust. The acid-induced weathering obscures the details of elaborate carvings on medieval cathedrals, ancient Greek columns, and Mayan ruins at alarming rates. Graffiti, pigeon excrement, and the growth of bacteria and fungi on rock surfaces may compound the damage.

Prevention Efforts

In the United States, the Acidic Deposition Control Program, Title IV of the Clean Air Act amendments of 1990, directed the Environmental Protection Agency (EPA) to reduce the adverse effects of acid rain. Public law mandated that the United States achieve 40 percent and 10 percent annual reductions in sulfur dioxide and nitrogen dioxide emissions, respectively, by the year 2000 from a 1980 base. The National Acid Precipitation Assessment Program coordinates interagency acid deposition monitoring and research, and assesses the cost, benefits, and effectiveness of acid deposition control strategies. This echoes the 1985 30 Percent Protocol of the Convention on Long-Range Transboundary Air Pollution. Twenty-one nations signed the protocol, thereby agreeing to reduce sulfur dioxide emissions 30 percent from 1980 levels by 1993.

Strategies to reduce acid deposition in the United States target large electric utilities responsible for 70 percent of the sulfur dioxide and 30 percent of the nitrogen oxide emissions. Utilities participate in a novel market-based emission allowance trading and banking system that permits great flexibility in controlling sulfur dioxide emissions. For example, a utility may choose to remove sulfur from coal by cleaning it, burn a cleaner fuel such as natural gas, or install a gas desulfurization system to reduce emissions. The $6 billion international Clean Coal Technology Demonstration Program, funded by governments and private industries, continues to develop technologies—such as catalytic conversion of nitrogen oxides to inert nitrogen—to radically decrease emissions of acid gases from coal-fired power plants.

Computer models of acid deposition in the northeastern United States predicted that a 50 percent reduction in sulfur dioxide emissions would decrease sulfur deposition by 44 to 48 percent. Between 1980 and 1996, U.S. electric utilities lowered annual emissions of sulfur dioxide by 30 percent from 17.5 million tons and nitrogen oxides by 14 percent from 7 million tons. Average ambient air concentrations of sulfur dioxide decreased 37 percent and nitrogen oxide concentrations 10 percent between 1987 and 1996. However, a similar trend in sulfate and nitrate deposition has not been observed.

Noreen D. Poor

SUGGESTED READINGS: For an interesting illustrated history of local acid deposition, read *Restoration and Recovery of an Industrial Region: Progress in Restoring the Smelter-Damaged Landscape Near Sudbury, Canada* (1995), edited by John M. Gunn. *Atlas of the Environment* (1992), by Geoffrey Lean and Don Hinrichsen, provides a color map and a descriptive text of regions sensitive to acid deposition. *Acid Rain and Our Nation's Capital* (1997), published by the U.S. Geological Survey, gives a photographic tour of acid-damaged buildings and monuments in Washington, D.C. Charles E. Little, in *The Dying of Trees: The Pandemic in America's Forests* (1995), explains his view of the causes and effects of forest decline across the United States. In "Acid Precipitation in Historical Perspective," *Environmental Science and Technology* 16 (1982), Ellis B. Cowling provides a concise history of acid rain. Magda Havas, Thomas C. Hutchinson, and Gene E. Likens discuss misconceptions about acid rain in "Red Herrings in Acid Rain Research," in *Environmental Science and Technology* 18 (1984).

SEE ALSO: Air pollution; Automobile emissions; Clean Air Act and amendments; Convention on Long-Range Transboundary Air Pollution; Sudbury, Ontario, emissions.

Acid mine drainage

CATEGORY: Water and water pollution

Acid mine drainage (AMD) occurs when acidified waters flow from mines and mine wastes. It can pollute groundwater, surface water, and soils, producing adverse effects on plants and animals.

During mining, rock is broken and crushed, exposing fresh rock surfaces and minerals. Pyrite, or iron sulfide, is a common mineral encountered in metallic ore deposits. Rainwater, groundwater, or surface water that runs over the pyrite leaches out sulfur, which reacts with the water and oxygen to form sulfuric acid. In addition, if pyrite is present in the mining waste materials that are discarded at the mine site, some species of bacteria can directly oxidize the sulfur in the waste rock and tailings, forming sulfuric acid. In either case, the resulting sulfuric acid may run into groundwater and streams downhill from the mine or mine tailings.

AMD pollutes groundwater and adjacent streams and may eventually seep into other streams, lakes, and reservoirs to pollute the surface water. Groundwater problems are particularly troublesome because the reclamation of polluted groundwater is very difficult and expensive. Furthermore, AMD dissolves other minerals and heavy metals from surrounding rocks, producing lead, arsenic, mercury, and cyanide, which further degrades the water quality. Through this process, AMD has contributed to the pollution of many lakes.

AMD runoff can be devastating to the surrounding ecosystem. Physical change and damage to the land, soil, and water from AMD directly and indirectly affect the biological environment. Mine water immediately adjacent to mines that are rich in sulfide minerals may be as much as 100,000 to 1,000,000 times more acidic than normal stream water. AMD water poisons and leaches nutrients from the soil so that few, if any, plants can survive. Animals that eat those plants, as well as microorganisms in the leached soils, may also die. AMD is also lethal for many water-dwelling animals, including plankton, fish, and snails. Furthermore, people can be poisoned by drinking water that has been contaminated with heavy metals produced by AMD, and some people have developed skin cancer by drinking groundwater contaminated with arsenic generated by AMD leaching.

Alterations in groundwater and surface water availability and quality caused by AMD have also indirectly impacted the environment by causing changes in nutrient cycling, total biomass, species diversity, and ecosystem stability. Additionally, the deposition of iron as a slimy orange precipitate produces an unsightly coating on rocks and shorelines.

AMD was so severe in the Tar Creek area of Oklahoma that the Environmental Protection Agency (EPA) designated it as the nation's foremost hazardous waste site in 1982. The largest complex of toxic waste sites in the United States was produced by the mines and smelters in Butte and Anaconda, Montana, with much of the pollution attributed to direct and indirect effects of AMD. AMD is also a widespread problem in many coal fields in the eastern United States.

Alvin K. Benson

SEE ALSO: Groundwater and groundwater pollution; Heavy metals and heavy metal poisoning; Water pollution.

Adams, Ansel

BORN: February 20, 1902; San Francisco, California

DIED: April 22, 1984; Carmel, California

CATEGORY: Preservation and wilderness issues

Ansel Adams is best known for his black-and-white photographs of the American West. However, he was also an environmental activist who served on the Sierra Club's board of directors from 1934 to 1971.

Ansel Adams was the only son of Charles and Olive Bray Adams. Although a gifted child, he detested public schools and received his education with tutors, graduating from a private school in 1917. His early interests centered on music and, after he acquired his first camera, photography. A visit to Yosemite National Park in California in 1916 sparked his interest in nature photography, and he made frequent return visits to Yosemite throughout his lifetime.

Adams was able to support himself with music and photography during his early working years. As his interest in photography increased, however, it became the dominant factor in his life. He traveled extensively throughout the American West photographing landscapes. His first published photograph appeared in the Sierra Club Bulletin in 1927, and the first of his numerous photographic collections was published in the same year. In 1928 he married Virginia Rose Best in Yosemite and held his first one-person exhibit in San Francisco. The Adamses had two children: Michael, born in 1933, and Anne, born in 1935.

In 1934 Adams was elected to the board of directors of the Sierra Club, a position he held until 1971. He was a cofounder of Group F/64, an organization dedicated to using photography to emphasize and preserve the natural beauty of the American West.

Adams's photographs, almost exclusively in black and white, were renowned for their sharp contrasts, detail, use of light and shadow, and ability to capture the beauty of their natural settings. Many of his exhibits were held in museums and at universities that recognized his artistic talent. In addition to this artistry, he developed several innovative techniques for developing photographs to enhance their contrast and embellish their appearance. He published several how-to books on photography and taught workshops on how to utilize his techniques. In addition, he established his own studio to exhibit and sell his works.

Largely self-taught, Adams was recognized as one of the leaders in nature photography. He not only photographed nature but also realized the need for environmental activism and lobbied extensively for conservation measures. As early as 1936 he approached the United States Congress to establish additional national parks in the western states. He was a member of President Lyndon B. Johnson's environmental task force and later met with Presidents Gerald Ford and Ronald Reagan regarding environmental issues.

Adams's work extended beyond nature photography to include architectural studies, portraits, and commercial photography. In 1943 he photographed the Manzanar Relocation Center, where Japanese Americans were interned by the U.S. government during World War II. In 1960 he was commissioned to photograph scenes from all nine campuses of the University of California. In his later years he received several honorary doctoral degrees and other awards, including one named in his honor by the Wilderness Society. During these years he traveled widely, giving lecturers and exhibiting his works.

Gordon A. Parker

SEE ALSO: National forests; National parks; Sierra Club; Wilderness Society; Yosemite.

Aerosols

CATEGORY: Atmosphere and air pollution

Aerosols are made up of aggregations of small particles, both liquid and solid, and the atmospheric gases in which they are suspended. Atmospheric aerosols diminish environmental quality throughout much of the world.

An aerosol is a multiphasic system consisting of tiny liquid and solid particles and the gas in which they are suspended. In unpolluted areas such as New Zealand, aerosols contain impurities from natural sources; acidity comes from carbonic acid (H_2CO_3). In central Europe and other industrialized areas throughout the world, fossil fuel combustion contributes large amounts of oxides of sulfur (SO_x) and oxides of nitrogen (NO_x) to the atmosphere, leading to the formation of sulfuric acid and nitric acid aerosols. In these polluted areas, acidity levels are much higher in fog than in rain, and dry deposition of sulfuric and nitric acid particulates from impactions of acid fog may be more damaging to buildings and the environment than acid rain. Forest canopies tend to scavenge acid aerosols; on conifer-covered mountains, cloud droplets are the major source of acid deposition. Dry deposition on canopies rapidly affects root systems and increases soil acidification over the long term. Soil acidification has been linked to a number of adverse effects on vegetation, especially tree dieback and forest decline in Europe.

Aerosols in industrialized areas often contain heavy metals—including chromium (Cr), iron (Fe), copper (Cu), cadmium (Cd), cobalt (Co), nickel (Ni), and lead (Pb)—latex, surfactants, and asbestos. When tetraethyl lead was used as a gasoline additive, inhalation of lead aerosols contributed a substantial fraction of the body burden of lead in urban dwellers. Recognition of this hazard led to a ban on leaded gasoline. Dry deposition of heavy metals in aerosols can have adverse effects on ecosystems; $CdSO_4$ and $CuSO_4$ are known to reduce root elongation in trees. Fluorides released by heavy industry contribute to tree dieback.

The adverse impact of aerosols on air quality has been noted by many writers since the time of John Evelyn, who, on January 24, 1684, recorded a marked decrease in atmospheric visibility and increased respiratory problems associated with London smog. During the late nineteenth and early twentieth centuries, the lethal effects of aerosols were evident in the greatly increased mortality that occurred during London smog episodes. A historic episode in 1952, during which smog in London reduced visibility to zero, caused an estimated four thousand excess deaths, doubling the normal death rate for children and for adults aged forty-five to sixty-four. This episode led to emission controls, and by the late 1960's, health effects related to coal burning were reduced to minimal levels.

Urban aerosols—with their mix of latex, soot, hydrocarbons, SO_x and NO_x, and other pollutants—have been implicated in the growing prevalence of asthma. In the United States, the National Institutes of Health (NIH) estimated that asthma prevalence rose 34 percent between 1983 and 1993. About 4.8 million children in the United States were estimated to suffer from asthma in 1993. Asthma deaths have rapidly increased in the United States, rising from 0.9 per 100,000 people in 1976 to 1.5 per 100,000 people in 1986. Deaths from asthma among African Americans of all ages rose from 1.5 per 100,000 in 1976 to 2.8 per 100,000 in 1986. African Americans between the ages of fifteen and twenty-four had an asthma death rate of 8.2 per one million in 1980; by 1993, the rate had increased to 18.8 per one million.

Aerosols also increase the absorption and scattering of light in the atmosphere, reducing visibility. When smoke particle concentrations exceed 80–100 mg/m^3, visibility falls below 1 kilometer (0.62 miles). Smoke palls can travel considerable distances; during the spring of 1998, agricultural burning in the Yucatán in Mexico created a pall that markedly diminished visibility in Dallas, Texas.

Particulates in aerosols serve as condensation nuclei and may enhance fog formation. Visibility reductions by hygroscopic air pollutants are noticeable at relative humidities of about 50 percent. A number of multicar chain-reaction accidents have occurred downwind of industrial plants when a hygroscopic plume has passed over cooling ponds adjacent to a highway, creating an abrupt reduction in visibility. Cars and trucks moving at a high speeds suddenly become engulfed in an area with visibility of about 1 meter (3 feet). Drivers, unable to see where the road is or the position of other vehicles, respond erratically, causing accidents.

Extremely low air temperatures can lead to the formation of ice fogs known as "arctic haze" over cities, which are fed by moisture and particles given off by combustion. Ice fogs are entirely anthropogenic, composed of minute ice crystals that substantially reduce visibility to the point where air travel is restricted. Ice fogs last for several days at a time and are frequent during winter in Fairbanks, Alaska, and many Canadian cities. Irkutsk, Siberia, Russia, reports an average of 103 days of fog yearly, all during winter.

Anita Baker-Blocker

SUGGESTED READINGS: Excess mortality resulting from particulates in Europe is reviewed in "Short-Term Effects of Ambient Sulphur Dioxide and Particulate Matter on Mortality in 12 European Cities: Results from Time Series Data from the APHEA Project," by K. Katsouyanni et al., in the *British Medical Journal* 314 (June 7, 1997). "Effects of Acidic Precipitation on Forest Ecosystems in Europe," by B. Ulrich, in *Acidic Precipitation Volume 2: Biological and Ecological Effects* (1989), edited by Domy C. Adriano and Arthur H. Johnson, presents data on dry deposition and its effects on vegetation.

SEE ALSO: Acid deposition and acid rain; Air pollution; Particulate matter.

Agent Orange

CATEGORY: Pollutants and toxins

"Agent Orange" was the name given to a powerful defoliant used extensively by the U.S. military during the Vietnam War. Spraying of Agent Orange deforested large sections of Southeast Asia, and exposure to the herbicide has been linked to the development of serious health problems in both military personnel and civilians.

Between 1962 and 1971, some 19 million gallons of herbicides were sprayed by the U.S. military over South Vietnam and Laos by airplane, helicopter, boat, truck, and manual sprayers in an effort to reduce ground cover for enemy troops and to destroy enemy crops. About 11 million gallons of this total was sprayed in the form of Agent Orange, a fifty-fifty mixture of the herbicides 2,4-D and 2,4,5-T combined with a kerosene-diesel fuel for dispersal. The principal components of both 2,4-D and 2,4,5-T decompose within weeks after application; however, 2,4,5-T contained between 0.05 and 50 parts per million of dioxin, primarily 2,3,7,8-tetrachlorodibenzo-para-dioxin (TCDD), which is among the deadliest chemicals known and which has a half-life of decades. Approximately 368 pounds of dioxin were released during the spraying, with profound effects on both the ecology of the region and the subsequent health of U.S. military personnel, the residents of Vietnam and Laos, and their offspring.

Agent Orange constituted about 60 percent of total volume of herbicides sprayed by U.S. personnel in the region; other substances used included dinoxol, trinoxol, bromacil, diquat, tandex, monuran, diuron, and dalapon as well as compounds known by such code names as Agent White, Agent Blue, Agent Purple, Agent Green, Agent Pink, and Agent Orange II ("Super Orange"). In addition to 2,4-D and 2,4,5-T, various mixes included picloram (a growth regulator similar to 2,4-D and 2,4,5-T) and cacodylic acid, an arsenic-containing organic that dehydrated and killed plants. Unlike 2,4-D and 2,4,5-T, which acted on plant surfaces, mobile picloram penetrated soil to be absorbed by roots and induce systemic effects.

Military Background

The U.S. military developed weapon herbicides during World War II at Fort Detrick, Maryland, and considered using them against Japanese food plots on Pacific islands. The British used herbicides in Malaya in the 1950's to destroy rebel food plots. Domestic tests of Agent Orange were conducted at Camp Drum, New York, and Eglin Air Force Base, Florida, in the 1950's. The first field tests were conducted in South Vietnam in 1961 and in Thailand in 1964 and 1965.

"Operation Ranch Hand" was the code name given to the U.S. Air Force program of herbicide application during the Vietnam War. The program involved a total of thirty-six aircraft that sprayed roughly 10 percent of the area of South Vietnam. The spraying and defoliation targeted jungles, inland forests, camp edges, roads, trails, railroads, and canals to make enemy movement conspicuous and easier to attack. Because of the possible risk to the crops of the United States' South Vietnamese allies, spraying of enemy food plots was not as extensive. In addition, the Army Chemical Corps conducted truck, helicopter, and manual spraying (particularly around base camps and transportation routes); the Navy, using small riverboats, sprayed edges of rivers and canals; and Special Forces troops conducted covert spraying operations.

Environmental Damage in Vietnam

Many of the trees in the tropical mangrove forests near the southernmost coasts of South Vietnam were killed by a single spraying. It is estimated that it will take up to a century for the mangrove forests to recover without reseeding. Since these and adjacent waters served as breeding and nursery grounds for wildlife, the area's ecology has been catastrophically affected.

Inland forests were less sensitive to Agent Orange. They would usually recover after single or double sprayings with only temporary foliage loss. However, three or more sprayings eventually induced tree death and converted forests to grasslands. Wildfires burned seeds and seedlings of native trees, further delaying recovery. Erosion washed sediment into deltas near river outlets, compounding environmental damage. Since approximately one-third of sprayed acreage was sprayed more than once, and 130,000 acres were sprayed more than four times, damage to Vietnam's ecology was extensive.

Elevated concentrations of toxins still affect Vietnam's population and environment. Because of persistent effects of dioxin, which is fat soluble and bioaccumulates up the food chain, health problems related to the spraying have affected not only Vietnamese alive during the war but also their offspring. Malformations and birth defects are common in Vietnamese children, and other maladies are suspected to be linked to dioxin exposure. It took more than two decades for a serious effort at characterizing and alleviating these problems to be begun, because it took years after the war for the problems to become recognized. In addition, Vietnam had other pressing economic, military, and rebuilding concerns. In the 1990's, the Vietnamese National Committee for the Investigation of the Consequences of the Chemicals Used During the Vietnam War, also known as the 10-80 Committee, solicited international help. In an effort to determine existing dioxin levels, from 1994 to 1998, Hatfield Consultants of West Vancouver, British Columbia, Canada, sampled soils, crops, fish, poultry, livestock, and human blood in the A Luoi Valley in central Vietnam near the old Ho Chi Minh Trail. The investigators also performed satellite characterization of topography changes in the Ma Da region of southern Vietnam caused by defoliation and sediment washing.

The Vietnamese Red Cross established a fund for Agent Orange victims in 1998; in April, 1998, Vietnam's prime minister ordered the first nationwide survey of Agent Orange-related problems. Vietnam has also established several "peace villages" that can accommodate up to five hundred patients and victims of Agent Orange.

Health Effects in Vietnam Veterans

Agent Orange and dioxin are now known to have caused comparable health problems in U.S. Vietnam veterans. Health problems linked to exposure include soft tissue sarcoma, non-Hodgkin's lymphoma, Hodgkin's disease, chloracne, and porphyria cutanea tarda (the first three are cancers, and the last two are skin diseases). There is a definite correlation with respiratory cancer, prostrate cancer, multiple myeloma, acute and subacute peripheral neuropathy, and spina bifida (abnormal spine development in children of veterans). Other suspected health effects include immune system disorders, reproductive difficulties and cancers, diabetes, endocrine and hormone imbalances, cancer in offspring, and malformations and defects (there is a much stronger evidence linking birth defects to dioxin in Vietnamese). It appears that many of these health problems (including spina bifida, birth defects, immune system problems, and cancer propensity) can be passed on to the children and even the grandchildren of those initially exposed.

These issues were controversial for years after the war, but as evidence accumulated, the Department of Veterans Affairs (VA) eventually addressed the issue. Free medical examinations and care were offered to Vietnam veterans with suspected Agent Orange-induced problems in 1978. By 1981, the VA had established a program providing follow-up hospital care to veterans with any health problem of which the cause was unclear. The VA now provides monthly compensation for those with the ten diseases for which there is a proven cause-and-effect or positive correlation. Also, compensation, health care, and vocational rehabilitation are provided to children of veterans with spina bifida. The VA now presumes that all military personnel who served in the Vietnam Theater were exposed to Agent Orange.

Overview

Other groups of people—including farmers, foresters, ranchers, chemical industry workers, incinerator workers, and paper-mill workers—are often exposed to trace dioxin and related chemicals such as polychlorinated biphenyls (PCBs) at levels in excess of those experienced by the typical Vietnam veteran. Workers in these areas occasionally exhibit increased frequencies of diseases (for example, prostrate cancer is especially common among farmers). Trace dioxins are commonly produced by the burning of chlorine-containing organics and other chemical processes. It is likely that nearly all people in the industrialized world has some amount of dioxin in their bodies. However, because of many epidemiological and exposure variables, symptoms vary widely, and it is difficult to establish conclusive links between symptoms and particular chemicals or activities.

Defenders of the spraying program point out that the use of Agent Orange and related herbicides was highly successful in meeting its military goals and, hence, likely saved the lives of many U.S. servicemen. Moreover, although the resulting environmental and health effects might have been less severe if another chemical had been used rather than the dioxin-containing 2,4,5-T, dioxin's long-term effects were not known at the time. The use of Agent Orange in Vietnam thus drives home the lesson that technology often has unanticipated consequences.

Robert D. Engelken

Suggested readings: An excellent overview of the Agent Orange problem can be found in *Veterans and Agent Orange: Health Effects of Herbicides Used in Vietnam* (1996), published by the National Academy Press. *Operation Ranch Hand,* published by the U.S. Government Printing Office in 1982, is a good description of the military operation responsible for the spraying of Agent Orange.

See also: Dioxin; Pesticides and herbicides.

Agricultural chemicals

Category: Pollutants and toxins

Chemicals are utilized by the agriculture industry to improve crop yield or the quality of produce. Agricultural chemicals generally fall into one of four categories: fertilizers, pesticides, harvest aids, or food additives. Of these, fertilizers and pesticides have the greatest environmental impact.

Plants require sunshine, water, carbon dioxide from the atmosphere, and mineral nutrients from the soil. Mineral nutrients may be subdivided into macronutrients (calcium, magnesium, sulfur, nitrogen, potassium, and phosphorus) and micronutrients (iron, copper, zinc, boron, manganese, chloride, and molybdenum). Plant growth and crop yields will be reduced if any one of these nutrients is not present in sufficient amounts. Micronutrients are required in small quantities, and deficiencies occur infrequently; therefore, the majority of agricultural fertilizers contain only macronutrients. Magnesium and calcium are utilized in large quantities, but most agricultural soils contain an abundance of these two elements, either derived from parent material or added as lime. Most soils also contain sufficient amounts of sulfur from the weathering of sulfur-containing minerals, the presence of sulfur in other fertilizers, and atmospheric pollutants. The remaining three macronutrients (nitrogen, potassium, and phosphorus) are readily depleted and are referred to as fertilizer elements. They must be added to most soils on a regular basis. Mixed fertilizers contain two or more nutrients. For example, a fertilizer labeled 10-10-10 contains 10 percent nitrogen, 10 percent phosphorus, and 10 percent potassium.

Fertilizers and Environmental Concerns

The application of fertilizer to agricultural soil is by no means new. Farmers have been applying manures to improve plant growth for more than four thousand years. For the most part, this practice had little environmental impact. Since the development of chemical fertilizers in the late nineteenth century, however, fertilizer usage has increased tremendously. During the second half of the twentieth century, the amount of fertilizer applied to the soil increased more than 450 percent. While this increase has more than doubled the worldwide crop production, it has also generated some environmental problems.

The production of fertilizer requires the use of a variety of natural resources, and some people have argued that the increased production of fertilizers has required the use of energy and mineral reserves that could have been used elsewhere. For every crop, there is a point at which the yield may continue to increase with the application of additional nutrients, but the increase will not offset the additional cost of the fertilizer. The economically feasible practice, therefore, is to apply the appropriate amount of fertilizer that produces maximum profit rather than maximum yield. Unfortunately, many farmers still tend to overfertilize, which wastes money and contributes to environmental degradation. Excessive fertilization can result in adverse soil reactions that damage plant roots or produce undesired growth patterns. Overfertilization can actually decrease yields. If supplied in excessive amounts, some micronutrients are toxic to plants and will dramatically reduce plant growth.

The most serious environmental problem associated with fertilizers, however, is their contribution to water pollution. Excess fertilizer elements, particularly nitrogen and phosphorus, are carried from farm fields and cattle feedlots by water runoff and are eventually deposited in rivers and lakes, where they contribute to the pollution of aquatic ecosystems. High levels of plant nutrients in streams and lakes can result in increased growth of phytoplankton, a condition known as eutrophication. During the summer months, eutrophication can deplete oxygen levels in lower layers of ponds and lakes. Excess nutrients can also be leached through the soil and contaminate underground water supplies. In areas where intense farming occurs, nitrate concentrations are often above recommended safe levels. Water that contains excessive amounts of plant nutrients poses health problems if consumed by humans and livestock, and it can be fatal if ingested by newborns.

Pesticides

Pesticides are chemicals designed to kill unwanted organisms that interfere, either directly or indirectly, with human activities. The major types of pesticides in common use are those designed to kill insects (insecticides), nematodes (nematocides), fungi (fungicides), weeds (herbicides), and rodents (rodenticides). Herbicides and insecticides make up the majority of the pesticides applied in the environment.

The application of chemicals to control pests is not necessarily a new technology. People have been using sulfur and heavy metal compounds as insecticides for more than two thousand years. Residues of toxic metals such as arsenic, lead, and mercury are still being accumulated in plants that are grown on soil where these materials were used. The commercial introduction of dichloro-diphenyl-trichloroethane (DDT) in 1942 opened the door for the synthesis of a host of synthetic organic compounds to be used as pesticides.

Chlorinated hydrocarbons such as DDT were eventually banned or severely restricted in the United States because of their low biodegradability and persistence in the environment. They were replaced by organophosphates, which rapidly biodegrade but are generally much more toxic to humans and other animals than chlorinated hydrocarbons. In addition, they are water soluble and, therefore, more likely to contaminate water supplies.

Carbamates have also been used in place of the chlorinated hydrocarbons, and while these compounds biodegrade rapidly and are less toxic to humans than organophosphates, they are also less effective in the killing of insects. A large number of herbicides have also been developed. These chemicals are generally classified as one of three types. Contact herbicides kill when they come in contact with the leaf surface. Systemic herbicides circulate throughout the plant after being absorbed and cause abnormal growth. Soil sterilants kill microorganisms necessary for plant growth and act as systemic herbicides.

Pesticides and Environmental Concerns

Approximately fifty-five thousand different pesticides are available in the United States, and Americans apply about 500 million kilograms of pesticides each year. Fungicides account for 12 percent of all pesticides used by farmers, while insecticides account for 19 percent and herbicides account for 69 percent. Approximately 2.5 tons of pesticides are applied each year throughout the world. While most of these hemicals are applied in developed countries, the amount of pesticides used in underdeveloped countries is rapidly increasing.

An agriculturalist uses a tractor to spray pesticides on crops. American farmers apply about 500 million kilograms of pesticides each year. (Ben Klaffke)

There is no doubt that pesticides have had a beneficial impact on the lives of humans by increasing food production and reducing food costs. In spite of these benefits, however, the use of pesticides has caused some environmental problems, including the development of pesticide resistance. When a new pesticide is released, it will be effective in reducing the number of target pests. Within a few years, however, a number of species will have developed genetic resistance to the chemical and will no longer be controlled with it. Other chemicals must then be developed to replace the one that no longer kills the pest. Many pests will also develop resistance to these newer pesticides. As a result, many synthetic chemicals have been introduced into the environment; however, the pest problem is still is great as it ever was.

Another problem with pesticides is that they do not just kill and go away; they can remain in the environment for varying lengths of time. For example, chlorinated hydrocarbons can persist in the environment for up to fifteen years. This can be beneficial from an economic standpoint because the pesticide has to be applied less frequently, but it can be detrimental from an environmental standpoint. When many pesticides are degraded, many of their breakdown products, which are often toxic to other organisms, may also persist in the environment for long periods of time.

A third problem with pesticides is their tendency to become more concentrated as they move up the food chain. A pesticide may not affect species at the base of the chain, but it may be toxic to organisms that feed at the apex because the concentration of the chemical increases at each higher level of the food chain. Many algal species can be sprayed with an insecticide without any apparent effect to the algae. However, the chemical can be detrimental to organisms such as birds that eat fish that have eaten insects that have fed on the algae.

A fourth problem is broad-spectrum poisoning. Few, if any, chemical pesticides are selective. They kill a wide range of organisms rather than just the target pest. Many insects are beneficial, but by using insecticides that kill these organisms along with the harmful insects, more damage than good may be done. Many pesticides, particularly insecticides, are also toxic to humans. Thousands of people, particularly children, have been killed by direct exposure to high concentrations of these chemicals. Many workers in pesticide factories have been poisoned through job-related contact with the chemicals. Numerous agricultural laborers, particularly in Third World countries where there are no stringent guidelines for the handling of pesticides, have also been killed as a result of direct exposure to these chemicals. While there is currently no direct evidence to show a cause and effect relationship between pesticides and cancer in humans, these chemicals have been suspected of being carcinogenic.

D. R. Gossett

SUGGESTED READINGS: *The Scientific Basis of Alternative Agriculture* (1986), by Miguel A. Altieri, describes alternative methods for restoring soil fertility, soil conservation, and the biological control of pests. A good discussion of environmental issues associated with the use of agricultural chemicals can be found in *Introduction to Environmental Study* (1980), by Jonathan Turk. *Silent Spring* (1962), by Rachel Carson, is a somewhat outdated but excellent dramatization of the potential dangers of pesticides. The book served as a major catalyst for bringing the hazards of using pesticides to the attention of the general public. Good discussions of environmental ethics, including the ethics of using agricultural chemicals, can be found in *People, Penguins, and Plastic Trees* (1995), by Christine Pierce and Donald VanDeVeer. Another good discussion of the environmental impact of agricultural chemicals can be found in *Environmental Issues in the 1990's* (1992), by Antionette M. Mannion and Sophia R. Bowlby. A very good discussion on human health issues associated with the use of pesticides can be found in *Man and Environment: A Health Perspective* (1990), by Anne Nadakavukaren.

SEE ALSO: Agricultural revolution; Dichlorodiphenyl-trichloroethane (DDT); Pesticides and herbicides.

Agricultural revolution

CATEGORY: Agriculture and food

Advances in agricultural technology during the twentieth century—referred to as the agricultural revolution—led to a dramatic increase in the worldwide production of food and fiber crops. This increase, however, did not occur without environmental consequences. Monocultures, fertilizers, irrigation, pesticides, and modern farm equipment have all impacted the environment.

Early agricultural centers were located near large rivers that helped maintain soil fertility by the deposition of new topsoil during each annual flooding cycle. Agriculture, however, eventually moved into other regions that lacked the annual flooding of large rivers, and humans began to utilize a technique known as slash-and-burn agriculture, in which land is cleared by burning existing vegetation. The ashes are then used to fertilize the land for growing crops. This type of agriculture is still practiced in many Third World countries and is one reason that the tropical rain forests are disappearing at such a fast rate. During the nineteenth century, the Industrial Revolution led the way for the development of many different types of agricultural machinery, which resulted in the mechanization of most farms and ranches.

The Green Revolution—as advances in agricultural science during the twentieth century came to be known—resulted in the development new, higher-yielding varieties of numerous crops, particularly the seed grains that supply most of the calories necessary for maintenance of the world's growing population. While higher-yielding crops, along with improved farming methods, resulted in tremendous increases in the world's food supply, they also led to an increased reliance on monoculture, the practice of growing only one crop over a vast number of acres. This practice has decreased the genetic variability of many agricultural plants, increased the need for commercial fertilizers, and produced an increased susceptibility to damage from a host of biotic and abiotic factors.

The agricultural revolution has resulted in the development of an agricultural unit that requires relatively few employees, is highly mechanized, devotes large amounts of land to the production of only one crop, and is highly reliant on agricultural chemicals such as fertilizers and pesticides. While this has led to tremendous increases in agricultural productivity, it has also heavily impacted the environment. During the past one hundred years, there has been a continual loss of good topsoil. Even under ideal conditions, the process of soil formation is very slow. Many agricultural techniques lead to the removal of trees and shrubs that provide windbreaks or to the depletion of soil fertility, which reduces the plant cover on the field. The result of these practices has been the exposure of the soil to increased erosion from wind and moving water to the extent that as much as one-third of the world's croplands are losing topsoil more quickly than it can be replaced.

Agriculture represents the largest single user of global water. Approximately 73 percent of all water withdrawn from fresh water supplies is used to irrigate crops. Some irrigation practices have actually been detrimental. Overwatering can waterlog the soil, and irrigation of crops in dry climates can result in salinization of the soil, which occurs when irrigation water rapidly evaporates from the soil, leaving behind the mineral salts that were dissolved in the water. The salts accumulate and become detrimental to plant growth. It has been estimated that as much as one-third of the world's agricultural soils have been damaged by salinization. In addition, there is a debate as to whether the increased usage of water for agriculture has decreased the supply of potable water fit for other human uses.

Nitrogen, phosphorus, and potassium are the nutrients that are most often depleted from agricultural soils, and it is necessary to apply them to the soil regularly in order to maintain fertility. The amount of fertilizer applied to the soil increased more than 450 percent during the second half of the twentieth century, which has caused environmental problems in some areas. Fertilizer elements, particularly nitrogen and phosphorus, are carried away by water runoff and are eventually deposited in rivers and lakes, where they contribute to the pollution of aquatic ecosystems. In addition, nitrates can accumulate in underground water supplies.

Agriculture is highly dependent on the use of pesticides to kill organisms such as insects, nematodes, weeds, and fungi that directly or indirectly interfere with crop production. The use of these pesticides has dramatically improved crop yields, primarily because they are designed to kill the pests before significant damage can occur to the crop. However, pesticides often kill nonpests, and evidence suggests that indiscriminate use of these chemicals can have detrimental effects on wildlife, the structure and function of ecosystems, and perhaps even human health. In addition, the overuse of pesticides can lead to the development of resistance in the target species, which can result in a resurgence of the very pest the pesticide was designed to control.

Modern agriculture also consumes large amounts of energy. Farm machinery requires large supplies of liquid fossil fuels. The energy required to produce fertilizers, pesticides, and other agricultural chemicals is another energy cost associated with agriculture. Energy used in food processing, distribution, storage, and cooking after the crop leaves the farm may be five times as much as the energy used to produce the crop. Most of the foods consumed in the United States require more calories of energy to produce, process, and distribute to the market than they provide when they are eaten.

D. R. Gossett

SUGGESTED READINGS: An excellent discussion of the environmental issues associated with modern agriculture can be found in *Living in the Environment: An Introduction to Environmental Science* (1992), by G. Tyler Miller, Jr. While *Silent Spring* (1962), by Rachel Carson, is somewhat outdated, it is an excellent dramatization of the potential dangers of pesticides and served as a major catalyst for bringing the hazards of using pesticides to the attention of the general public. Very good discussions of environmental ethics, including the ethics of land use for agriculture, can be found in *People, Penguins, and Plastic Trees* (1995), by Christine Pierce and Donald VanDeVeer.

SEE ALSO: Agricultural chemicals; Green Revolution; Pesticides and herbicides.

Air pollution

CATEGORY: Atmosphere and air pollution

Air pollution is the atmospheric presence of materials or energy in such quantities and of a duration sufficient to cause harm to living organisms, effect weather and climate change, or damage human-made materials and structures. Its effects on metals, fabrics, and materials used by humans and its biological effects on humans and other organisms make air pollution a significant environmental problem that must be faced by local and global communities.

The atmosphere is a mixture of two types of gases: those whose concentrations are constant over a long period of time and those whose concentrations are variable. Among the former are nitrogen (N_2), which makes up approximately 78 percent of the atmosphere, and oxygen (O_2), which constitutes about 21 percent of the atmosphere. Along with these two are argon (Ar), almost 1 percent of the atmosphere, and trace amounts of neon (Ne), helium (He), krypton (Kr), hydrogen (H_2), and xenon (Xe), none of which seems to have any major effect on the atmosphere. Among the gases whose concentrations vary are carbon dioxide (CO_2), water vapor (H_2O), methane (CH_4), carbon monoxide (CO), ozone (O_3), ammonia (NH_3), hydrogen sulfide (H_2S), and several oxides of nitrogen and sulfur. Water vapor has the highest degree of variability and has a significant effect on the atmosphere because of its ability to change phase readily, absorbing or emitting energy as it does so. The trace components of the atmosphere are continually redistributed by the fairly complex circulation patterns of the air known as winds.

In addition, the atmosphere is characterized by the various physical effects acting on it or taking place within it. The most important of these are solar radiation and thermal energy. The sun acts as an almost perfect black body radiator at an effective temperature of 3,316 degrees Celsius (6,000 degrees Fahrenheit). Some of the incoming solar radiation is absorbed by atmospheric gases such as oxygen, ozone, carbon dioxide, and water vapor, allowing about 80 to 85 percent to reach the ground under clear sky conditions. Cloud cover ensures that only about 50 percent of the solar radiation reaches the earth's surface on the average.

This incident solar radiation is absorbed by the atmosphere and the earth's surface and reradiated at longer wavelengths, mostly infrared. The amount of this radiation that reaches space is affected by atmospheric concentrations of carbon dioxide and water, both of which absorb some infrared radiation. This absorption produces the "greenhouse effect," an important process that keeps the lower atmosphere at a higher temperature than the upper atmosphere, supporting life on the planet.

Sources of U.S. Air Pollutant Emissions, 1994
In Thousands of Tons

Source	Particulates	Sulfur Dioxide	Nitrogen Oxides	Volatile Organic Compounds	Carbon Monoxide
Total	**45,431**	**21,118**	**23,615**	**23,174**	**98,017**
Fuel combustion, stationary sources	1,033	18,497	11,728	886	4,884
Industrial processes	621	1,986	798	2,695	5,355
Solvent utilization	2	1	3	6,313	2
Storage and transport	59	5	3	1,773	58
Waste disposal and recycling	250	37	85	2,273	1,746
Highway vehicles	311	295	7,530	6,295	61,070
Off highway	411	283	3,095	2,255	15,657
Miscellaneous	42,743	14	374	685	9,245

Source: U.S. Department of Commerce, *Statistical Abstract of the United States, 1996*, 1996.

Note: Lead emissions are not included in the table. Lead emissions for 1994 were about 4,956 tons, mostly from industrial processes and highway vehicles.

The atmosphere is a chemical system that is not in equilibrium, mainly because of the activities of living organisms. The respiration of many organisms produces carbon dioxide, which, by means of photosynthesis, produces oxygen. The huge amount of oxygen in the atmosphere is almost entirely the result of photosynthesis. Methane, the main hydrocarbon in the atmosphere, is produced by microbial degradation of organic matter in marshes, paddy fields, and the digestive systems of animals. Microorganisms that degrade nitrogen compounds in animal urine create ammonia. Forests are a large source of more complex hydrocarbons such as alkanes, alkenes, and esters (the odors of flowers and fruits).

These natural processes have resulted in an atmosphere that has, since the formation of the earth, reached a steady-state composition. Since the nineteenth century Industrial Revolution, however, human activities have significantly changed the amounts of some of the compounds found naturally in the atmosphere and have introduced some new compounds that have the potential to affect seriously the delicate balance that exists among the earth, its atmosphere, and living organisms.

Acid Rain

The acidity of precipitation is usually expressed in terms of its pH values, where pH represents the concentration of hydrogen ions present. The pH scale generally extends from 0 to 14, with pH 7 representing neutral solutions, pH values greater than 7 basic solutions, and less than 7 acidic solutions. The pH scale is logarithmic so that an increase of one pH unit corresponds to a tenfold increase in hydrogen ion concentration. Unpolluted precipitation has a pH value of around 5. This acidity is mainly from the atmospheric presence of carbon dioxide, which forms carbonic acid, and of chlorine, from salt, which forms hydrochloric acid. Other natural contributions that affect pH, regionally or globally, are ammonia, soil particles, sea spray, and volcanic emissions of sulfur dioxide and hydrogen sulfide. Precipitation that predates the Industrial Revolution has been preserved in glaciers and generally has a pH of more than 5, sometimes as high as 6.

Precipitation with a pH value of less than 5 is termed acid rain. Precipitation with pH values averaging less than 4.5 are common in large areas of Europe and eastern North America. Individual storms may be much more acidic, such as a rainstorm in West Virginia in 1978 that had an unofficial pH of 2.0. The major causes of this increase in acidity seem to be sulfates and nitrates. Ice cores from Greenland covering a period of 115 years showed threefold increases in sulfate concentrations since the beginning of the twentieth century and twofold increases in nitrate concentrations since 1955.

The fact that acid rain is largely a regional phenomenon gives clues to the sources of the major contributors. Much acid precipitation results from the combustion of fossil fuels, especially high-sulfur coal. The sulfur is oxidized by burning into sulfur dioxide (SO_2), which is released into the atmosphere with the main combustion products, carbon dioxide and water. High concentrations of sulfur dioxide in the air produce a grayish haze known as "London fog," similar to that which affected London in December of 1952. Many people experienced respiratory difficulty, and the number of deaths from respiratory causes directly paralleled measured average smoke and sulfur dioxide concentrations. In the atmosphere, sulfur dioxide reacts with water to produce sulfuric acid. This in turn falls to the earth's surface as both wet and dry deposition.

Nitrogen and oxygen in the air do not react at any significant rate but readily combine to form nitrogen oxides in the high-temperature combustion processes found in power plants, smelters, steel mills, and internal combustion engines. In the atmosphere gaseous nitrogen oxides produce the brownish haze often seen over Los Angeles during the summer months. The nitrogen oxides go through various reactions in the atmosphere, some of which result in nitric acid (HNO_3), which eventually reaches the earth's surface.

Among the effects of these acids are the corrosion of human-made objects such as metallic structures, stone buildings, and statues. Acid rain falling on lakes increases their pH, which alters various salt concentrations on which fish depend for health. The effect of acid rain on vegetation is unclear and remains an important area of research. Increased pH of some soils seems to allow the release of metals such as aluminum, manganese, zinc, nickel, lead, mercury, and cadmium, some of which have toxic effects on humans.

U.S. Air Pollution Trends, 1950-1994

In Thousands of Tons

Year	PM-10	PM-10, Fugitive Dust	Sulfur Dioxide	Nitrogen Dioxides	Volatile Organic Compounds	Carbon Monoxide
1950	17,133	(NA)	22,358	10,093	20,936	102,609
1960	15,558	(NA)	22,227	14,140	24,459	109,745
1970	13,044	(NA)	31,161	20,625	30,646	128,079
1980	7,050	(NA)	25,905	23,281	25,893	115,625
1985	4,094	40,889	23,230	22,860	25,798	114,690
1990	3,882	39,451	22,433	23,038	23,599	100,650
1994	3,705	41,726	21,118	23,615	23,174	98,017

Source: U.S. Department of Commerce, *Statistical Abstract of the United States, 1996*, 1996.

Note: PM-10 emissions consist of particulate matter smaller than 10 microns in size. Lead emissions are not included in the table. Lead emissions in 1970 were 219,471 tons; in 1980, 74,956 tons; in 1990, 5,666 tons; in 1994, 4,956 tons.

OZONE

Ozone (O_3) is a form of oxygen found in small quantities throughout the earth's atmosphere. In the troposphere (the lowest layer of the atmosphere), ozone is of interest for a number of reasons. First, it plays an important role in the control of photochemistry. This is a group of processes in which compounds produced in the reduced state from natural or anthropogenic sources are oxidized to chemically inert materials such as carbon dioxide or to materials that can be precipitated from the atmosphere, such as nitric acid. Photochemical reactions in the troposphere provide the chief cleansing mechanism by which some materials are removed from the atmosphere. The importance of ozone to this process arises from its dissociation by ultraviolet radiation to produce reactive atomic oxygen. Some of the atomic oxygen reacts with water to produce hydroxyl (OH) radicals, which are responsible for the oxidation of most trace gases.

In addition, ozone in the troposphere is an important pollutant. It is implicated in the breakdown of natural polymers such as rubber, cotton, leather, cellulose, some paints, plastics, nylon, and fabric dyes. Since ozone is a very strong oxidant, it is a potential irritant to the lungs of humans and animals. Finally, tropospheric ozone, because of its oxidizing ability, is involved in global climate control because of its ability to influence concentrations of such greenhouse gases as carbon dioxide and methane. In addition, ozone is itself a greenhouse gas.

In the stratosphere, ozone provides an essential umbrella that partially shields the earth's surface from ultraviolet radiation. In the upper atmosphere, ozone formation involves oxygen and ultraviolet radiation. The reaction is $O_2 + h\upsilon \rightarrow O + O$, where $h\upsilon$ is ultraviolet energy. The oxygen atoms then react with oxygen molecules to produce ozone according to the reaction $O_2 + O \rightarrow O_3$. This photochemical process by which ozone is produced is balanced by the photochemical process that destroys it: $O_3 + h\upsilon \rightarrow O_2 + O$. Both processes involve the absorption of ultraviolet radiation, and the dynamic chemical equilibrium that exists between them removes a portion of the ultraviolet energy as it travels toward the earth's surface.

Knowledge of the equilibrium chemistry between oxygen and ozone allowed prediction of the equilibrium concentration of ozone in the upper atmosphere. Since the early 1970's measurements of stratospheric ozone concentrations over the Antarctic continent have pointed to concentrations much lower (sometimes by as much as 50 percent) than expected. Since then, similar reductions in ozone concentrations have been seen at other locations.

The materials chiefly responsible for this disruption of the ozone layer seen to be a group of compounds called chlorofluorocarbons (CFCs). They were found useful in air conditioning and refrigeration systems, and as blowing agents in plastic forming processes, solvents in the electronics industry, and propellants in spray cans. Since they were believed to be free of side effects, large amounts of these chemicals were expelled into the atmosphere. The very inertness, which is so advantageous at the earth's surface, becomes a great disadvantage as CFC molecules begin their long journey to the stratosphere.

Once in the stratosphere, CFC molecules absorb ultraviolet radiation and break down, yielding chlorine atoms. These chlorine atoms catalyze the conversion of ozone to oxygen. Since the chlorine atom is a catalyst in the process, it is released to continue its destructive activity for many years. CFC molecules have projected lifetimes as long as one hundred to two hundred years, so the problem will remain for a long time even though suitable substitutes are being developed for CFCs.

GLOBAL CLIMATE CHANGE

When energy in the form of electromagnetic radiation strikes a molecule, the energy may be reflected, absorbed, or transmitted. Solar energy striking the earth's surface is absorbed and heats the land and water, which in turn radiate energy in the form of infrared radiation back toward space. Eventually an equilibrium state is reached in which the amount of energy absorbed and radiated by earth is equal. In the absence of an atmosphere, the equilibrium temperature of the earth would be about -21 degrees Celsius (-5.8 degrees Fahrenheit).

The atmosphere contains gases that transmit ultraviolet and visible radiation but absorb infrared wavelengths. Therefore, the infrared energy radiated by earth toward space is trapped in the air layer, increasing its temperature and that of the earth's surface. The equilibrium temperature of the earth because of this phenomenon is about 12 degrees Celsius (53.6 degrees Fahrenheit), 33 degrees Celsius (59.4 degrees Fahrenheit) warmer than it would be without those gases. These gases do for the earth what glass walls and roofs do for the temperature of a greenhouse; therefore, they are known as greenhouse gases.

Any molecule with two or more atoms that has no center of symmetry is a potential greenhouse gas. The important greenhouse gases in the earth's atmosphere are carbon dioxide, methane, nitrous oxide, ozone, and CFCs. This collection of gases absorbs radiation across the infrared range of wavelengths so there are no windows for reflected infrared radiation to escape back into space. Concentrations of all of the gases are increasing at rates that vary from 0.25 percent per year for nitrous oxide to 5 percent per year for CFCs.

Water vapor is the greatest contributor to the greenhouse effect, but its concentration is generally considered to be unaffected by human activities. After water the most important of the greenhouse gases is carbon dioxide, which seems to be responsible for about 50 percent of the overall warming. The concentration of carbon dioxide in the atmosphere has increased by almost 30 percent since the beginning of the Industrial Revolution. This increase has been caused primarily by the burning of fossil fuels. Among the products resulting from any hydrocarbon combustion are water and carbon dioxide. All processes that depend on energy from coal, oil, or natural gas are contributing to the total amount of greenhouse gases in the atmosphere.

Whether weather phenomena such as frequent serious storms, El Niño, droughts, and floods are directly related to the greenhouse effect is a topic of heated debate among scientists and nonscientists. Cores covering thousands of years of accumulation taken from the Antarctic ice pack were examined for clues about concentrations of atmospheric gases and average temperatures. An almost direct correlation was found between carbon dioxide concentration and surface temperature. The historical evidence seems to point to potentially serious consequences if alternate forms of energy are not quickly developed.

Other Pollutants

In addition to gases, air contains suspended particulate matter. The particles are collections of molecules, sometimes similar, sometimes different. The constituents of particulate matter differ over time and space. In urban areas, particulate matter often contains sulfuric acid and other sulfates, carbon, or higher molecular weight hydrocarbons that result from incomplete combustion of fossil fuels. Particulate matter and sulfur dioxide are the common pollutants found in urban smog. Over time, suspended particles tend to increase mass by combining or acting as nuclei on which vapors condense. Eventually these fall to the ground or are washed out by precipitation.

The greatest concern over particulate matter in the atmosphere is the fact that often the particles are small enough to be inhaled and retained in the respiratory system. Vegetation is affected when the particles coat the leaves and plug stomata, reducing the absorption of carbon dioxide and suppressing photosynthesis and hence plant growth. Particulate matter adhering to painted surface and buildings reduces the lifetimes of materials and coatings and often causes corrosion, especially in moist atmospheres.

Other pollutants in the atmosphere include radioactive materials, carbon monoxide (CO), lead, and hydrocarbons. Radioactive materials result from natural processes (including the decay of materials such as uranium) or human technology. Radioactive nuclides produce ionizing radiation, which has the potential for long-term effects on cells, including cell death, genetic mutations, or malignant tumor formation. Carbon monoxide results from the incomplete combustion of hydrocarbons. It is an unstable compound that quickly oxidizes to carbon dioxide. It is absorbed through the lungs and forms a complex with hemoglobin that is more tightly bound than oxygen. In this way carbon monoxide prevents oxygen from reaching individual cells, eventually resulting in death.

Eight metals—beryllium, cadmium, chromium, lead, manganese, mercury, nickel, and vanadium—may be found in the air. Although all are potentially harmful, only lead is widely dispersed throughout the environment, mainly because of its use as an additive in gasolines. Most countries have taken major steps to reduce this use of lead, but residual concentrations still affect humans, especially children in urban areas. Hydrocarbons and their derivatives may be found as solids, liquids, or gases. Although some are the result of natural processes, most are by-products of combustion processes. Some of the hydrocarbons are toxic even in small concentrations, but the major contribution of hydrocarbons is their involvement in atmospheric photochemistry.

Grace A. Banks

Suggested readings: In *Fundamentals of Air Pollution* (1994), Richard W. Boubel, Donald L. Fox, D. Bruce Turner, and Arthur C. Stern give an introduction to the topic from the perspective of the effects of air pollution. Louise B. Young gives an easy-to-read presentation of scientific knowledge on such issues as ozone depletion, acid rain, and global warming in *Sowing the Wind: Reflections on the Earth's Atmosphere* (1990). Derek M. Elsom discusses national and international approaches to atmospheric pollution control in *Atmospheric Pollution: A Global Problem* (1992). Slightly more technical but still readable are works by Alan Wellburn, *Air Pollution and Climate Change: The Biological Impact* (1994), and Peter Brimblecombe, *Air Composition and Chemistry* (1996). *Air Quality* (1997), by Thad Godish, provides readers with a comprehensive overview of air quality, its science, and management practices.

See also: Acid deposition and acid rain; Air pollution policy; Automobile emissions; Clean Air Act and amendments; Convention on Long-Range Transboundary Air Pollution; Indoor air quality; Particulate matter; Smog.

Air pollution policy

Category: Atmosphere and air pollution

Air pollution policy is legislation initiated by the federal government that establishes standards for individual air pollutants and requires businesses whose activities negatively affect air quality to alter their operations.

The Clean Air Act of 1963, the original legal basis for all air pollution control efforts throughout the United States, laid the foundation for what some consider to be the most progressive, wide-reaching, and complicated environmental cleanup legislation in the world. Implementation of the earliest federal, state, and local clean air laws were considered by many to be ineffective, leading to several sweeping amendments.

The first set of amendments to the Clean Air Act in 1970 resulted in emissions standards for automobiles and new industries, in addition to establishing air-quality standards for urban areas. Devised through an exceptionally cooperative bipartisan effort, the 1970 amendments were proclaimed by President Richard M. Nixon to be a "historic piece of legislation" that put Americans "far down the road" toward achieving cleaner air. Specific concentration levels for several hazardous substances were established, with the individual states required to develop comprehensive plans to implement and maintain these standards.

Tightly controlled scientific methodology was used for the first time to assess and determine acceptable levels for public and environmental health for six major pollutants: carbon monoxide, sulfur dioxides, nitrogen oxides, particulate matter, photochemical oxidants, and hydrocarbons. Emission standards for air pollution sources such as automobiles, factories, and power plants were established that also limited the discharge of air pollutants in geographical areas where air quality was already acceptable, thus preventing their deterioration.

The major amendments of 1970 also stimulated many states to pass regional and local air pollution legislation, with some areas eventually passing laws that later proved to be even tighter than federally established guidelines. During this period, the Environmental Protection Agency (EPA) began strongly suggesting the tightening of rules regulating the amount of lead that could be added to gasoline, a significant source of lead poisoning in urban children and young adults, thus laying the groundwork for the future elimination of all leaded gasolines. Many sectors of the business community challenged the wording of some of the 1970 amendments, arguing that the language was vague and required clarification, particularly regarding the deterioration of air quality in areas that were already meeting federal standards.

1977 AMENDMENTS

The 1977 amendments to the Clean Air Act were stimulated by growing public and civil awareness of the necessity for further clarification of standards and the increased knowledge that came from a decade of scientific pollution control research. Industrial areas that were in violation of air-quality standards, called nonattainment areas, were only allowed to expand their factories or build new ones if the new sources achieved the lowest possible emission rates. Additionally, other sources of pollution under the same ownership in the same state were required to comply with pollution-control provisions, and unavoidable emissions had to be offset by pollution reductions by other companies within the same region. These emissions-offset policies forced new industries within a geographical region to request formally that existing local companies reduce their pollution production, often resulting in the new companies paying the considerable expense of new emissions-control devices for the existing companies.

Protection of air quality in regions that were already meeting federal standards sparked considerable congressional debate as many environmentalists claimed that existing air-quality standards gave some industries a theoretical license to pollute the air up to permitted levels. Rules for the "prevention of significant deterioration" within areas that already met clean-air standards were set for sulfur oxides and particulates in

Milestones in Air Pollution Policy

YEAR	EVENT
1963	The Clean Air Act sets aside $95 million to reduce air pollution in the United States.
1970	The Environmental Protection Agency is established to enforce environmental legislation.
1970	Clean Air Act amendments establish stricter air-quality standards.
1977	Additional Clean Air Act amendments extend compliance deadlines established by the 1970 amendments and allow the EPA to bring civil lawsuits against companies that do not meet air-quality standards.
1988	The United Nations sponsors the Convention on Long-Range Transboundary Air Pollution, which is designed to reduce acid rain and air pollution.
1987	The Montreal Protocol is signed by twenty-four nations pledging to reduce the output of ozone-depleting chlorofluorocarbons.
1990	Clean Air Act amendments increase regulations on emissions that cause acid rain and ozone depletion and also establish a system of pollution permits.
1997	The Environmental Protection Agency issues updated air-quality standards.

1977, and many individual experts and organizations lobbied for the inclusion of other pollutants such as ozone.

A final major change mandated by the 1977 amendments was the strengthening of the authority of the EPA to enforce laws by allowing it to use civil lawsuits in addition to the criminal lawsuits that were previously required. Civil lawsuits have the advantage of not carrying the burden-of-proof requirements needed for criminal convictions; this legal dilemma previously motivated violating companies to take part in lengthy legal battles wherein the legal costs were less than the purchase and maintenance of the necessary pollution-control devices. The EPA was also empowered to levy noncompliance penalties without having to file a lawsuit, using the argument that violators have an unfair business advantage over competitors who are currently complying with established legislation. Additionally, several "right-to-know" laws went into effect beginning in 1985 that required manufacturing plant managers to make health and safety information regarding toxic materials available to current and prospective employees, business partners, and sponsors.

1990 AMENDMENTS

In 1990 the Clean Air Act was further amended to address inadequacies in previous amendments, with major changes including the establishment of standards and attainment deadlines for 190 toxic chemicals. The amendments were approved with the same bipartisan efforts as the 1970 amendments, prompting President George Bush to state that the new legislation moved society much closer toward the clean air environment that "every American expects and deserves."

The 1990 amendments established a market-based measure for pollution taxes on toxic chemical emissions, thus enhancing the incentive for businesses to comply as quickly as possible. Emission standards were tightened for automobiles, and mileage standards for new vehicles were raised, which attacked the pollution problem at its center by prompting numerous significant steps toward improved fuel efficiency. Notable results of these measures include significant reductions in vehicular emissions of sulfur dioxide and nitrogen oxide (50 percent), carbon monoxide (70 percent), and other harmful substances (20 percent).

The 1990 amendments also established market-based incentives to reduce nitrogen and sulfur oxides because of their role in the growing controversy regarding acid deposition within rainwater. The EPA was empowered to create tradable permits that stipulated permissible emission levels for nitrogen and sulfur oxides. The permits were issued to U.S. companies whose rates were lower than those set by current requirements for the improvement of air quality. This landmark legislation enabled companies that implemented innovative and cost-effective means to reduce air pollution to sell their unused credits to other companies.

Other significant legislation passed within or assisted by the 1990 amendments included the beginning of the phasing out of numerous ozone-depleting chemicals and the implementation of strategies that would help sustain the environment. Many businesses complained that the considerable additional expenses associated with implementing these new laws created unnecessary burdens for industry that in many cases outweighed the potential environmental benefits. President Bill Clinton, however, continued tightening acceptable levels of smog and soot but did begin allowing flexible methods for reaching these improved goals over a ten-year period. This marked a significant change from the earlier administration of President Ronald Reagan, which proposed a relaxation of environmental standards to favor industrial and technological interests. Clinton is credited with associating the problem of controlling fossil fuel emissions with the threat of global warming, an issue that has undergone considerable debate within the United Nations (U.N.). A 1990 amendment requiring the use of gasoline containing 2 percent oxygen by weight in regions classified as being in severe or extreme nonattainment for the federal ozone standard was followed by several state-level requirements.

The clean air changes of 1990 led the California Air Resources Board to introduce the most stringent vehicle-emissions quality controls to date later that year. By 1998 2 percent of all new cars sold in California were required to have pollution-control devices that released no environmentally harmful emissions at all, and the figure was required to rise to 10 percent by 2003. These monumental state laws also dictated that the hydrocarbon emissions of all new cars sold in California be at least 70 percent less than those sold in 1993 by the year 2003. Thirteen northeastern states later passed similar, but somewhat less rigorous, laws. Only New York retained the 2 percent law. As of 1998, California still faced the greatest air-quality challenges in the nation, with seven of the top ten metropolitan smog areas located in that state.

A final important clause in the 1990 amendments was Section 129, which required the EPA to regulate emissions coming from solid waste incinerators, including incinerators used for disposal of medical waste. Medical waste incinerators are among the largest sources of dioxin and airborne mercury, which are widely believed to contribute to serious health problems.

Indoor Air Quality

Indoor air pollution is a relatively new environmental problem that has only recently been brought to public attention, as evidenced by the fact that by the end of the 1990's no significant legislation had been passed to address it. Some research studies have suggested that indoor air pollution may contribute to more cases of lung cancer than outdoor pollutants, since most people spend more than 90 percent of their time indoors. Sick building syndrome, which causes an assortment of debilitating health problems—such as chronic respiratory problems, sinus infections, sore throats, and headaches—has been estimated to affect between 10 million and 20 million Americans. Some influential lobbyists have called for an entire new set of amendments to the Clean Air Act to regulate indoor pollutants, while some lawyers have unsuccessfully argued that the Clean Air Act as it now stands could also apply to indoor air pollutants.

The EPA has identified asbestos, formaldehyde, radon, and cigarette smoke as the four most serious sources of indoor air pollution. Steps toward the regulation of asbestos have been taken, but they have created numerous technical and legal problems for the EPA, causing many to complain that the organization is overworked and underfunded. The 1976 Toxic Substances Control Act gives the EPA broad authority to control the production, distribution, and disposal of potentially hazardous chemicals, with bans on products that contain formaldehyde, a preservative for biological specimens, receiving past consideration.

Other legislation, such as the 1976 Consumer Product Safety Act, has granted federal and state authority over consumer products that are potentially dangerous to public health and the environment, with many products that generate indoor air pollution arguably falling under that jurisdiction. For example, emission standards for stoves could regulate the output of carbon monoxide, whereas standards for plywood and textiles could set standards for formaldehyde emissions. Indoor air pollution is a much greater problem in developing countries where environmental standards for factories are considerably less restrictive than in the United States. In such countries wood, dung, and crop residues are primary sources of cooking and heating fuels. Both of these issues are of concern to the United Nations.

The 1969 National Environmental Policy Act was the first legislation to introduce environmental impact statements, which involved a formal process designed to predict how a development project or proposed legislation would affect natural resources in a given area. The sequence of steps in these important statements includes screening to decide if a project requires assessment and, if so, what level of detail is required; preliminary assessment to identify the magnitude, significance, and importance of the project's impact; scoping to ensure that the statement focuses on key issues and to determine where more detailed information is needed; and implementation, which involves detailed investigation to predict impacts and assess consequences. Environmental audits—the voluntary regulation of a business or organization's practices with respect to environmental health—have also proven beneficial and are becoming more commonplace.

Children in Eastern Europe wear filter masks to reduce the harmful effects of sodium dioxide emitted by nearby factories. The United Nations is working to establish international guidelines for the reduction of such air pollution. (Reuters/David Brauchli/Archive Photos)

Air Pollution Policy and Air Quality

Air pollution remains a critical environmental risk despite the initiation of several improvements since the 1970 amendments. An important national health objective for the year 2000 was to reduce exposure to air pollutants to a level where 85 percent of Americans lived in environments that met EPA standards. Between 1970 and 1991, the Clean Air Act and its supporting legislation enabled a 61 percent drop in particulate emissions, a 50 percent drop in carbon monoxide emissions, a 27 percent drop in sulfur emission, and a 38 percent drop in hydrocarbons. These substantial improvements all occurred at a time when the population grew by 28 percent, and the gross national product nearly doubled in size. However, the EPA also announced in 1991 that more than 50 percent of Americans were breathing air that violated the ozone standard and that nearly one hundred metropolitan areas failed to meet federal air-quality regulations, even though many of the deadlines for meeting those standards had been repeatedly extended. The American Lung Association conservatively estimates that breathing polluted air costs Americans more than $50 billion annually in health care and related costs.

In July, 1997, the EPA issued updated air-quality standards following the most complete scientific review process in the history of the organization. Following review by hundreds of internationally recognized scientists, industry experts, and public health officials, major steps were taken toward the improvement of environmental and public health by revising ozone standards for the first time in twenty years. In addition, the first standards for particulate matter were introduced. The EPA's 1997 study concluded that many standards did not protect the environment enough and placed too many

Americans at risk. Data indicate that repeated exposure to pollutants at levels previously considered to be acceptable can cause permanent lung damage to children and those who regularly exercise and work outdoors in many urban environments. The American Cancer Society found that the risk of early death is more than 15 percent higher in areas where levels of fine particulates are the highest, whereas another study revealed that expected life span can be shortened by up to two years in heavily polluted cities.

Lung disease has become the third leading cause of death in the United States, with deaths from asthma attacks among children and young people more than doubling between the years of 1980 and 1993. Laws were passed that directed the EPA to review the public health standards for the six major air pollutants at least every five years to ensure that they reflect the most recent scientific data, in addition to laying out specific procedures to obtain these results and appropriately revise the standards. The amendments to the Clean Air Act require that pollution limits be based solely upon health, risk, exposure, and damage to the environment as determined by the best available scientific data.

Congress deliberated the issue of cost versus the technical feasibility of meeting current clean air standards in 1970, 1977, and 1990, with both the legislative and executive branches firmly siding each time with putting public health and the environment ahead of industry. The EPA has assembled an implementation package for updates in air-quality standards to give to states, local governments, and businesses to assist them in meeting rigorous standards in a timely and economical fashion. In 1997 the President's Council of Economic Advisors estimated that the annual cost of reducing ozone alone would range from $11 billion to $60 billion.

On the international level, a 1987 U.N. convention in Canada saw twenty-four nations agree to guidelines established to protect the ozone layer via the Montreal Protocol. The Montreal Protocol was renegotiated in 1990, with important adjustments being the phase-out of certain chlorocarbons and fluorocarbons by the year 2000 and the provision of aid to developing countries to assist in making this transition. In 1988, as part of the U.N.-sponsored Convention on Long-Range Transboundary Air Pollution, the United States and twenty-four other nations ratified a protocol that froze the rate of nitrogen oxide emissions at 1987 levels.

Daniel G. Graetzer

SUGGESTED READINGS: Christopher J. Bailey's *Congress and Air Pollution: Environmental Policies in the USA* (1998) gives a thorough review of the environmental politics behind the legislative battles and laws regarding air pollution. John O'Neill's *Ecology, Policy, and Politics: Human Well-Being and the Natural World* (1993), M. S. Greve's *The Demise of Environmentalism in American Law* (1996), and Nigel Haigh and Frances Irwin's *Integrated Pollution Control in Europe and North America* (1990) describe environmental degradation and actions taken by individuals and societies active in nature preservation, environmentalism, and ecology. Robert Jennings Heinsohn and Robert Lynn Kabel, *Sources and Control of Air Pollution* (1999), and Kenneth Wark, Cecil F. Warner, and Wayne T. Davis, *Air Pollution: Its Origin and Control* (1998), provide excellent reviews of urban air pollution and the effects of the numerous amendments to the United States Clean Air Act. R. Shep Melnick's *Regulation and the Courts: The Case of the Clean Air Act* (1983) and William H. Rodgers's *Environmental Law* (1986) give detailed accounts of the numerous precedent-setting arguments that immediately followed the first amendments to the Clean Air Act.

SEE ALSO: Acid deposition and acid rain; Air pollution; Automobile emissions; Clean Air Act and amendments; Montreal Protocol; Ozone layer and ozone depletion.

Alar

CATEGORY: Agriculture and food

The use of the growth regulator alar to improve the quality and appearance of apples became controversial in the late 1980's, when a debate arose over the chemical's carcinogenic properties.

Alar, also known as daminozide, is a growth regulator manufactured by Uniroyal Chemical Company. In the late 1960's, farmers began using the product to improve the quality and appearance of apples. The use of alar by apple growers to preserve the crispness of their fruit, especially Delicious, Staymen and McIntosh apples, as the fruit was sent to market was a common practice for more than twenty years. In 1989, a controversy over the potential harmful effects of the chemical erupted, and alar was accused—erroneously, according to many experts—of being the most potent carcinogen in the food supply.

The first questions about the chemical were addressed by Dr. Bela Toth of the University of Nebraska, whose research claimed that alar created tumors in mice but not in rats. Toth's discovery did not stress the fact that the rodents were fed massive amounts of the chemical, far exceeding the maximum tolerated dose used in cancer testing.

At the time, the Environmental Protection Agency (EPA) disregarded the study. In 1983, however, the same organization, under attack by environmental and media groups critical of the environmental policies of the Ronald Reagan Administration, began its questioning of alar. Steve Schatzow, a lawyer, was appointed to lead the Office of Pesticide Programs (OPP); in conjunction with other organizations, including the Natural Resources Defense Council (NRDC) and the American Council on Science and Health (ACSH), the OPP began the fight against alar by proclaimimg it to be the most potent cancer-causing substance in the food industry.

The effect of OPP's announcement was dramatic. Consumers poured apple juice down drains, stores pulled apple products from their shelves, and farmers suffered losses estimated in the hundreds of millions of dollars. The anti-alar campaign became more aggressive when such celebrities as *Sixty Minutes* newsman Ed Bradley, activist Ralph Nader, and actress Meryl Streep—who set up a group called "Mothers and Others for Pesticide Limits"—expressed fears about the chemical.

Despite the anti-alar fanfare, subsequent studies on the consumption of traces of the chemical, whether in apple juice or any other form of apples, proved to be negative. After 1990, no mainstream, peer-reviewed research demonstrated any linkage between the chemical or its breakdown product, UMDH, and cancer. For example, Dr. Jose R. P. Cabral, an investigator with the International Agency for Research on Cancer (IARC), declared that alar was safe to use and that his group's experiments had not found tumors in rodents that had consumed reasonable quantities of the chemical. In the wake of such findings, the U.S. Food and Drug Administration issued a statement affirming that eating apples that had been treated with alar posed no health threat.

Soraya Ghayourmanesh

SEE ALSO: Agricultural chemicals; pesticides and herbicides.

Alaska Highway

DATE: completed 1943
CATEGORY: Land and land use

Originally a military highway built during World War II, the Alaska Highway has become an important transportation and tourist corridor through northwest Canada and Alaska. The presence of the highway has caused environmental change in the regions through which it passes in conjunction with land-use development associated with transportation and tourism.

In the early 1940's, during World War II, fear that Japan would invade Alaska prompted the construction of the Alaska Highway for military use. Begun in 1942 and completed in 1943, the Alaska Highway was an all-weather gravel road that connected Dawson Creek, British Columbia, Canada, with Fairbanks, Alaska. The highway was initially a military road only and, as such, was designed for military transport; little attention was given to environmental considerations in deference to the war effort. The road had to be constructed for ease and speed of movement of military goods, and engineering efforts were aimed at meeting these needs. Grades, road cuts, and stream crossings were made quickly without much consideration of the environmental impact of constructing a 2,400-kilometer-long (1,500-mile-long) highway through previously untraversed wilderness.

For a few years following World War II, civilian travel on the highway was limited and carefully controlled. With the removal of travel restrictions in 1947, however, traffic increased, and environmental degradation inevitably followed. While early travelers had to attend to most of their needs themselves, the flow of traffic proved an incentive for the provision of services. Service stations, restaurants, and lodging facilities all grew to meet the needs of the traveling public on the Alaska Highway. The presence of a well-maintained highway and available public facilities fostered additional growth. The new facilities themselves created, in turn, a need for waste removal, storage areas, and underground fuel tanks. Passing, as it does, through wilderness areas, historically important settings such as the Klondike, and active mining areas, interest in and use of the highway has grown accordingly.

Regular improvements have been made to the highway. It has been upgraded from a gravel to a paved surface, curves have been straightened, and grades have been improved. These changes themselves have imposed an additional imprint on the environment adjacent to the Alaska Highway. The successive efforts at widening and straightening the route have increased the direct physical impact on the surrounding landscape.

Jerry E. Green

SEE ALSO: Conservation; Road systems and freeways.

Alaska National Interest Lands Conservation Act

DATE: 1980
CATEGORY: Preservation and wilderness issues

The Alaska National Interest Lands Conservation Act of 1980 was intended to resolve conflicts regarding land ownership in Alaska between Native Alaskans and the federal and state governments.

The 1959 Alaska Statehood Act allowed the state government to choose 102.5 million acres of federal land from that part of the public domain not reserved for parks or other designated use. The land selection process proceeded slowly, complicated by such issues as Native Alaskan claims, which were not clarified in the statehood bill. The discovery of oil fields in northern Alaska prompted passage of the 1971 Alaska Native Claims Settlement Act (ANCSA). The bill extinguished native land claims and allowed for construction of the Trans-Alaskan Pipeline. In order to secure support for the ANCSA from conservation groups, Congress included a provision in the bill that allowed the secretary of the interior to withdraw up to 80 million acres from the public domain for the establishment of conservation units, including national parks and wilderness areas. Congress would make the final determination regarding the final disposition of withdrawn lands. In order to provide a forum for cooperative planning on the issue, Congress created the Joint Federal-State Land Use Planning Commission.

An intense political battle ensued. Federal agencies, including the Forest Service and the National Park Service, competed for the authority to manage withdrawn lands. Environmentalists favored agencies that would limit development, while Alaska state officials threw their support behind agencies that might prove willing to allow multiple use and the exploitation of natural resources for economic gain. After years of contentious debate, in 1979 the House of Representatives passed a bill establishing 127 million acres of conservation units, 65 million of which were designated wilderness areas. The Senate version of the bill significantly reduced the acreage, and the legislation did not become law. The following year, the election of Ronald Reagan to the presidency and a Republican majority to the Senate convinced House members that they had no alternative but to accept the Senate version of the bill, which passed in 1980 and became known as the Alaska National Interest Lands Conservation Act (ANILCA). The bill set aside 104 million acres of land as conservation units, with 57 million acres of wilderness. The National Park Service received 44 million acres, the Fish and Wildlife Service 50 million acres, and the Forest Service 2.5 million acres. The Bureau of Land Management, the federal agency most amenable to development, received a patchwork of marginal lands that no other agency desired.

ANILCA more than tripled the amount of land protected as wilderness in the United States, and its provisions placed 75 percent of national park land in Alaska. However, it did not solve the conflicts over Alaska land use. The state contested some land withdrawals. The act allowed for the construction of pipelines and roadways through conservation units. Moreover, ANILCA contained no measures that protected the habitats of migratory animals. Finally, the bill included provisions that allowed for the future exploitation of mineral resources, including oil, should both the Congress and the president deem it necessary.

Thomas Clarkin

SEE ALSO: Bureau of Land Management, U.S.; Forest Service, U.S.; National parks; Trans-Alaskan Pipeline; Wilderness areas.

Alternative energy sources

CATEGORY: Energy

The world's dependence on nonrenewable fossil fuels—such as coal, oil, and natural gas—for energy has led to environmental problems that will continue to compound themselves until clean, renewable energy resources are put into use. Such alternative sources include solar, tidal, geothermal, and wind energy.

The ultimate source of energy on earth is radiant energy from the sun, or solar energy. On average, the earth receives approximately 2 calories of energy from the sun per square centimeter per minute at the outer atmosphere. One-third of this energy is reflected away from the earth by the atmosphere, clouds, and light-colored surfaces on earth. The remaining two-thirds are converted into different forms of energy by the natural processes of chemistry, heat, and motion. If nonpolluting, sustainable alternative sources of energy are to be developed, they must be linked, either directly or indirectly, to solar energy.

Although solar energy appears to be derived from nuclear fusion, nuclear energy from chemical elements on earth is derived from nuclear fission. When the atoms of heavy radioactive elements such as uranium or plutonium are split (fission), or when two light atomic nuclei from elements such as helium or hydrogen are fused together (fusion), heat energy is produced. The heat energy generated by fission or fusion can also be used to generate electrical power. Although there is a tremendous amount of energy in a small amount of matter, nuclear energy is a highly inefficient conversion process under existing technology because a tremendous amount of waste heat is passed to the environment.

ENERGY CONVERSION

An outstanding property of energy is its ability to be converted from one form to another. Some transformations occur as a result of natural processes, while others are caused by human intervention. The development of fossil fuels is an example of natural transformations. Radiant energy from the sun is converted to chemical energy by plant photosynthesis. Some of it is then passed on to animals in the food chain and stored in favorable environments where fossil fuels develop from plant and animal remains. This transformation takes millions of years to complete, so such sources are considered nonrenewable. When chemical energy is burned, it is converted to heat energy that may be used to power machines, heat space, and produce electricity. Gravitational energy, also called potential energy, is converted to kinetic energy when matter is put into motion. Examples of kinetic energy include the power harnessed from winds, streams, and tides. These highly efficient energy transformations result in only low levels of pollution.

When energy is converted from one form to another, the net gain or loss of energy is negli-

Windmills along Interstate 10 near Palm Springs, California. Wind is a clean, renewable source of energy. (AP/Wide World Photos)

gible. However, some of the energy is degraded or converted into forms that cannot be used efficiently. In order for energy to perform useful work, it must be concentrated at levels adequate to perform a desired task. Some diffused heat energy is always released to the environment during energy conversion processes.

Energy is unevenly distributed over the earth in many different forms. More solar energy is received in equatorial regions than at higher latitudes. It is redistributed, however, by hydrological and atmospheric systems. Kinetic energy from rivers and streams is confined to locations having adequate precipitation and the proper geographic relief. Energy from tides is available only in areas near lakes and seashores. On the other hand, energy from wind is potentially available everywhere, depending upon prevailing winds and local pressure gradients. The highly desirable fossil fuels produced by chemical processes are unevenly distributed over the earth because geologic processes make some environments more conducive to their development. Environmental degradation is associated with extracting and transporting energy to areas of consumption.

Energy begets energy because one or more usable energy sources must be employed to extract another source from the natural environment, transport it to desirable locations, and convert it to the desired form. Some energy sources are more accessible than others because their extraction from the natural environment requires less usable energy. For example, the energy required to drill the average inland oil well is less than that needed to drill a deep offshore oil well. An understanding of this concept is necessary for an objective assessment of the impact on the environment. Environmental degradation increases in response to the amount of energy necessary to extract energy from the environment.

More than enough energy exists on earth to supply human needs, but converting the available supply to usable forms and transporting it to places of need is an expensive process. By conserving present supplies and developing technology that will allow the use of alternative energy resources, humans can develop cleaner sources of energy and become self-sufficient energy consumers.

ENERGY CONSUMPTION TRENDS

The world's dependence on fossil fuels has increased steadily since 1950, from 1,715 million tons of oil equivalent in 1950 to 7,856 million tons of oil equivalent in 1995. The increasing use of oil to fuel motor vehicles is largely responsible for the growth in the dependence on fossil fuels. Coal use is increasing at a less dramatic rate. Worldwide use of natural gas is experiencing the fastest rate of increase in consumption of the fossil fuels. Environmental concerns and new technology are allowing natural gas to replace coal and oil in many applications, reducing the growth of gaseous pollutants in the atmosphere.

Though world nuclear generating capacity had by 1999 reached an all-time high, its growth rate had declined dramatically since the 1980's. Nuclear power is not likely to contribute significantly to energy needs in the foreseeable future, although several new plants are being constructed in Asian countries and France. Public opposition to waste disposal has reduced the acceptance of nuclear energy in Asian countries and should cause further declines in the future.

Although geothermal energy (heat energy from the earth's interior) makes up a very small percentage of total energy consumption, its electrical generating capacity is on the increase. Although the United States is the leading user, electricity from geothermal energy plays a much more important part in the energy mix of island countries such as Iceland, the Philippines, and Indonesia.

The use of wind power is rapidly growing. Like geothermal energy, it is used primarily to produce electricity and makes up a very small percentage of the world's energy consumption. Much of the growth in wind energy is occurring in Europe and Asia.

Worldwide Alternative Energy Consumption Trends, 1950-1995

YEAR	POWER CONSUMPTION (MEGAWATTS)			
	NUCLEAR	GEOTHERMAL	WIND	PHOTOVOLTAIC
1950	—	200	—	—
1955	—	262	—	—
1960	1,000	374	—	—
1965	5,000	556	—	—
1970	16,000	711	—	—
1975	71,000	1,287	—	1.8
1980	135,000	2,471	10	6.5
1985	250,000	4,414	1,020	22.8
1990	328,000	5,832	1,930	46.5
1995	344,000	6,798	4,821	78.6

Source: Data adapted from Lester R. Brown et al., *Vital Signs, 1997: The Trends That Are Shaping Our Future.* New York: W. W. Norton, 1997.

The use of photovoltaic (PV) cells is dramatically increasing. This technology is used to convert sunlight into electricity. PV cells are used to provide a direct source of electricity for homes, to power communications equipment, to run water pumps, and to generate electricity linked to electrical grid systems. The capacity to produce less costly PV cells is increasing, making them more affordable.

Alternative Sources of Energy

The generation of electricity from nuclear energy is a controversial alternative to the use of fossil fuels. Proponents of nuclear energy argue that it is the safest, cheapest, and cleanest of available alternatives in meeting projected energy demands. Opponents question the desirability of nuclear power because of the likelihood of severe health and environmental hazards caused by the production of nuclear energy and its wastes. In the wake of the 1986 Chernobyl accident in Ukraine and the Monju accident in Japan during start-up testing in 1995, this alternative has fallen out of favor with the public.

The development of geothermal power holds more promise for the future, especially in the regions where heat energy stored in the earth's crust can be extracted. Technical problems that may hinder the development of geothermal energy are hydrogen sulfide emissions from geothermal power plants and the large initial investment required. Simpler and cleaner methods to exploit geothermal energy for space heating and cooling, as well as generating electricity, are in the early stages of development. These methods involve systems of fluids that extract and store energy below the earth's surface.

The future development of hydroelectric power is limited in developed countries because many of the most efficient sites are already being used, and many other rivers are being protected from development by legislation. Though there is more potential for developing hydropower, it floods large amounts of arable land and destroys wildlife habitats.

Like hydroelectric power, the energy from winds and tides appears to be endless. Electricity from tidal power is confined to coastal locations and will have only local significance. The energy output from tides depends on major differences between high and low tides and can therefore supply only a fraction of the potential energy output of rivers. Wind power can be harnessed almost anywhere on earth and has great potential, despite its unpredictable strength in generating electricity. Energy from windmills and tides is clean, but the systems are usually noisy. Coastal beaches and ecosystems are also disturbed.

Energy from the sun has the greatest advantages with the least effect on the environment. Although solar energy is being used to heat and cool homes through the use of rooftop solar collectors, large-scale use poses technical difficulties because the conversion of solar energy into electricity is not yet an efficient process. Through photosynthetic processes, energy from the sun is stored as chemical energy. This is referred to as biomass. Biomass conversion includes energy from organic materials such as animal waste, municipal waste, and agricultural and wood products. Widespread mass production of energy from these sources is possible and appears promising because of the renewable, nonpolluting nature of discarded waste products. Natural gas (methane) from landfills is an example of biomass conversion. Another alternative energy source is the fuel cell, a device that efficiently converts a fuel such as hydrogen into electricity. These battery-like devices produce very little air pollution and are quiet.

Of the many available alternative energy sources, conservation is the cleanest and most practical approach. Even with an aggressive conservation program, no single alternative can provide future energy needs as less developed countries increase their demand for energy. However, a judicious mix of conservation and alternative energy sources may ultimately decrease environmental pollution along with dependence on the earth's limited fossil fuels. The complete production system, from material inputs to final disposal of waste products, of alternative energy sources must be assessed when evaluating their impact on the environment.

Jasper L. Harris

SUGGESTED READINGS: An easy-to-read review of energy development trends is provided in *Vital Signs 1997* (1997), by Lester Brown, Michael Renner, and Christopher Flavin of the Worldwatch Institute. Global data tables on energy consumption, energy production, carbon dioxide emissions from fossil fuels, and atmospheric concentrations of greenhouse and ozone-depleting gases are presented in *World Resources 1998-1999: A Guide to Global Environment* (1998), a joint publication by the World Resources Institute, the United Nations Environment Programme, the United Nations Development Programme, and the World Bank. A fundamental discussion of fossil fuels and alternative energy sources is presented in *Earth, Resources, and the Environment*, by Douglas G. Brookins (1981). A readable discussion of alternative energy sources and their influence on the environment is presented by G. Tyler Miller in *Energy and Environment: The Four Energy Crises* (1975). Present and future energy problems associated with fossil fuels and alternative energy sources are addressed in a readable form in *Environment: 98/99* (1998), by John L. Allen.

SEE ALSO: Geothermal energy; Hydroelectricity; Nuclear power; Solar energy; Tidal energy; Wind energy.

Alternative fuels

CATEGORY: Energy

Fuel or energy usage refers to the transformation of energy from one form to another form that is more useful to humankind. The direct burning of fossil fuels such as gas, oil, or coal releases stored energy, which can be used for such purposes as heating homes and propelling vehicles. However, the consumption of fossil fuels leads to air and thermal pollution. Therefore, researchers have begun to develop alternative fuels that are likely to produce minimal environmental pollution.

Air pollution can result from the burning of fossil fuels in such sources as industrial furnaces and electric generating plants. The engines of automobiles are one of the largest sources of pollution because the burning takes place too quickly for complete combustion to take place, resulting in the production of large amounts of noxious gases. Even when combustion is complete, the carbon dioxide that is thrown into the atmosphere absorbs some of the natural infrared radiation emitted by the earth that would otherwise escape. Many researchers have predicted that the buildup of atmospheric carbon dioxide gas and the consequent heating of the atmosphere, referred to as the greenhouse effect, will raise the average temperature of the atmosphere by several degrees by the end of the twenty-first century, causing shifts in rainfall patterns, melting of polar icecaps, and widespread desertification. An alternative scenario predicts that a layer of carbon dioxide will block some of the sun's rays and cause a new ice age. In either case, experts agree on the importance of limiting the burning of fossil fuels.

Another type of environmental pollution is thermal pollution. Every heat engine, from automobiles to power plants, exhausts a certain amount of heat into the environment. Most electricity-producing power plants make use of a heat engine to transform thermal energy into electricity, and the exhaust heat is generally absorbed by a coolant such as water. If the engine is run efficiently (30 to 40 percent at best), the temperature at the exhaust end must be kept as low as possible. One way to keep temperatures low in power plants is to use large amounts of water as a coolant. The water is usually obtained from a nearby river, lake, or ocean, and it is returned to the source after it runs through the power plant. However, the transfer of heat to water raises the temperature of the water. A rise in temperature of only a few degrees can cause significant damage to aquatic life when the water is returned to its source, in large part because the warmed water holds less dissolved oxygen. The lack of oxygen can adversely affect fish and other organisms and at the same time may encourage excessive growth of other organisms, such as algae, thus disrupting the ecology of an area.

Most of the electricity produced in power plants makes use of a heat engine coupled with an electric generator. Electric generators are devices that transform mechanical energy into electric energy. At a steam plant, for example, coal, oil, or natural gas is burned to boil water and produce high-pressure steam, which turns a turbine. The advantages of such plants are that the technology needed to build them is well known, and they are not expensive to run. The disadvantages are that the products of combustion create air pollution, their efficiency is limited, and the waste heat produces thermal pollution. Alternative fuels are sought to avoid or minimize those disadvantages.

Advantages of Alternative Fuels

About 90 percent of all energy in the United States is produced with fossil fuels, which also account for about 80 percent of the world's annual energy use. The remaining 20 percent comes from alternative fuels, which include a contribution of 6 percent from nuclear power and 14 percent from other sources. According to United States federal law, alternative fuels include alcohol fuels such as methanol (methyl alcohol), denatured ethanol (ethyl alcohol) and other alcohols, either in pure form (called "neat" alcohols) or mixed with unleaded gasoline (85 percent or more of the mixture must consist of alcohol); compressed natural gas (CNG) and liquefied natural gas (LNG), mainly methane; liquefied petroleum gas (LPG, one of the many component molecules found in gasoline); hydrogen; coal-derived liquid fuels; fuels other than alcohols derived from biological materials, such as soy bean, rapeseed, or other vegetable-oil-based fuels; and electricity.

Before the introduction of gasoline as a motor fuel in the late nineteenth century, vehicles were often powered by what are now considered alternative fuels. For example, illuminating or coal gas (a form of methane or natural gas) was used in early prototype internal combustion vehicles in the 1860's. Electricity, stored in lead acid batteries, was a popular energy source for vehicles from as early as the 1830's until the 1920's. During the 1880's, Henry Ford built one of his first automobiles and fueled it on ethanol, which was often called "farm alcohol" because it was made from corn.

After the oil crisis of the 1970's, many of these "alternatives" to gasoline began returning to the transportation fuel market. Alternative fuels are needed for two main reasons: energy security and air quality. Energy security comes from the fact that the use of alternative fuels reduces reliance on oil imports from nations in the Middle East. Air-quality benefits are derived from the fact alternative fuels are inherently cleaner than gasoline because they are chemically less complex. The longer and the more chemically complex a molecule is, the less likely it is to be completely burned. This incomplete combustion of the molecule in an internal combustion engine causes the release of carbon monoxide, nitrogen oxide, and other molecules in the exhaust. Alternative fuels, because of their simpler chemical makeup, release fewer emissions from incomplete combustion.

Use of compressed natural gas shows carbon monoxide levels of 65 to 76 percent less than regular gasoline. Ozone reactivity for CNG is 89 to 96 percent cleaner than gasoline. Methanol vehicles have 37 percent less carbon monoxide emissions than gasoline and up to 56 percent less ozone reactivity. Methanol is an excellent automotive fuel and has long been used for racing cars. Propane-powered vehicles have 43 to 46 percent less carbon monoxide emissions and 57 to 61 percent better ozone reactivity.

"Gasohol" is a term used for the mixture of 10 percent ethyl alcohol with gasoline. It raises the octane rating of fuel and significantly reduces the carbon monoxide released from the tailpipes of automobiles. However, it raises the vapor pressure of gasoline, increasing the release of evaporative volatile hydrocarbons from the fuel system and oxides of nitrogen from the exhaust. These substances are components of urban smog. Methane in the atmosphere traps the heat from the earth and thus may contribute to global warming.

Electric vehicles, which have no internal combustion engine, offer an even better alternative because they do not burn fuel. Their only sources of pollution are the power plants that create the electricity, which can be regulated more closely than cars for their sources of pollution.

Another advantage of alternative fuels is that they evaporate less readily than gasoline. Evaporation from a car's fuel tank contributes to smog. However, petroleum-based fuels (gasoline and diesel) offer excellent energy content by volume. By comparison, alternative fuels have less energy density. Hence, alternative fuels may power a vehicle for fewer miles on a gallon-to-gallon basis than gasoline.

NUCLEAR AND GEOTHERMAL ENERGY

Three nuclear processes may provide nuclear power as an alternative fuel: fission, fusion, and radioactive materials. The fission process is used in all nuclear power plants, since fusion has not been economically favorable. Radioactive materials naturally release energy, which is not suitable for large-scale fuel production. Nuclear fuel is used to heat steam in a manner similar to fossil fuels. A nuclear power plant is thus essentially a steam engine using uranium as its fuel. It suffers from the low efficiency characteristic of all heat engines and the accompanying thermal pollution. However, nuclear power plants in normal operation produce practically no air pollution. Nuclear processes release no carbon dioxide gas and thus do not contribute to the greenhouse effect.

Both fossil-fuel and nuclear plants heat water to steam for the steam turbine. Natural steam, as geothermal energy, can be obtained from the earth itself. In many places, water beneath the ground is in contact with the hot interior of the earth and is raised to high temperature and pressure. Energy can be tapped from steam that has risen to the earth's surface in the form of hot springs, geysers, or steam vents, or it can be tapped from subterranean sources by drilling down to trapped steam beds. In some areas it is possible to drill two parallel wells down to hot, dry rock in contact with the earth's interior and pass cold, pressurized water down one well. The heated water or steam will return up the other well.

City buses powered by natural gas, such as this one, emit less carbon monoxide than those that use gasoline for fuel. (Ben Klaffke)

The largest geothermal plant in the world is located in Northern California. A similar plant has been successfully operating in Italy since the beginning of the twentieth century. A number of others are functioning in various parts of the world. Geothermal energy produces little air pollution, although there is some gas emission. In addition, the spent hot water causes thermal pollution, and the often high mineral content of the water may pollute the environment and corrode the power plant apparatus. Nonetheless, geothermal energy is a reasonable, inexpensive means of fuel production and holds considerable promise, although naturally produced steam can, presumably, run out, leaving a site dry.

Ocean thermal energy conversion (OTEC) is a process that can be applied primarily in tropical seas. In the tropics, the difference in temperature between water at the surface of the ocean and water 1 kilometer deep is about 20 degrees Celsius. This temperature difference could be used to drive a heat engine. The working fluid would have to be a substance with a lower boiling point than water that could drive the turbines. The capital costs are high for such plants because the structure is made in the sea. The efficiency would be low because of the small temperature difference, at best about 7 percent. The ocean environment may cause problems such as corrosion and fouling by biological organisms caught in the intake water. However, there would be little problem with air pollution or radioactive disposal, so the advantages might outweigh the disadvantages.

WATER, WIND, AND SUN POWER

Hydroelectric power plants were in use long before the advent of fossil fuels. Hydroelectric plants use falling water instead of steam to turn the turbines. They produce practically no air or water pollution, and they are nearly 100 percent efficient, since little waste heat is produced. However, they are not always reliable because of drought. Another form of alternative fuel obtainable from water power is tidal energy. A basin behind a dam is filled at high tide, and the water is released at low tide to drive turbines. At the next high tide, the reservoir is filled again, and the inrushing water also turns turbines. Good sites for tidal power are not plentiful and would require large dams across natural or artificial bays.

Windmills were once widely used to harness wind energy as fuel. Their comeback, on a much grander scale, began late in the twentieth century. Windmills of various sizes are being produced, from small, 3 kilowatt models (for remote houses, for example) to large models that produce several megawatts each. Windmills are generally clean, although a large array might be considered an eyesore and may cause noise pollution.

The direct use of solar energy is also possible. For example, a solar heating system may be made by placing water-carrying tubes in contact with a large, black surface that absorbs the sun's radiant energy and heats the water. The surface is covered with a piece of glass to prevent loss of heat. The heated water is circulated to a large, well-insulated reservoir, where it is stored and recirculated to heaters. The reservoir can also serve as the source for the hot-water supply. Natural convection or pumps can be used to move the water in the two parts of the system. However, some form of backup system is needed when there are prolonged periods of heavy clouds. Although such systems would cause thermal pollution and might affect the local climate, there would be virtually no air or water pollution.

Another direct user of sunlight is the photovoltaic (PV) cell, which converts sunlight directly into electricity without the need for a heat engine. Thermal and other pollution are very low, since no heat engine is involved. Chemical pollution produced in their manufacture in large numbers could, however, be serious. Using PV cells on a large scale would also require a large land area, since the sun's energy is not very concentrated. Another possibility would be to place solar cells and concentrators in orbit to direct solar energy to the earth as a relatively environmentally safe fuel.

M. A. K. Lodhi

SUGGESTED READINGS: A study of the complex problem of fuel use and environment is provided in John M. Fowler's *Energy and Environment*

(1975). For an understanding of energy problems, see Earl Cook's *Man, Energy, Society* (1976). For a scientific approach to the problem of alternative fuels and environmental issues, see D. C. Giancoli's *Physics: Principles with Applications* (1995). Worldwide environmental damage has been estimated by F. Barfir, T. N. Vaziroglu, and H. J. Plass, Jr., in their article in *International Journal of Hydrogen Energy* 15 (1990). "Hydrogen City," *International Journal of Hydrogen Energy* 12 (1990), by M. A. K. Lodhi, discusses alternative fuels and environmental issues and lists environmentally safe sources and devices.

SEE ALSO: Alternatively fueled vehicles; Synthetic fuels.

Alternative grains

CATEGORY: Agriculture and food

Agricultural researchers have made efforts to find alternatives to high-yield grain crops, the harvest of which has led to severe erosion and increased use of fertilizers and pesticides.

More than one-half of the calories consumed daily by the world's human population comes from grains. Most of these grains are produced by plants of the grass family, Poaceae. Major cereal plants domesticated many centuries ago include rice (*Oryza sativa*), wheat (*Triticum aestivum*), and maize or corn (*Zea mays*). Other important grain crops, also plants of the grass family, include barley (originating in Asia), millets and sorghum (originating in Africa), and oats and rye (originating in Europe).

Since the early twentieth century, the scientific principles of genetics have been applied to improvements of crop plants. The most notable improvements occurred between 1940 and 1970. As a result of irrigation, improved genetic varieties, and the use of large amounts of fertilizers and pesticides, yields of major crops greatly increased. Norman Borlaug received a Nobel Prize in 1970 for his contributions to these developments, which came to be called the Green Revolution. However, it soon became apparent that the Green Revolution was not the boon first envisioned. For maximum yield, large-scale farming involving huge investments of capital is required. Also, environmentalists became concerned over the resulting erosion and the environmental damages caused by the use of large amounts of fertilizer and pesticides.

Various alternatives have been proposed. In the case of grain crops, several approaches offer promise, including more widespread use of minor cereals, especially those tolerant of unfavorable growing conditions; development of new cereal plants by hybridization or other genetic manipulations; and utilization of pseudocereals, nongrass crop plants that produce fruits (grains) similar to those of cereal plants.

Most sorghum (*Sorghum bicolor*) grown in the United States is used for silage (milo) or molasses (sweet sorghum). In Africa and India, various grain sorghums are grown in regions where rainfall is too low for most other grain crops. Well adapted to hot, dry climates, their grains are used to make a pancakelike bread. "Millet" refers to several grasses that are useful cereal plants because they also tolerate drought well. In Africa the most important are pearl millet (*Pennisetum glaucum*) and finger millet (*Eleusine coracana*). Grains of both species can be stored for long periods and are used to make bread and other foods. Other, perhaps less important, grain plants also called millet include foxtail millet (*Setaria italica*), native to India but now grown in China; proso millet (*Pamicum milaeceum*), native to China but grown in Russia and central Asia; sanwa millet (*Echinochloa frumentacea*), cultivated in East Asia; and teff (*Eragrostis teff*), an important food and forage plant of Ethiopia. Such grain sorghums and millets have the potential to grow in areas with hot, dry climates far beyond the regions where they are now being utilized.

In a distinct category is wild rice (*Zizania aquatica*). Native to the Great Lakes region of the United States and Canada, it has been, and still is, harvested by American Indians. Like the common (but unrelated) rice, it grows in flooded fields. Attempts to cultivate wild rice since the 1950's have been somewhat successful as the result of the development of nonshattering varieties.

However, it remains an expensive, gourmet item. Two cereal plants have promise because of the high protein content of their grains. Wild oat (*Avena sterilis*) is a disease-resistant plant with large grains. Job's tears (*Coix lachryma-jobi*), native to Asia, is now planted throughout the tropics. Research on these and related species continues.

Although all important cereal plants have been improved, either by genetic engineering or by more conventional genetic techniques, the most notable new alternative grain plant is triticale (*Triticosecale* sp.). The first human-made cereal, it is the result of crossing wheat with rye. The sterile hybrid from such a cross was made fertile by doubling its chromosomes. Thus, triticale varieties produce viable seeds. Triticale combines the superior traits of each its parents: the cold tolerance of rye and the higher yield of wheat. The protein content of triticale compares favorably with that of wheat, and its quality, as measured by lysine content, is higher. However, flour made from triticale is inferior for making bread unless mixed with wheat flour.

Pseudocereals are plants that are not in the grass family but that produce nutritious hard, grainlike fruits that can be stored, processed, and prepared for food much like grains. They belong to several plant families. Many grow under conditions not suitable for the major cereal crops. Buckwheat (*Fagopyrum esculentum*), of the buckwheat family, Polygonaceae, probably originated in China. It tolerates cool conditions and is adapted to short growing seasons, thus permitting it to be grown in the temperate regions of North America and Europe. In the United States, it is often associated with pancakes but is used in larger quantities for livestock feed. In Eastern Europe, the milled grain is used for soups.

Quinoa (*Chenopodium quinoa*) of the goosefoot family, Chenopodiaceae, has been cultivated by Indians of the Andes Mountains for centuries. The leafy annual produces grainlike fruits (actually achenes) with a high protein content and exceptional quality (high in lysine and other essential amino acids). After its bitter saponins have been removed, it can be cooked and eaten like rice or made into a flour. Quinoa has been cultivated in the Rocky Mountains of Colorado since the 1980's and has become a gourmet food in the United States. Most amaranth (*Amaranthus* sp.) plants are New World weeds. They belong to the amaranth family, Amaranthaceae. A few species were used by Aztecs and other native peoples, but their use was banned by the Spanish. Since the late 1970's, plant breeders have targeted several species for improvement. The results are highly nutritious grains rich in lysine that are suitable for making flour. Research in Pennsylvania and California has resulted in improved varieties.

Thomas E. Hemmerly

Suggested readings: Among the books that discuss cereal and related crop plants are *Economic Botany: Plants in Our World* (1995), by Beryl B. Simpson and Molly C. Ogorzaly, and *Plants and Society* (1999), by Estelle Levetin and Karen McMahon. Although older, *Plants for Man* (1972), by Robert W. Schery, is also useful.

See also: Food and Agriculture Organizations; Genetically engineered foods; High-yield wheat; Sustainable agriculture.

Alternatively fueled vehicles

Category: Energy

Researchers have attempted to design and build cars that can burn alternative fuels that would reduce or eliminate harmful pollutants emitted by traditional fuels, such as gasoline.

Traditional vehicle fuels—gasoline and diesel fuel—contribute numerous harmful pollutants to the environment, including reactive organic gases, carbon monoxide, methanol, carbon dioxide, sulfur oxides, and particulate matter. Many scientists believe that these pollutants are at least partly responsible for such problems as ozone depletion and global warming. In response to potential air pollution dangers, the U.S. Congress passed the Clean Air Act in 1990, which mandated measures to decrease the levels of environmentally harmful emissions. Therefore, the development of alternative fuels has

The General Motors EV1, a mass-produced electric car. (General Motors)

become an important focus in both governmental and industrial circles. Among researchers, however, there are opposing views regarding the seriousness of the problems caused by vehicular emissions.

Five types of alternative fuels are considered most likely to replace or supplement gasoline and diesel fuel in vehicles: alcohol, natural gas, liquefied petroleum gas (LPG; also known as propane), vegetable oil, and hydrogen.

The similarities of gasoline and diesel fuel to the methanol and ethanol make these two alcohols attractive alternative fuel options. Already liquids, they are familiar substances that can be easily produced using existing technology. Methanol does not produce soot or smoke when it burns, does not contribute to ground-level ozone, and can refuel a car just like conventional gasoline. Historically, methanol has been less expensive than gasoline. It also has a higher octane rating than gasoline, and a spill is much less damaging than a gasoline leak because methanol is highly water soluble and quickly disperses. Disadvantages include cost of production, safety concerns caused by the invisibility of its flame and its corrosiveness, cold-weather intolerance, and limited range of travel per tankful of fuel. Also, methanol is not as clean burning as electricity or hydrogen. Ethanol is used most commonly as an enhancer, an additive to conventional or reformulated gasoline. The oxygen content of ethanol works through a "blend-leaning" effect by which the engine fuel and emission systems are "fooled" into reacting as if air were being added. This results in about 30 percent fewer carbon emissions than in gasoline-powered engines.

Natural gas is abundant and cost-effective since little processing is needed. Its high octane rating provides emissions advantages. Because natural gas cannot be stored in sufficient amounts for use in vehicles, it is more often compressed or liquefied. However, the liquefied form contains only 60 percent of the energy that the same volume of gasoline affords. Liquefied petroleum gas (LPG, or propane) has the advantage of already being produced and sold within the existing infrastructure: It uses a familiar and well-established pipeline and storage system. It is nontoxic, and the odorant added to propane enables easy detection of excessive concentrations that could cause asphyxiation. However, rapid evaporation can damage skin on contact, and propane is more highly explosive than gasoline.

Biodiesels, which are fuels made partly from vegetable oils such as soybean, sunflower, peanut, or rapeseed oil, have good emission properties and a low sulfur content. They are nontoxic and biodegradable. Biodiesels require little modification to fuel systems and supply networks, but small-scale production has kept the cost high. Also, cold weather intolerance and limited shelf life are drawbacks. Hydrogen can be combusted directly in internal-combustion engines and can be used in fuel cells to produce electricity with high efficiency and with water vapor as the only emission. The problem of storage can be solved by storing it as a compressed gas. However, it has a limited operating range.

Despite the potential of these alternative fuels, certain environmental concerns must be addressed. The emissions from methanol occur in the form of formaldehyde, classified by the U.S. Environmental Protection Agency (EPA) as a likely carcinogen. However, the EPA considers the health risk from ground-level ozone and carbon monoxide to be greater. Ethanol is toxic only in high concentrations and is less so than gasoline, diesel, and methanol. While natural-gas-powered vehicles are effective in reducing emissions of carbon monoxide and ozone, natural gas is made of methane, a greenhouse gas; thus, it may contribute to the global warming problem. Although biodegradable, the greatest obstacle of biodiesel is that it tends to increase nitrogen oxide, which, in the presence of solar radiation, reacts to form ground-level ozone.

From an environmental standpoint, propane and hydrogen are quite promising as alternative fuels. Propane does not cause environmental contamination and compares favorably with other alternative fuels in its levels of greenhouse gases. Hydrogen is almost ideal as a low-emissions fuel, producing little nitrogen oxide in internal-combustion engines and only water and heat in fuel cells.

Victoria Price

SUGGESTED READINGS: Richard L. Bechtold's *Alternative Fuels Guidebook* (1997) provides an easy-to-read discussion of major alternative fuels and their advantages and disadvantages with respect to the environment. *Alternative Motor Fuels* (1996), by Maureen Sheilds Lorenzetti, follows a historical overview with information about clean gasolines and diesels, alternative fuels, and appendices on federal and state legislation impacting alternative fuels. *The Energy Sourcebook* (1991), edited by Ruth Howes and Anthony Fainberg, offers a guide to energy technology, resources, and policy. *The Environment* (1996), edited by A. E. Sadler, offers opposing viewpoints on various environmental issues.

SEE ALSO: Alternative fuels; Automobile emissions; Synthetic fuels.

American alligator

CATEGORY: Animal and endangered species

The American alligator inhabits coastal areas, swamps, ponds, and marshes of the southeastern United States, from North Carolina to Florida and along the Gulf Coast to Texas. Once an endangered species, the alligator made a striking comeback in the last few decades of the twentieth century thanks to conservation efforts.

The American alligator (*Alligator mississippensis*) is a surviving member of the crocodilians—a family of reptiles that roamed the earth along with dinosaurs 230 million years ago. Ranging between 1.8 and 4.6 meters (6-15 feet) in length and weighing up to 454 kilograms (1,000 pounds), it is the largest reptile in North America.

Spanish explorers were the first Europeans to come across the American alligator, which they named *el lagarto* ("the lizard"), an expression Americans later turned into "alligator." Various eighteenth century descriptions of alligators by trappers, explorers, adventurers, and naturalists quickly earned the reptile a place in legend and folklore. Americans and Europeans alike became fascinated with erroneous, but entertaining, portrayals of the reptile as an almost mythological, smoke-breathing dragon.

Though white southerners and Native Americans hunted the alligator for centuries, the reptile faced no serious, widespread threat as a species until the 1870's, when a worldwide demand arose for its soft hides, which were turned into belts, hats, shoes, and handbags. Their value grew in the following decades, and by the 1940's alligator populations were so dangerously reduced that southern states outlawed hunting and trapping them. These actions, however, stimulated illegal poaching during the 1960's and drove the alligator to the brink of extinction.

In 1973 the federal government put the alligator—along with its cousin, the American crocodile (*Crocodylus acutus*)—on the endangered species list and banned the trafficking of its hides. The protection worked so well that alligators are no longer regarded as endangered. Open hunting of alligators is still illegal, though Florida and Louisiana allow permitted hunts to control alligator populations and protect furbearing animals.

New problems, however, confront the American alligator. Land development continues to destroy its natural habitat. However, human activity also creates new artificial living spaces for alligators in canals and drainage ditches. These new environs often put alligators in close proximity to humans. As a result, alligators commonly appear in private swimming pools, docks, farms, driveways, toll booths, and even schools and shopping centers. Their encounters with humans are on the rise. Though alligators feed mainly on fish, snails, crabs, amphibians, and small mammals, they also occasionally consume dogs, cats, and even calves. Attacks on humans—especially small children—do occur, though they are rare.

Pollution also threatens alligators. In 1996 University of Florida researchers reported serious reproductive problems among alligators in Lake Apopka, Florida's third-largest lake. The researchers suspect a 1980 chemical spill and agricultural pesticide runoff into the lake of causing the fertility problems. According to one University of Florida study, Lake Apopka lost 90 percent of its alligator population in a recent twenty-year period. More research is needed to determine whether the Lake Apopka situation is an isolated case or an indicator of a wider problem that could again put alligators in grave danger.

John M. Dunn

SEE ALSO: Endangered species; Endangered Species Act.

Amoco Cadiz oil spill

DATE: March 16, 1978
CATEGORY: Water and water pollution

On March 16, 1978, the tanker Amoco Cadiz *grounded off the coast of Brittany, France, spilling its cargo of crude oil into the sea.*

The four-year-old very large crude carrier (VLCC) *Amoco Cadiz* had been built in Spain. At 331 meters (1,086 feet) long and 68.6 meters (225 feet) wide, and with a draft of 19.8 meters (65 feet), it was one of the largest ships afloat. The vessel was American owned (by Standard Oil of Indiana), Liberian flagged, and crewed by Italians.

The *Amoco Cadiz* was in the final stages of its voyage from the Persian Gulf to Rotterdam in the Netherlands with 223,000 tons of mixed Kuwaiti and Iraqi crude oil. The vessel was northbound along the coast of Brittany, France, at about 10:00 A.M. when a steering failure occurred. The ship was about 24 kilometers (15 miles) off the French coast. Within two hours, the tugboat *Pacific* was alongside the *Amoco Cadiz* connecting a towline. The tug ran out about 914 meters (3,000 feet) of steel towing wire in an attempt to keep the large tanker off the rocks. After only two hours of towing, the towline broke. During this time, shipboard engineers had attempted to fix the damaged rudder, but the system was beyond repair. The ship was 10.5 kilometers (6.5 miles) off the coast of France.

Wreckage of the Amoco Cadiz *oil tanker after it ran aground off the coast of Brittany, France. The oil spill was comparable in magnitude to the* Torrey Canyon *disaster, which had occurred eleven years before near England.* (AP/Wide World Photos)

By 9:00 P.M., the tug had reattached a towline to the stern of the *Amoco Cadiz.* Yet shortly thereafter, the large tanker grounded on the Roches de Portsall. It immediately began leaking its cargo of crude oil over the coast of Brittany. This area accounts for almost 40 percent of the marine life and 7 percent of the oysters harvested in France. Within three days, the slick covered almost 80 kilometers (50 miles) to the north of the ship and 32 kilometers (20 miles) to the south. The French government implemented their oil-spill cleanup plan. Within six days of the spill, all of the vessel's tanks were open to the sea, and the slick measured 129 kilometers (80 miles) by 29 kilometers (18 miles). The French were unable to pump oil off the grounded vessel because of both poor weather and poor charts of the area.

Ten days after the ship grounded, the highest tide of the period occurred, and beach cleaning began in earnest. By the beginning of April, the French had mustered 5,800 military personnel, 3,000 civilians, 28 boats, and more than 1,000 vehicles for the cleanup operation. By the end of May, 206,000 tons of material had been cleaned off the shores of Brittany. Only 25,000 tons of this was actually oil. The rest was sand, rock, seaweed, and other plant life. The rescue efforts found ten thousand dead fish and twenty-two thousand dead sea birds.

The cleanup effort was declared a success even though many criticized its slow response and fragmented efforts. The bulk of the oil cleanup was attributed to the sea itself. The relatively deep water and fast current along the shore helped the sea both disperse and dissipate the 223,000 tons of crude oil that had spilled into it over a six-day period.

Robert J. Stewart

SEE ALSO: *Argo Merchant* oil spill; *Braer* oil spill; *Exxon Valdez* oil spill; Oil spills; *Sea Empress* oil spill; Tobago oil spill; *Torrey Canyon* oil spill.

Amory, Cleveland

BORN: September 2, 1917; Boston, Massachusetts
DIED: October 14, 1998; New York, New York
CATEGORY: Animals and endangered species

Cleveland Amory founded the Fund for Animals and established the Black Beauty Ranch as a refuge for unwanted or abused domestic and wild animals. His decades of activism for animal rights and protection have saved thousands of animals from extermination and helped bring the issue of cruelty to animals into the public spotlight.

From the time he was a young child, Amory harbored a dream to create a sanctuary for animals where they would be protected from harm and allowed to roam free. To this end, he established the nonprofit Fund for Animals in 1967 and opened the fund's Black Beauty Ranch in Murchison, Texas. In 1977 the first major rescue of animals was initiated by the fund. The U.S. Park Service had scheduled the extermination, by shooting, of all the wild burros living in the Grand Canyon in Arizona. Over two years, the Fund for Animals orchestrated a helicopter airlift of 577 burros from the floor of the Grand Canyon, 2,134 meters (7,000 feet) below its rim. An intensive, countrywide adoption campaign was conducted for the burros; those animals not adopted found homes at Black Beauty Ranch. Two years later, a total of five thousand burros also earmarked for destruction were rescued from Death Valley National Monument and the Naval Weapons Center at China Lake, both in California.

Cleveland Amory founded the Fund for Animals in 1967 to rescue animals from abusive situations. His efforts helped spark public debate about animal cruelty issues. (Archive Photos)

In the early 1980's the fund rescued over sixty wild Spanish Andalusian goats from San Clemente Island off the coast of California. The United States Navy had decided to eradicate the animals that populated the island, saying that it was mandated by the Endangered Species Act (1973) because the goats were eating the native, endangered species of vegetation on the island. The Navy intended to allow hunters to kill the animals for a fee. The Fund for Animals again arranged a helicopter airlift of the animals to temporary quarters in San Diego, California, and another adoption campaign was successfully conducted.

Black Beauty Ranch has been home to thousands of domestic and exotic animals rescued from neglectful or abusive situations. One famous resident is Nim, a chimpanzee that communicates with sign language. Rescued from a research lab, Nim was scheduled to be used for medical research after his perceived usefulness as a subject in animal communication studies ended. The fund has also rescued wild horses from U.S. Bureau of Land Management lands, buffalo from private hunting grounds, and elephants from circuses and zoos.

Amory and the Fund for Animals, along with generous donors to the fund, have developed the resources to conduct high-profile rescues of large numbers of animals and have received much national and international publicity in the process. This publicity has helped raise the public's consciousness about issues of cruelty and neglect toward animals. An in-depth discussion of the activities of the Fund for Animals and a history of animal abuses is contained in Amory's book *Ranch of Dreams* (1997).

Karen E. Kalumuck

See also: Animal rights; Animal rights movement; Pets and the pet trade.

Animal rights

Category: Animals and endangered species

The animal rights philosophy promotes a higher valuing of animals, sometimes to the same legal status as humans. Animal rightists vary in beliefs from pet owners merely interested in the welfare of cats and dogs to purists who believe that all animals should be left in nature and beyond the reach of human interference.

Standard animal rights concerns focus most heavily on the use of animals in biomedical research and education, and secondly on the use of animals as a source of food. Protests are also staged against the fur industry, zoos, circuses, the pet industry, and hunting. This contrasts with actual usage, where food, hunting, and pets make up nearly 99 percent of animals used. While some animal rightists seek allies within certain environmentalist groups that share a similar counterculture philosophy, many other environmental groups believe that animals are needed for pollution monitoring, ecosystem management, and the perpetuation of natural food chains, and thus reject the core animal rights tenets.

Animal Cruelty

All states in the United States have animal abuse laws to protect most birds and mammals from cruel treatment, abandonment, poisoning by nonowners, and failure to be provided adequate food and water. Most laws do not include the larger numbers of invertebrates, fish, reptiles, and amphibians, and most states exclude coverage of animals used in science research. Such laws are directed toward human behavior and do not establish "rights" of animals any more than laws concerning land use establish "land rights." States define wildlife as property of either the landowner or the state.

Cruelty includes direct acts of physical abuse—such as beating a horse—as well as neglect, as when an animal starves to death. Cruelty usually involves intentional, malicious, or knowingly committed acts and excludes instances of accident or ignorance. Many states focus on unjustified infliction of pain in defining cruelty. Therefore animal breeders, racetracks, and other facilities that confine animals usually must meet minimal requirements for exercise, space, light, ventilation, and clean living conditions. This long tradition of U.S. animal cruelty law reflects widespread public acceptance of animal welfare principles, which animal rightists wish to expand to challenge animal use for food, research, education, fur, hunting, pets, and zoos.

Until the 1980's most animal rights organizations in the United States were not activist, but instead focused on animal welfare and providing shelters. The public was aware of the role that animal research had played in the earlier development of antibiotics, insulin, surgical techniques, and vaccinations, and most adults had recent memories of many major diseases. However, protocols for Federal Drug Administration (FDA) drug approval, as well as agricultural and environmental safety standards, based on animal testing were criticized by a new generation that had little experience with the effects of widespread disease.

Two books—*Animal Liberation* (1977), by Peter Singer, and *The Case for Animal Rights* (1983), by Tom Regan—provided an updated philosophy for activists who questioned the human use of animals for medical research, food, fur, and education. New animal rights organizations were formed, and old organizations shifted their mission, including People for the Ethical Treatment of Animals (PETA), New England Antivivisection Society (NEAVS), Ethical Science Education Committee, and the underground Animal Liberation Front (ALF).

Animals as a Food Source

Farmers own agricultural animals as property. Meat processing is a well-established industry, and a large majority of the American population traditionally eats meat. Animal rights efforts have therefore concentrated on attacking specific farm practices and portraying a meat diet as unhealthy and antienvironmental. The confinement of animals to produce veal, poultry, and eggs is pictured as inhumane.

Animal rightists also leverage nutritional concerns among some members of the medical

Animal Rights Milestones

YEAR	EVENT
1977	Peter Singer publishes *Animal Liberation,* which elaborates an animal rights philosophy and gains a substantial following.
1983	Tom Regan's *The Case for Animal Rights* reinforces the message of Singer's books and adds more people to the movement.
1988	The first public school dissection opt-out law in the United States is passed in California; Florida, New York, Maine, Pennsylvania, and Illinois soon follow.
1993	The First Congress in Alternatives in Animal Usage provides a forum for discussion and debate between animal research advocates and animal rights groups.

community into a broad condemnation of consuming meat. Meat, cheese, and milk comprise a major food group, although in the mid-1990's the United States Department of Agriculture (USDA) began promotion of a food pyramid representing a smaller meat diet near the pyramid peak. Humans demonstrate an evolutionary history of eating both plant and animal products, in tooth structure, digestive anatomy, and the need for some amino acids found originally in animals. However, new plant breeds can provide some of these nutrients, and a vegetarian diet is now less risky.

Animal rightists contend that vegetarianism is an ecological necessity. The primary argument is based on energy flow. On average, only 10 percent of the chemical energy of one food level is stored in the next food level; for example, one hundred units of grain sustain ten units of cow, which in turn produce one unit of human. Using this logic, the human population could increase tenfold if it switched from eating meat to eating grain directly. However, critics point out that most of a cow's diet is grass that is indigestible to humans and is often not grown on land suitable for crops. The majority of meat animals worldwide, from pigs to fish, are not fed on grain and may even be critical to recycling wastes. Some animal rightists claim a "hamburger connection" between tropical deforestation and local hamburger chains. Although there is no export of rain forest beef to the United States, the destruction of South and Central American rain forests to make room for cattle ranches and farms continues at a rapid rate.

BIOMEDICAL RESEARCH

Extensive animal research was utilized in the development of germ theory and the conquest of most major infectious diseases. Nearly all Nobel prizes in medicine awarded from 1901 onward involved animal research and, until the 1980's, met with widespread public approval. Antivivisection and humane societies spoke with a weak voice and addressed standard animal cruelty issues until the publication of *Animal Liberation* and *The Case for Animal Rights,* which provided a more detailed rationale for their cause. During the 1970's and 1980's many young people—the first full generation raised in the absence of major infectious diseases, away from contact with farm animals, and in a climate that was increasingly antiscience—joined animal rights organizations.

Animal rightists have four major objections to animal research. They believe that organisms have an intrinsic right to live free of human interference, violation of which is called "speciesism." A second set of concerns centers on the suffering and pain caused to research animals. A third concern is use of pets acquired through shelters and the pet industry. Finally, new technologies have been developed that many claim make direct research on animals unnecessary.

While more than 85 percent of research is conducted with rodents, most protests are targeted toward research on dogs and cats (fewer than 2 percent of research animals) and nonhuman primates (fewer than 0.5 percent). Some activists contend that all animal research can be abandoned and replaced with tissue culture techniques and computer simulations. Activists often complain that current animal-care regulations are not being enforced. Some even contend that there is a major criminal industry that steals pets to supply research animals. The Draize test for establishing the safety of cosmetics on rabbit eye tissue is considered frivolous research.

Animal rights groups seek to control local animal shelters and restrict transfer of animals to research and education. Ongoing attempts are made to gain legislation to restrict use of pound animals; eliminate safety testing; allow private citizens to sue for enforcement of the Animal Welfare Act (AWA); include farm animals, rats, and mice under the AWA; and include research usage under state animal cruelty laws.

Many scientists respond by claiming that biological systems are far more complex than simple computer models and that drugs do not always respond in tissue culture as they would in a whole organism. Most research organizations follow the "three R's" of replacement, reduction, and refinement. Tissue cultures are one method of replacing some animals, reduction uses fewer animals, and refinements reduce the pain or distress of lab animals. Animals taken from shelters for research constitute fewer than 1 percent of the dogs and cats that would be otherwise be euthanized. Research facilities are inspected by the USDA's Animal and Plant Health Inspection Service (APHIS), which enforces AWA criteria. The Food and Drug Administration (FDA) and the Environmental Protection Agency (EPA) also have laboratory practice regulations. Advocates of animal testing claim that 95 percent of lab animals are not subjected to pain; the remaining animals, involved in studies of pain itself, are provided pain-relieving drugs or anesthetics as soon as the study permits.

Medical research using animals is defended by the National Association for Biomedical Research (NABR) and the Incurably Ill for Animal Research (IIFAR). This issue reportedly generates the third-largest amount of legislative mail (following Social Security and the federal deficit). Backlash to activist raids on research facilities led to the passage of the Animal Research Facility Protection Act in the United States.

Education and Other Issues

The first state law enacted to allow a student to opt out of lessons using animals was passed in California in 1988 in response to a student who protested dissection. Although she failed to win her federal court case to establish a right to avoid such coursework, the California law made the issue moot in that state. Florida, New York, Maine, Pennsylvania, and Illinois also eventually passed laws that allowed students to opt out of classroom dissection.

Animal rights organizations protest against supply companies that provide animals to schools. Some pamphlets allege that dissection contributes to the decline in amphibians and endangerment of other species. They contend that dissection and animal experimentation desensitizes students and that only a small percentage of students need to learn lab skills. Computer simulations, stuffed animals, plastic models, and other alternatives are promoted as effective and economical alternatives. One organization maintains a toll-free hotline for students who do not wish to dissect and provides materials written for elementary, secondary, and college levels.

Public schools are adverse to entering into public debates and controversy concerning such allegations. However, many scientists and educators have responded by claiming that alternatives fail to provide the multisensory experiences that make labs meaningful and that genuine lab results are critical in teaching the reality base of science. Laboratory work continues to be the main incentive for many students to decide to enter medical and health careers. Field biologists claim that nearly all dissection animals come from food harvesting, captive-rearing, or animal shelters. They also claim that dissection, rather than desensitizing students, normalizes attitudes toward organs, blood, and feces. While not all students will become doctors, all will become patients and need to have a mental image of their own internal anatomy.

Although more furs are coming from captive breeding, such as mink farms, protests continue to target fur-trapping techniques. Demonstrators who splash paint on expensive fur coats are usually given news coverage. Zoos are generally portrayed as prisons. This contrasts with most environmental groups, who value maintaining and reintroducing animal populations that would otherwise have gone extinct. Circuses are portrayed as promoting both slavery and abuse. Animal training for circuses, shows, and film production may be subject to scrutiny and harassment. Hunting is a rural tradition with widespread support in the American West and Midwest. Protests may involve "shadowing" hunters to drive away game animals.

Animal rights organizations vary widely in their philosophy toward pet ownership. Pet owners are major contributors to these organizations, and the extent to which organizations are willing to advocate pure animal freedom defines the sometimes volatile differences between the groups. Analysis of animal rights membership reveals a higher number of females with higher education and includes an increasing number of pet veterinarians but not farm veterinarians. Although they constitute fewer than 0.5 percent of the population, members of animal rights organizations have succeeded in securing substantial state and federal legislation and policies and are expected to grow in number.

John Richard Schrock

SUGGESTED READINGS: *Animal Liberation* (1977), by Peter Singer, and *The Case for Animal Rights* (1983), by Tom Regan, revived the animal rights movement in the Western world. *Science, Medicine, and Animals* (1991), by the National Academy of Sciences and Institute of Medicine, provides a defense for animal use in research. *They Threaten Your Health: A Critique of the Antivivisection/Animal Rights Movement* (1985), by Ernest Verhestel, is a strongly written, heavily annotated, detailed response to animal rights arguments against animal research. *Animals and Their Legal Rights: A Survey of American Laws from 1641 to 1990* (1990), published by the Washington, D.C.-based Animal Welfare Institute, is a readily available compilation of state and national animal cruelty laws. *Animal Experimentation: The Moral Issues* (1991), edited by Robert M. Baird and Stuart E. Rosenbaum, provides a weak defense of science research, but a reader can find most issues addressed. *Education and Training in the Care and Use of Laboratory Animals: A Guide for Developing Institutional Programs* (1991), by the National Research Council, describes the extent research laboratories are held to humane standards of animal care. *In the Name of Science: Issues in Responsible Animal Experimentation* (1993), by F. Barbara Orlans, discusses a wide array of research issues in detail from an animal protectionist viewpoint.

SEE ALSO: Animal rights movement; Pets and the pet trade; Singer, Peter.

Animal rights movement

CATEGORY: Animals and endangered species

People involved in the animal rights movement share philosophical beliefs based on the idea that all animals are entitled to an equal claim on life and liberty and possess the same rights to existence as humans. Animal rightists oppose those who believe that animals are available for human use as food, beasts of burden, amusement, recreation, and for study and experimentation.

The philosophical concept of animal rights arose during the seventeenth and eighteenth centuries along with the development of biological science. The growing interest in biology gave rise to a sort of sideshow in which living, conscious dogs were cut open to display the animal's internal organs to a crowd. A variety of blood sports were popular as well, including bullbaiting and bearbaiting. In these, a bull or bear was chained in a ring along with one or more dogs that were trained to attack the larger animals. Dog fights, in which various terrier breeds were encouraged to attack each other, were also popular.

Birth of Animal Rights

In 1824 the Royal Society for the Prevention of Cruelty to Animals (RSPCA) was founded in Great Britain to enforce new anticruelty laws. However, the laws and their enforcement had little, if any, effect in rural areas that were far from the watchful eyes of the police or RSPCA agents. On many farms, animals were still kept in filthy conditions and beaten if they balked at hauling overloaded wagons. The slaughter of animals for market was carried out as simply and quickly as possible.

Biomedical research, especially in human anatomy and physiology, rapidly advanced in Europe during the early to mid-nineteenth century. While anatomical study could be satisfied with human corpses, physiologists required living material. Animals became the targets. Although many of the animals used in medical research were rats and mice, dogs were also frequently used. Many of the animals were stolen pets, while others were strays that were found roaming on the streets. The treatment of the dogs in medical laboratories varied, but, for the

Memberships of Animal Rights Organizations, 1997

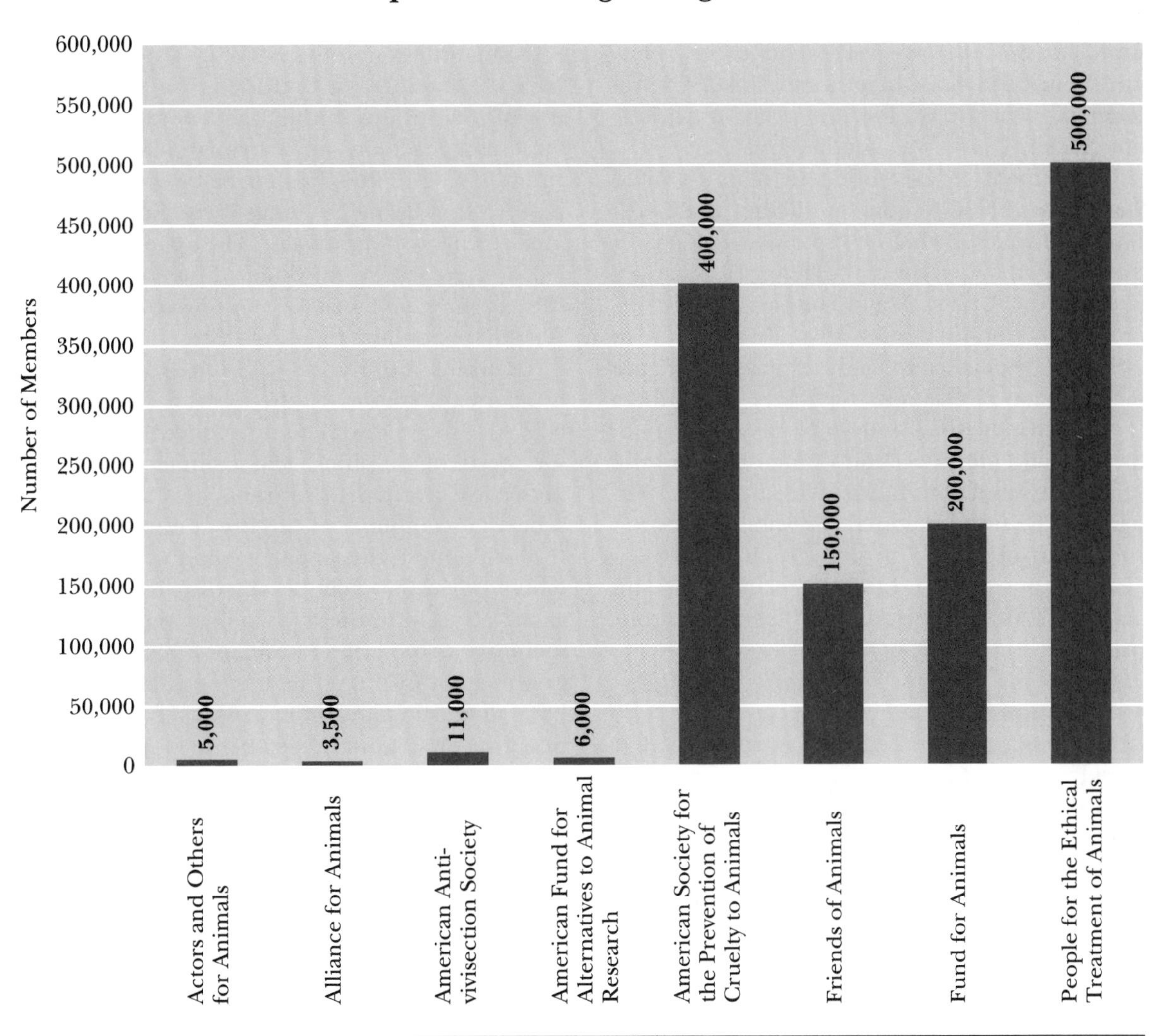

Source: Data are extracted from *Gale Encyclopedia of Associations.*

purposes of good science, the animals had to be maintained and treated in clean, sanitary, and relatively stress-free environments. Many dogs died at the hands of medical researchers, and opposition to the practice quickly grew. Many of the objections came from the public, who had heard stories of both real and imagined horrors suffered by animals in the experiments.

The British government sought to quiet the complaints by passing the Cruelty to Animals Act of 1876. The act did not prohibit the practice of experimenting on live animals. Instead, it set regulatory procedures that had to be followed in the laboratories. Animal rights were no less an issue in the United States. The first documented humane society in the United States was the American Society for the Prevention of Cruelty to Animals (ASPCA), incorporated in 1866. Another pioneer group, the American Antivivisection Society (AAS), was founded in 1883.

The animal rights movement in the United States was relatively quiet until the first Earth Day in 1970, after which it rapidly expanded. Of the more than eighty animal rights organizations in the United States, fifty-seven (70 percent) were founded after 1970. In addition to the older groups, among the organizations are such diverse associations as Actors and Others for Animals, the Coalition for Non-Violent Food, the Animal Political Action Committee, and the American Fund for Alternatives to Animal Research. The list also contains several adversarial and confrontational groups, such as Greenpeace, People for the Ethical Treatment of Animals (PETA), and the Animal Liberation Front (ALF).

Animal Rights Issues

The plight of the whales was an issue that quickly attracted animal rights proponents. For centuries whaling was conducted from sailing ships with handheld harpoons in the manner made famous in Herman Melville's novel *Moby Dick* (1851). Even with such crude equipment and methodology, whalers reduced the whale population in the Atlantic Ocean and turned their attention to the Pacific Ocean. In the late nineteenth century, steam (and later, diesel) vessels and cannon-fired harpoons increased the whalers' efficiency. The methods of the whalers aroused the ire of many people. Frequently, the harpooned whale was forced to tow the steel "catcher" ship for hours until the animal succumbed to the injuries from the explosive-headed harpoon. Another whaling technique was to harpoon and kill whale calves. The mother and other adults hovered around the injured or killed calf and were in turn harpooned.

Economics rather than animal rights brought about the formation of the International Whaling Commission (IWC) in 1946. The commission was established to manage the whale stocks. It had no regulatory authority. Norway, Iceland, and Japan continued to hunt whales despite the recommendations of the IWC. These nations have defended their activities as a sustainable use of a natural resource. Opponents view it as an archaic activity and a violation of animal rights. In 1972 the United States government passed the Marine Mammal Protection Act as a move to protect the whales. The apparent recovery of the Pacific gray whale stocks suggests that the act may have been a step toward achieving the goals of animal rights activists.

Dolphins are also at risk, but from indirect human exploitation. In the Eastern Tropical Pacific Ocean, yellowfin tuna frequently swim below pods of dolphins, and commercial fishermen have learned to set their nets around the schools of dolphins to capture the tuna. As the fish are netted, the trapped dolphins drown. As many as 132,000 may be killed in this manner each year. The protests of animal rights groups included a boycott of canned tuna. The U.S. government instituted regulations that required both domestic and foreign fishermen to follow practices that would release the dolphins and still retain most of the tuna. In five years the accidental catch and kill of dolphins was reduced to 25 percent of what it had been. The reduced tuna catches, however, forced many U.S. fishermen out of the industry, leaving a void that was quickly filled by foreign fishermen.

Animal rights activists have turned their attention to circuses, zoos, animal theme parks, the fur trade, and any activity in which live animals are used. For example, in February, 1995, when

Animal rights activists protest at a Harvard University primate research center in Massachusetts. Anticruelty groups have been trying to stop the use of animals in medical testing since the nineteenth century. (AP/Wide World Photos)

the Ringling Brothers and Barnum & Bailey Circus was preparing to visit Richmond, Virginia, the management asked the news media not to mention the time of arrival. The circus management wanted to move their animals from the rail yard to the show grounds with as little fanfare as possible. The circus had received threats from animal rights groups and sought to avoid any confrontation and possible risks to animals and the public. Despite the changes that have taken place, however, it is unlikely that the controversy about animal rights will soon disappear.

Albert C. Jensen

SUGGESTED READINGS: Joy Williams's "The Inhumanity of the Animal People," *Harpers* 295 (August, 1997), suggests that animal cruelty stems from the fact that animals are voiceless and thus unable to complain about exploitative treatment. Peter Singer's struggle to protect apes from further exploitation is described by Daniel W. McShea in "On the Rights of an Ape," *Discover* 15 (February, 1994). *Animal Rights: Opposing Viewpoints* (1996), edited by Andrew Harnack, presents essays that explore the animal rights debate. "A Sporting Chance," *Sierra* 81 (May/June 1996), by Charles Pope, describes the alliance between hunters and animal protectionists against land developers, oil companies, large-scale mining operations, and the timber industry. The dissolution of an alliance between PETA and the Nature Conservancy is described by Bill Preen in "PETA's Pig Fight," *Garbage* 6 (Summer, 1994). The removal of native prairie dogs from proposed baseball diamonds in Hutchinson, Kansas, is described by Kevin Fedarko in "Please Don't Shoot the Prairie Dogs," *Time* 150 (July 7, 1997). James B. Whisker's *The Right to Hunt* (1981) offers a detailed, well-supported defense of hunting. *Animal Rights: The Changing Debate* (1996), edited by Robert Garner, brings together essayists with different viewpoints to debate animal rights issues.

SEE ALSO: Animal rights; Greenpeace; People for the Ethical Treatment of Animals; Singer, Peter.

Antarctic Treaty

DATE: 1961
CATEGORY: Preservation and wilderness issues

The 1961 Antarctic Treaty is an international agreement in which signatory nations have agreed to set aside the Antarctic region for scientific and peaceful pursuits.

Antarctica, the large body of land and ice that surrounds the South Pole, is among the earth's most unique and wild places. The only continent with no indigenous human population, Antarctica exceeds 12 million square kilometers (5 million square miles) in size, almost 1.5 times larger than the continental United States. An ice layer averaging more than 1.6 kilometers (1 mile) in thickness covers approximately 95 percent of the land area. Although very few terrestrial species are found on the continent, the surrounding waters are rich in marine life and support large populations of marine mammals, birds, fish, and smaller creatures, some of which are found nowhere else on earth. Antarctica and its surrounding waters play a key but not yet fully understood role in the planet's weather and climate cycles.

In the early twentieth century seven nations asserted territorial claims on Antarctica, which persisted unresolved for decades. International scientific cooperation among twelve countries during the 1957 International Geophysical Year (IGY) led to the establishment of sixty research stations on the continent. As the IGY drew to a close, the scientific community argued that Antarctica should remain open for continuing scientific investigation and should be unfettered by national rivalries over territory.

This led to the negotiation of the Antarctic Treaty, which entered into force in 1961. The treaty prohibits military activity on the continent and promotes scientific cooperation among the parties. Signatory nations agree to freeze existing territorial claims and make no new ones. Initially, twelve nations signed and became consultative parties, agreeing to hold regular consultative meetings to discuss implementation of the treaty. Fifteen more countries have since been granted consultative party status. Seventeen additional nations have acceded to the treaty but are not full parties and participate as observers only. The Antarctic Treaty was also the first major arms control agreement among the nuclear weapons states and provided a model for several subsequent agreements, including the Limited Test Ban Treaty (1963). The treaty specifically prohibits nuclear test explosions in Antarctica.

Consultative meetings have resulted in more than 150 implementation recommendations to national governments and have led to two additional treaties: the 1978 Convention for the Conservation of Antarctic Seals and the 1982 Convention on the Conservation of Antarctic Marine Living Resources (CCAMLR), which addresses fishery management. An unusual feature of the latter is that the applicable territory is defined by ecosystem criteria rather than by political boundaries. The Antarctic Treaty, related agreements, and recommendations form what is known as the Antarctic Treaty System.

Mineral discoveries in the 1970's led some nations and private companies to contemplate plans to exploit mineral resources. Environmental advocates and the scientific community, sharing a concern about potential impacts, led a long fight to prevent such activities. In 1991 the historic Antarctic Environmental Protocol was finally adopted, which banned mineral and oil exploration for a minimum of fifty years. Annexes to the protocol contain legally binding provisions regarding environmental assessments, protection of indigenous plants and animals, waste disposal, marine pollution, and designation of protected areas. The protocol entered into force in January, 1998, after ratification by all consultative parties.

Environmental issues that still posed challenges to the protection of Antarctica in the late 1990's included threats of overexploitation of Antarctic fisheries, impacts of expanding tourism, and the need for an agreement among the treaty parties regarding liability for environmental damages.

In 1977 environmental organizations interested in Antarctica formed the Antarctic and Southern Ocean Coalition (ASOC), which by 1998 included 230 member organizations from fifty countries. The ASOC has been accorded status as an expert observer to the Antarctic

Treaty System and represents member group interests. Environmentalists who have endorsed the concept of Antarctica as a "world park" lobbied for and strongly support the environmental protocol.

Phillip Greenberg

SEE ALSO: Antarctica Project; Limited Test Ban Treaty.

Antarctica Project

Date: founded 1982
Category: Ecology and ecosystems

The Antarctica Project was founded in 1982 as a nonprofit, nonpartisan, nongovernmental organization dedicated to the protection of Antarctica.

The Antarctica Project was created to coordinate both domestic and international environmental efforts affecting policy in Antarctica and helped to ensure ratification of the Antarctic Environmental Protocol on January 14, 1998, after six years of intense lobbying. This strengthens the environmental mission of the Antarctic Treaty System, a group of international treaties that set policy and determined the state activities that are permitted in the Antarctic region for member governments. Through the Antarctic Environmental Protocol the entire continent of Antarctica was declared off-limits to exploitation of its oil, gas, and mineral resources for a minimum of fifty years. The protocol also protects Antarctic wildlife and habitat, and controls pollution and waste management.

The Antarctica Project encourages scientists to study the Antarctic ecosystem in an environmentally friendly manner and ensures that their research activities comply with the protocol's guidelines. The Antarctica Project is the secretariat for the Antarctic and Southern Ocean Coalition (ASOC), founded in 1977, which has more than 230 member organizations in fifty countries. The ASOC was instrumental in keeping Antarctica free of oil and mineral exploitation until the signing of the protocol.

ASOC members work closely with international governmental agencies, tourists, and scientists studying the Southern Ocean, which is home to abundant seals, fish, whales, penguins, and a multitude of other seabirds. Members participate in the meetings of the Convention on the Conservation of Antarctic Marine Living Resources (CCAMLR) and work to protect the Southern Ocean Whale Sanctuary, established in 1994 to protect the feeding grounds for more than 90 percent of the world's great whales.

Through the protocol, Antarctica was declared a "natural reserve, dedicated to peace and science." Human activities are firmly regulated to minimize impacts on the fragile environment. Antarctica plays a primary role in regulating global environmental processes; therefore, the Antarctica Project supports research on the continent that will lead to an understanding of global climate change, atmospheric and oceanic systems, global sea level and tidal fluctuations, atmospheric pollution, and ozone destruction.

Future goals of the Antarctica Project and the ASOC include the proper implementation of the Antarctic Environmental Protocol by each country, protection of the biological diversity of the Southern Ocean, and the guarantee that Antarctic visitors act in an environmentally sensitive manner in accord with the protocol.

The Antarctica Project is the primary nongovernmental source of information regarding Antarctica and the Antarctic Treaty. It is responsible for numerous publications—including books, slides, videos, and posters—for educational and advocacy purposes. A quarterly newsletter is also available.

Diane Stanitski-Martin

SEE ALSO: Antarctic Treaty; Climate change and global warming; Ozone layer and ozone depletion.

Antienvironmentalism

Category: Philosophy and ethics

The emergence of antienvironmentalism reflects a long-standing impasse between environmental-

ists, who are concerned with the fate of all species and the environment in the wake of humankind's destructive activities, and those who consider humankind's immediate economic and lifestyle needs to be more important than all other concerns.

In the early twentieth century, environmentalism in the United States was largely fostered by wealthy sportsmen who saw the need to protect the outdoors in order to maintain satisfactory areas for hunting, fishing, and camping. The movement got a populist boost from the publication of Rachel Carson's *Silent Spring* in 1962, which presented an easily understood account of the dangers of toxic substances in the environment. For the first time, the public began to demand that laws be enacted to protect the environment and clean up land, water, and air that had already been polluted.

Growing Environmental Movement

For several years the environmental movement gathered strength as the public voted politicians with environmental orientations into office. Public outcry surged against polluting companies, leading to boycotts of products. Grassroots, citizen-led efforts such as recycling programs and litter patrols gained support as the public became more educated and concerned about environmental issues. Among the issues that pitted environmentalists against the government and industry were toxic waste incineration, habitat destruction by logging and mining companies, and use of public lands, including national parks.

Two oil crises during the 1970's served to focus awareness on energy conservation and the need to develop alternatives to energy derived from fossil fuels. Many feared that oil supplies were dwindling, while others wished to end U.S. reliance on oil-exporting nations in the Middle East. One important result was a general reduction in the size of motor vehicles. This, along with other technological advances, helped make cars more fuel efficient. However, such research came at great cost to automakers.

The 1970's were also characterized by landmark legislation that imposed strict limits on

Logging companies continue to fight government restrictions on timber harvests that were enacted during the 1960's and 1970's by proenvironment politicians. (Ben Klaffke)

pollution output and resource use, and also provided for the remediation of polluted land and water. Large fines were imposed for violations of the new laws, which were enforced by the newly formed Environmental Protection Agency (EPA). One of the most important and pivotal developments was the passage of the Comprehensive Environmental Response, Compensation, and Liability Act, or Superfund, which provided vast sums of public money for the cleanup of designated industrial and military waste dumps and other degraded sites. Signed into law by the U.S. Congress in 1980, Superfund's provisions allowed the government to bring lawsuits against the responsible parties, requiring them to help pay cleanup costs. In order to avoid fines, many industries were forced to develop and implement costly waste-processing technologies.

The political and economic situation began to change in the late 1970's as industry mounted a counteroffensive against environmental laws. Businesses contended with the burgeoning number of environmental regulations by finding and exploiting loopholes in legislation. A growing number of industries used stalling tactics and countersuits to delay or eliminate the need to implement required changes. Meanwhile, in the western United States, a coalition of loggers, miners, cattle ranchers, farmers, and developers demanded that the federal government transfer control of large tracts of federally owned land to individual states. Members of the so-called Sagebrush Rebellion felt that state ownership would give them more power to exploit the natural resources on the land.

Antienvironmentalism and Wise Use

A severe backlash against environmentalism began to occur when Ronald Reagan replaced Jimmy Carter as president of the United States in early 1981. Many environmental laws and regulations, which were seen as barriers to economic progress, were weakened or abolished. A large number of federal judges who had started their careers during the 1960's retired, and they were replaced by politically conservative judges who began to interpret existing laws in favor of industry. The office of the EPA was weakened, and funding for environmental enforcement and remediation was slashed. Secretary of the Interior James Watt, who had been a leader of the Sagebrush Rebellion, promoted legislation to open previously protected areas to mining and oil exploration. The general public, experiencing growth and prosperity for the first time in many years, began to favor short-term economic gains and turned a blind eye to news of the weakening environmental movement.

The late 1980's saw the birth of the wise-use movement, which appeals to humankind's pragmatic and optimistic side by asserting that some optimal balance of resource use and restoration is practicable and that technology, given time and funding, will develop workable solutions to existing environmental problems. This position assumes that the complex ecosystems involved can be understood well enough to know what these balances should be. Advocates of wise use believe that all public land, including national parks, should be opened to mining and drilling. Like the Sagebrush rebels, they also promote the strengthening of the rights of states and property owners to exploit resources with minimal federal regulation.

According to the tenets of wise use, the harvesting of timber from ancient forests would be followed by the planting of an equivalent acreage of saplings. Logging would be timed according to growth rates, and technology would produce fast-growing varieties of trees that would furnish adequate ecosystems for wildlife in the new forests. However, environmentalists argue that ancient forests represent complex, irreplaceable ecosystems that cannot be substituted with new forests planted by logging companies. Another example involves coastal wetlands, which provide vital habitat to numerous species and contain a high degree of biodiversity. Wetlands are frequently located in areas that are desired by real estate developers wishing to build vacation homes or resorts. Environmental protection laws based on the tenets of wise use mandate that destroyed wetlands must be replaced by a new wetland of equivalent size. Again, environmentalists worry that too many threatened species would be lost in the process of destroying and replacing wetland areas.

SCIENTIFIC CONTROVERSY

Another significant trend involved public confusion over scientific debates concerning such topics as the ozone hole and global warming. Industry and government scientists often questioned and condemned dire predictions advanced by other scientists. The government frequently responded by requesting additional research before requiring vast, expensive reductions of known pollutants. Rather than believe the frightening scenarios painted by some scientists, many people sided with scientists who questioned the validity of these and other threats to the global environment.

Mainstream environmental organizations that had evolved from small groups of fervent individuals were now led by full-time professional lobbyists based in Washington, D.C. Details of new legislation were negotiated among industry, government, and environmental leaders. National environmental organizations grew increasingly cumbersome and expensive to run. Many began accepting large donations from the same industries over which they were trying to watch, which created serious conflicts of interest. Top industry executives became members of the boards of directors for environmental organizations. At the same time, these corporations also made large donations to elected officials and thus gained access and influence in government. As the 1990's progressed, membership in the large environmental groups began to decline as "donor fatigue" set in, questions persisted about the true urgency of environmental issues, and cynicism arose about the possibility of environmental progress under such circumstances.

A relaxation of concern about environmental problems came about in the late 1990's in the wake of encouraging news about improvements of environmental indicators such as air-pollution levels of certain gases in the aftermath of implementation of cleaner energy production. For example, atmospheric levels of sulfur dioxide, which leads to acid rain, decreased in the United States and Europe after cleaner coal- and oil-burning technologies were implemented. The air-quality goals of many cities were met by a combination of fuel efficiency and "scrubber" smokestacks.

A controversial strategy advanced by a coalition of industry, government, and environmental leaders involves tradeable pollution permits. According to the plan, the government assigns utilities a certain number of pollution units per year. An especially clean-running plant will not need all of its units and will be able to sell them to plants that exceed their allotments. The system has been criticized by many environmentalists who contend that some utilities are able to buy their way out of the need to reduce pollution. The position thought to be antienvironmentalist in this context would maintain that the plan is a realistic method of controlling overall levels of pollution without putting older utilities out of business while they endeavor to upgrade their performance.

The revelation that environmental degradation can, in certain cases, be reversed over a fairly short span of time led to the argument on the part of antienvironmentalists that nature is surprisingly resilient, and therefore environmental protection does not need to be so stringent, costly, and regressive. Environmentalists counter such arguments with a call to remain vigilant and to include the health of the environment in national and global visions of the future.

Wendy H. Hallows

SUGGESTED READINGS: For a detailed study of both sides of a large number of environmental questions, see *A Moment on the Earth: The Coming Age of Environmental Optimism* (1995), by Gregg Easterbrook. The book concludes with the notion that nature is resilient and that humankind is best served by "ecorealism." The following books present readable histories of environmentalism in the twentieth century and conclude that the movement has been damaged by the antienvironmental countermovement: *Losing Ground: American Environmentalism at the Close of the Twentieth Century* (1995), by Mark Dowie, and *A Fierce Green Fire: The American Environmental Movement* (1993), by Philip Shabecoff. *A Conspiracy of Optimism: Management of the National Forests since World War Two* (1994), by Paul W. Hirt, details the struggles among government, environmentalists, and the logging industry. Bill McKibben's *The End of Nature* (1989) presents a compelling account of a future in which climate patterns have been altered by global warming, along with a thoughtful philosophical discussion of the meaning of humankind's impact on the earth. *Sustaining the Earth: The Story of the Environmental Movement—Its Past Efforts and Future Challenges* (1990), by John Young, offers a blueprint for sustainable use of the earth's resources.

SEE ALSO: Sagebrush Rebellion; Watt, James; Wise-use movement.

Antinuclear movement

CATEGORY: Nuclear power and radiation

The antinuclear movement consists of a loose collection of organizations and individuals whose goal is to rid the earth of nuclear weapons. Most would also eliminate nuclear power reactors.

Antinuclear activists believe that if nuclear weapons are available, they will eventually be used; therefore, they seek to rid the earth of all such armaments. By organizing seminars, rallies, and protests, activists seek to force public debate on issues previously left to insiders, and they help shape the political climate. Activists from the environmental organization Greenpeace probably helped persuade the French government to reduce the number of underground nuclear tests in their 1995-1996 series. Activists have often been successful in capturing favorable television news coverage and mobilizing hundreds of thousands of citizens to protests. However, translating this success into practical results has

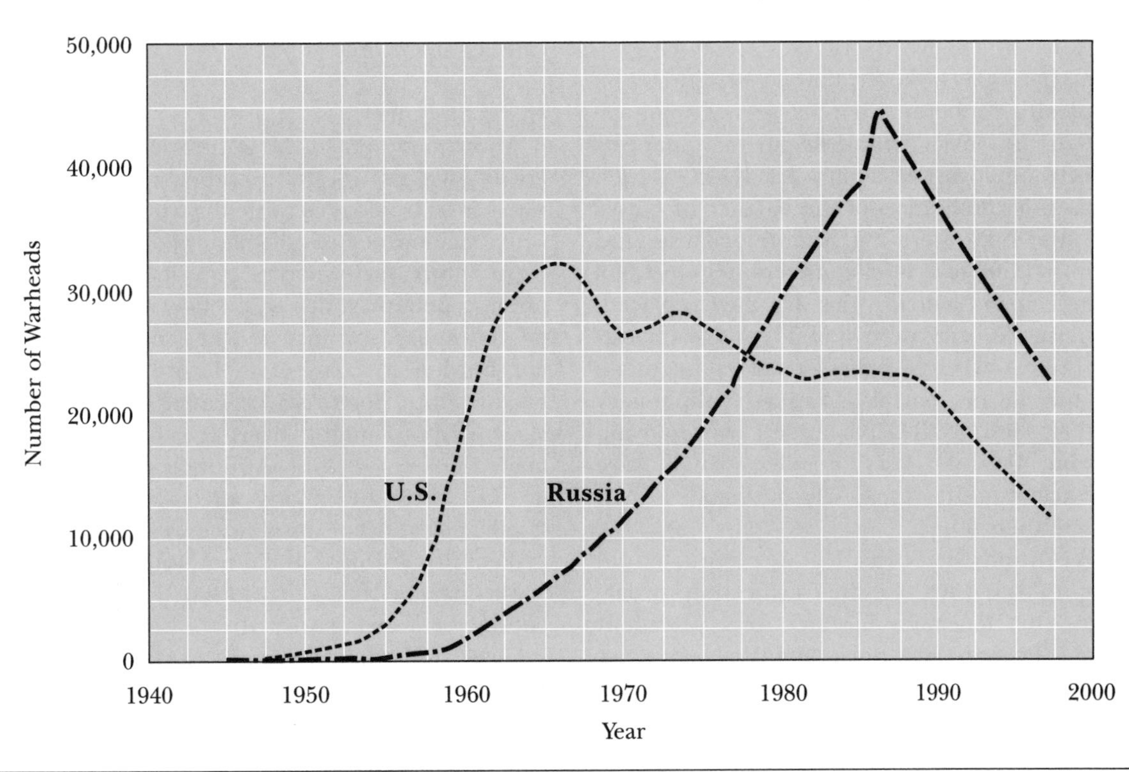

Source: Data adapted from *The Bulletin of the Atomic Scientists* (November-December, 1997).
Note: For the years following 1987, approximately 50 percent of the Russian warheads and 75 percent of the U.S. warheads shown were operational. The remaining warheads were held in reserve or had been retired but not yet destroyed.

proven elusive, and antinuclear referendums are generally defeated. At a conference on abolishing nuclear weapons held at Boston College in October of 1997, David McCauley offered a possible explanation. He suggested that the movement is fundamentally anarchistic and that distrust of people in power stifles cooperation with established political practices.

Nuclear Freeze

Antinuclear activists claimed that arms negotiators were allowing themselves to be defeated by complexities and suggested a mutual and verifiable freeze by the United States and the Soviet Union on the testing, production, and deployment of all nuclear weapons. The campaign attracted sympathetic television news coverage and gained momentum, until, by the late 1970's, hundreds of thousands of people were attending massive rallies in the United States and Europe to protest the planned deployment of the ground-launched cruise missile (GLCM) and the Pershing II missile.

Leaders of the North Atlantic Treaty Organization (NATO), however, dared not ignore the complexities inherent in nuclear negotiations. In 1976 the Soviet Union began to update its forces by deploying SS-20 missiles. NATO viewed this as a destabilization of the balance of power since the SS-20s were much more accurate and could fly 1,500 to 3,000 kilometers (930 to 1,860 miles) farther than the missiles they replaced. Furthermore, they were mobile, and each carried three warheads. NATO tried to negotiate with the Soviets, but all attempts at diplomacy failed. In 1983 NATO responded by deploying Pershing II and GLCM missiles, which were more accurate than the SS-20s but had less than one-half their range.

NATO and the Soviet Union eventually did agree to intrusive, on-site verification procedures. On December 8, 1987, U.S. president Ronald Reagan and Soviet general secretary Mikhail Gorbachev signed the Intermediate-Range Nuclear Forces Treaty to completely eliminate an entire class of weapons: SS-20s, Pershing IIs, GLCMs, and all other ground-launched nuclear missiles with ranges between 500 and 5,500 kilometers (310 and 3,420 miles), along with their launchers, support facilities, and bases. The treaty also banned flight testing and production of these missiles. The antinuclear movement deserves credit for helping educate the public and adding to the pressure that pushed politicians to the bargaining table. However, based on history, it seems unlikely that the treaty would have come about had NATO not proceeded with deployment of its own weapons.

An antinuclear initiative called Abolition 2000 was proposed in 1995. Within a few years, more than 1,200 nongovernment organizations (NGOs) on six continents had voiced support for it. Abolition 2000 called for movement toward "clean, safe, renewable forms of energy production that do not provide the materials for weapons of mass destruction and do not poison the environment for thousands of centuries." It also called for negotiations on a nuclear weapons abolition convention to be concluded by the year 2000. The convention would establish a time table for phasing out all nuclear weapons and would include provisions for effective verification and enforcement.

Nuclear Power

Many members of the antinuclear movement are also against nuclear power. They believe that nuclear technology is too dangerous and see reactor accidents caused by natural disaster, equipment failure, or human error as inevitable. They do not believe that radioactive waste can be safely disposed of; instead, they see it as an unfair burden to pass on to future generations. Furthermore, terrorists or nations without nuclear weapons might divert reactor material to make nuclear weapons. Antinuclear activists are not interested in making nuclear power plants safer. They want to shut down current plants and make it impossible to build new ones.

During the 1960's the antinuclear movement became concerned with the possible effects of low-level radiation from nuclear power plants. Several scientists have produced studies that purport to show that low levels of radiation cause an increased incidence of cancer in communities located near nuclear power plants. Many scientists believe these studies are flawed and find considerable evidence that the human body can

Police try to remove an antinuclear activist who lies on railroad tracks to protest the transport of nuclear waste in Germany. Such actions often gain wide media coverage, which brings more people to antinuclear rallies. (AP/Wide World Photos)

repair damage caused by sufficiently low levels of radiation.

The antinuclear movement was far more successful when it turned to issues of safety regulations and environmental law. They lobbied for stricter regulations, and they went to court to force nuclear plants to follow safety regulations that they had previously been allowed to bypass. The effect was to make nuclear plants safer, but it also added to their cost and delayed licensing. Using the courts to delay construction became a powerful tool. In 1967 construction time for a nuclear power plant in the United States averaged 5.5 years, but by 1980 it had reached twelve years. Increased construction time, added safety features, and the tendency of nuclear power companies to try new designs instead of settling on a standard design raised the price of electricity from nuclear plants. In 1976 the price of electricity from coal-fired and nuclear plants was nearly the same, but by 1990 nuclear power was twice as expensive as coal power in the United States. With this, the antinuclear movement's goal was reached in the United States: With economics against them, planners ceased construction of new nuclear power plants.

Nuclear Power in Asia and Europe

Nuclear power provides 19 percent of the electricity in the United States, 28 percent in Japan, 35 percent in Germany, 50 percent in Sweden, and 79 percent in France. Countries that must import much of their fuel, such as Japan and France, find nuclear power particularly attractive. Nuclear plant construction time in Japan has remained about four years since the early 1960's, and Japan became engaged in a vigorous expansion program during the 1990's.

Antinuclear sentiment led Sweden to schedule a phase-out of nuclear power. The first plant was to be shut down by July, 1998. However, lack of replacement power, fears by workers and industry that the plant's shutdown would lead to higher energy prices and exacerbate unemployment, and a lawsuit by the plant's owners seeking indemnification delayed the shutdown. Polls showed that 60 percent of the Swedes favored keeping the nuclear plants open until they reached their planned lifetimes rather than shutting them down early. As part of the price of forming a ruling coalition with Germany's Social Democratic Party, the environmentalist Green Party extracted a promise that Germany's nuclear power reactors would eventually be shut down. Germany's government faced problems similar to Sweden's in trying to phase out nuclear power and felt additional pronuclear pressure from England and France, who held contracts worth $6.5 billion to reprocess German nuclear fuel.

One of the great challenges faced the antinuclear movement is finding alternative sources of energy. Many nonnuclear power plants also produce serious environmental effects. According to the National Resources Defense Council, sixty-four thousand people in the United States may die prematurely each year from cardiopulmonary causes linked to air pollution. Power plants are major contributors to such pollution, with older, coal-fired plants being among the worst offenders. Burning coal not only produces copious amounts of carbon dioxide but also releases particulates, sulfur compounds, lead, arsenic, naturally occurring radioactive elements, and other harmful elements. Pollution-control equipment accounts for 40 percent of the cost of a new coal-fired plant and 35 percent of its operating costs. This equipment reduces the pollution but does not eliminate it. Many segments of the antinuclear movement actively support the development of alternative energy sources such as solar, geothermal, wind, and biomass power. The United States leads the world in the use of alternative energy; however, such sources account for only 2.7 percent of the electricity generated in the nation.

Scientific Groups

A few scientific groups have been especially influential in arms control. Many scientists who developed the atomic bomb were against leaving decisions about its uses to a few elite government and military officials. Immediately after World War II, some of them formed the Atomic Scientists of Chicago and began publication of the *Bulletin of the Atomic Scientists.* Another group formed the Federation of Atomic Scientists (FAS). Efforts of these scientists contributed to the founding in 1946 of the Atomic Energy Commission (AEC), a civilian agency that took control of the materials, facilities, production, research, and information relating to nuclear fission from the military.

In response to the escalating arms race, in 1955 Albert Einstein and Bertrand Russell published a manifesto calling upon scientists to assemble and appraise the perils of nuclear weapons. Cyrus Eaton, a wealthy industrialist and admirer of Russell, invited twenty-two scientists from both sides of the Iron Curtain to a conference in July, 1957. It was held in Eaton's summer home in the small village of Pugwash, Nova Scotia. The scientists were able to function as icebreakers between governments; in fact, some were government advisers. More Pugwash conferences followed, providing invaluable contacts, networks, and facts for those involved in arms control. The antinuclear movement in general, and scientific organizations in particular, were instrumental in bringing about several treaties, including the Limited Test Ban Treaty of 1963. Signatory nations agreed to end above-ground nuclear testing.

The Union of Concerned Scientists (UCS) was formed in 1969 to combat the establishment of an antiballistic missile (ABM) defense system in the United States. It also played a key role in defeating a scheme to rotate two hundred MX missiles among 4,600 protective silos. The plan would have cost $37 billion and swallowed vast tracts of the western desert of the United States. The UCS has also worked on renewable energy options and other environmental issues.

Charles W. Rogers

SUGGESTED READINGS: *The Antinuclear Movement* (1982), by Jerome Price, provides a good history and analysis of the antinuclear movement. *Nuclear Power: Both Sides* (1982), edited by Michio Kaku and Jennifer Trainer, and *Nuclear Energy:*

Principles, Practices and Prospects (1996), by David Bodansky, present both the successes and problems of nuclear energy. *The Nuclear Freeze Debate* (1983), edited by Christopher A. Kojm, and *Seeds of Promise: The First Real Hearings on the Nuclear Arms Freeze* (1983), by Randall Forsberg, Richard L. Garwin, Paul C. Warnke, and Robert W. Dean, are excellent sources of information on the freeze campaign of the early 1980's. The latter book was sponsored by the FAS. Hugh Gusterson, *Nuclear Rites: A Weapons Laboratory at the End of the Cold War* (1996), bridges the gulf between weapon scientist and antinuclear activist.

See also: Green movement and Green parties; Limited Test Ban Treaty; Silkwood, Karen; Union of Concerned Scientists.

Aquifers and aquifer restoration

Category: Water and water pollution

An aquifer is a water-bearing geological formation that can store and transmit significant amounts of groundwater to wells and springs. Since groundwater supplies a substantial amount of water in many localities, any contamination of the source severely reduces its resource value. Consequently, a variety of restoration techniques have been developed to clean contaminated aquifers.

All rocks found on or below the earth's surface can be categorized as either aquifers or confining beds. An aquifer is a rock unit that is sufficiently permeable to allow the transportation of water in usable amounts to a well or spring. (In geologic usage, the term "rock" also includes unconsolidated sediments such as sand, silt, and clay.) A confining bed is a rock unit that has such low hydraulic conductivity (or poor permeability) that it restricts the flow of groundwater into or out of nearby aquifers.

There are two major types of groundwater occurrence in aquifers. The first type includes those aquifers that are only partially filled with water. In those cases, the upper surface (or water table) of the saturated zone rises or declines in response to variations in precipitation, evaporation, and pumping from wells. The water in these formations is then classified as unconfined, and such aquifers are called unconfined or water-table aquifers. The second type occurs when water completely fills an aquifer that is located beneath a confining bed. In this case, the water is classified as confined, and the aquifers are called confined or artesian aquifers. In some fractured rock formations, such as those that occur in the west-central portions of New Jersey and eastern Pennsylvania, local geologic conditions result in semiconfined aquifers, which, as one might expect, have hydrogeologic characteristics of both unconfined and confined aquifers.

Wells that are drilled into water-table aquifers are simply called water-table wells. The water level in these wells indicates the depth below the earth's surface of the water table, which is the top of the saturated zone. Wells that are drilled into confined aquifers are called artesian wells. The water level in artesian wells is generally located at a height above the top of the confined aquifer but not necessarily above the land surface. Flowing artesian wells occur when the water level stands above the land surface. The water level in tightly cased wells in artesian aquifers is called the potentiometric surface of the aquifer.

Water flows very slowly in aquifers, from recharge areas in interstream zones at higher elevations along watershed boundaries to discharge areas along streams and adjacent floodplains at lower elevations. Thus, aquifers function as pipelines filled with various types of earth material. Darcy's law governing groundwater flow was developed in 1856 by Henry Darcy, a French engineer. In brief, Darcy's law states that the amount of water moving through an aquifer per unit of time is dependent on the hydraulic conductivity (or permeability) of the aquifer, the cross-sectional area (which is at a right angle to the direction of flow), and the hydraulic gradient. The hydraulic conductivity depends upon the size and interconnectedness of the pores and fractures in an aquifer. It ranges through an astonishing twelve orders of magnitude. There are very few other physical parameters that ex-

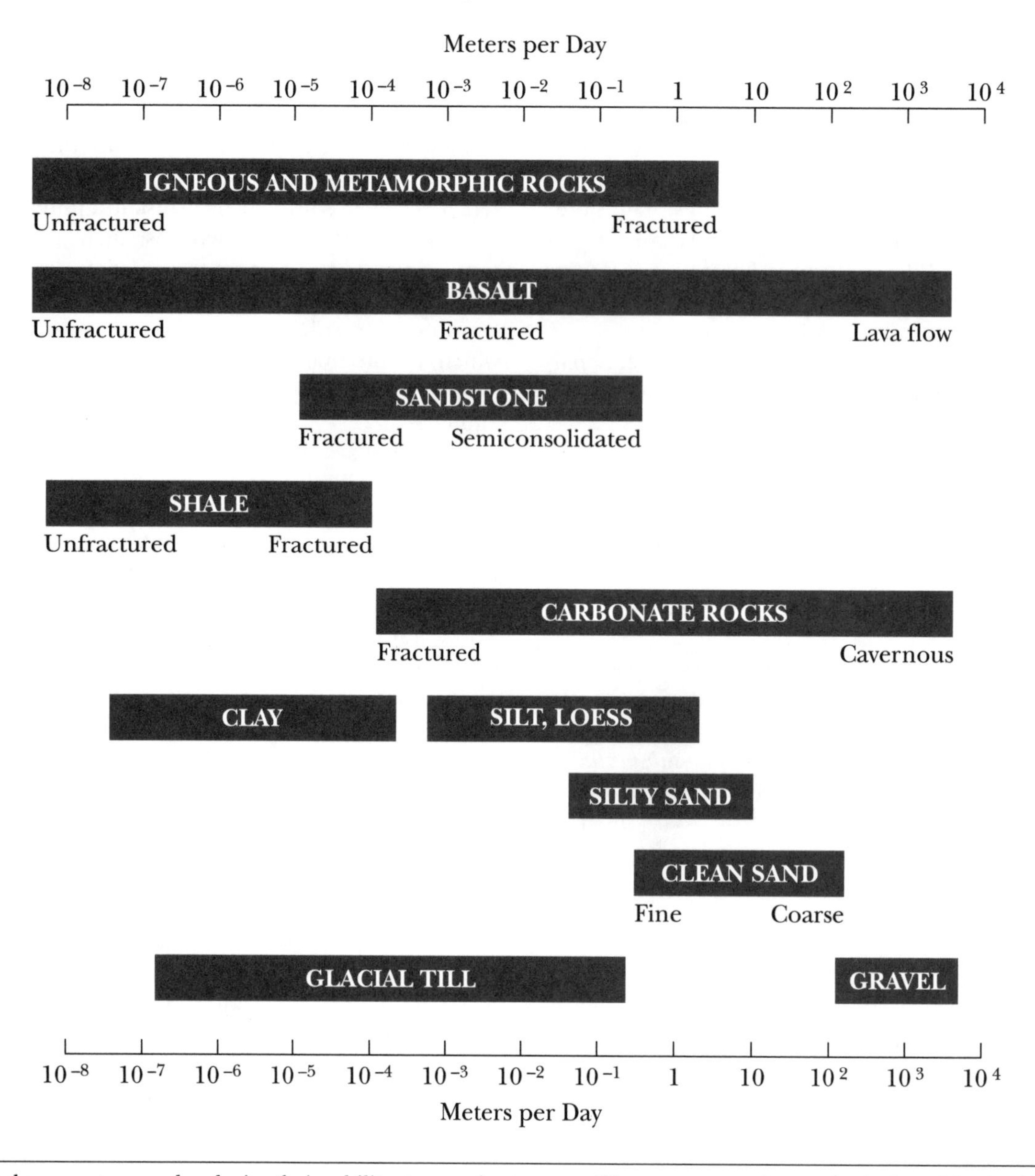

Rocks vary tremendously in their ability to conduct water. The meters-per-day scale is logarithmic: Each increment to the right and left of 1 indicates a change by a power of 10. To the right, 10 meters, 1,000 meters, and 100,000 meters; to the left, 0.1 meter, 0.01 meter, 0.001 meter, and so on.

Source: Ralph C. Heath, *Basic Ground-Water Hydrology*, U.S. Geological Survey Water-Supply Paper 2220, 1983.

hibit such a wide range of values. For example, the hydraulic conductivity ranges from an extremely low 10^{-7} to 10^{-8} meters per day in unfractured igneous rock such as diabase and basalt to as much as 10^3 to 10^4 meters per day in cavernous limestone and coarse gravel. Typical low permeability earth materials include unfractured shale, clay, and glacial till. High permeability earth materials include lava flows, coarse sand, and gravel.

In addition to this wide range of values, hydraulic conductivity varies widely in place and directionality within the same aquifer. Aquifers are isotropic if the hydraulic conductivity is about the same in all directions and anisotropic if it is different in different directions. As a result of all of these factors, groundwater yield is extremely variable both within the same aquifer and from one aquifer to another when they are composed of different rocks.

Since groundwater flows slowly compared to surface water, any contaminant that gets into the groundwater could be around for a long time, perhaps hundreds or thousands of years. Thus, prevention of groundwater contamination is simpler and much more cost-effective than trying to correct a problem that has been in existence for years.

Restoration of a contaminated aquifer may be accomplished, albeit at a price, by one or more of the following procedures: providing inground treatment or containment, providing aboveground treatment, or removing or isolating the source of contamination. The first approach involves natural or in situ treatment based on physical, chemical, or biological means, such as adding nutrients to existing subsurface bacteria to help them break down hazardous organic compounds into nonhazardous materials. The second approach uses engineered systems such as pumping wells or subsurface structures, which create hydraulic gradients that make the contaminated water stay in a specified location, facilitating removal for later treatment. Regardless of the restoration method selected, the source that is continuing to contaminate the aquifer should be removed, isolated, or treated.

Robert M. Hordon

SUGGESTED READINGS: A detailed and well–illustrated review of the various aquifers and hydrologic regions in North America is contained

A flowing artesian well. (Ben Klaffke)

in the Geological Society of America publication *Hydrogeology* (1988), edited by William Back, Joseph S. Rosenshein, and Paul R. Seaber. Aquifer characteristics are thoroughly covered in such standard textbooks as *Applied Hydrogeology* (1994), by Charles W. Fetter, and *Groundwater Hydrology* (1980), by David K. Todd. Aquifer restoration techniques are discussed in a "cook book" style by Jeff Kuo in *Practical Design Calculations for Groundwater and Soil Remediation* (1998). The treatment procedures to remedy contamination by petroleum and other dangerous organic compounds is contained in *Remediation of Petroleum Contaminated Soils* (1998), edited by Eve Riser-Roberts.

See also: Drinking water; Groundwater and groundwater pollution; Water pollution; Water quality; Wells.

Aral Sea destruction

Category: Ecology and ecosystems

The Aral Sea lies in Central Asia, within the former Soviet republics of Kazakhstan and Uzbekistan. Diversion of water from the sea by the Soviet government beginning in the 1950's led to drastic social, economic, and environmental changes in the region.

Prior to 1960, the Aral Sea was a freshwater lake, the fourth-largest in the world, covering 67,300 square kilometers (26,000 square miles). For thousands of years, the Aral has been fed by two streams, the Amu Darya and the Syr Darya, which annually carried fresh water more than 2,400 kilometers (1,500 miles) across largely desert terrain. With no natural outlet, the water balance of the Aral Sea was maintained by extremely high losses to evaporation.

During the 1950's the government of the Soviet Union undertook a massive project to greatly increase the amount of land under cultivation for cotton in Central Asia. To accomplish this, large amounts of water were diverted from the Amu and Syr into irrigation canals. Construction on the largest of these, the 845 kilometer (525 mile) Kara Kum Canal, was begun in 1954 and completed in 1975. Since the development of the canal system, inflow to the Aral has dropped dramatically. The deltaic wetlands of both rivers have largely disappeared, and the bed of the Amu is essentially dry where it finally reaches the Aral Sea.

By the mid-1990's, as a result of water diversion, the Aral Sea had been reduced in surface area by more than 50 percent and in volume by more than 70 percent (an amount equal to one and one-half times that of Lake Erie). In addition, the water level had dropped by 18.2 meters (60 feet). Towns that were once ports at the water's edge became landlocked many miles from the sea. Projections suggested that if the Aral continued to shrink, the sea might cease to exist by the year 2010.

The shrinking of the Aral Sea has brought a number of serious social, economic, and environmental consequences. The Aral was the site of a major fishing industry with twenty-four native species of fish. All are gone now, victims of water salinity that has increased to three times greater than that of the oceans. The island of Muynak, which once lay within the delta of the Amu Darya, now lies more than 64 kilometers (40 miles) from the sea. The thriving cannery industry that existed in Muynak persisted for a few years by processing frozen fish imported from the Baltic Sea and the North Atlantic. With the breakup of the Soviet Union in 1991, however, other former Soviet republics ceased shipping their fish to Muynak, and the industry died.

The region surrounding the Aral Sea has become the site of one of the world's most dramatic examples of environmentally induced disease. The 28,500 square kilometers (11,000 square miles) of exposed former lake bed are covered with precipitated salts. Each year some 43 million tons of dust and dried salt are picked up by the winds and carried as massive dust storms. This wind-borne dust has caused high incidences of emphysema and chronic bronchitis. Arthritic diseases have increased sixtyfold. Other health problems that have shown dramatic increases in the region are throat, intestinal, and liver cancers. The region around the Aral Sea has the highest maternal mortality (120 per 100,000 live births) and infant mortality (60 per 1,000 live births) levels in the former Soviet Union. More than 20 percent of the women between the ages of thirteen and nineteen have kidney diseases, and 23 percent have thyroid dysfunctions. It is estimated that more than 80 percent of the women of the Karakalpakia region suffer from anemia. Pesticides are found in mothers' milk, and high levels of heavy metals, salts, and pesticides occur in the drinking water.

Although the diversion of water for irrigation initially resulted in a tripling of cotton production, the wind-deposited salts have gradually accumulated in the soils, and productivity is declining. Cattle have died in large numbers from eating salt-poisoned vegetation. In addition to containing high levels of salt, the soils are waterlogged, the result of poor drainage.

High evaporation rates over the Aral Sea had, in the past, formed a column of humid air that had deflected and diverted the hot, dry desert winds in the summer and the cold Siberian winds in the winter. This no longer occurs. Summer temperatures in the region are now hotter and winter temperatures are colder. Both seasons show marked decreases in precipitation, and the summer growing season has shortened. The water evaporated from the Aral no longer falls as precipitation on the mountains to the south, the source of the river water, thus disrupting the local hydrologic cycle.

The Aral Sea can never be restored to its former state. Even to stabilize it in its present position would require massive funding, which is not available. Further, this position could only be maintained by cutting the irrigation acreage in half and returning the water flow to the Aral. Some measures that could be taken, however, include improved drainage systems and lining of canals to eliminate seepage into the soil and groundwater. Dumping of untreated municipal and industrial wastes into the Amu must be curtailed. Solar and wind energy, which exist in abundance in the region, might be used to power desalinization and water transfer projects. Proposed schemes that would transfer water from Siberian rivers, however, are impractical.

In 1993 all five of the states that occupy the Aral Sea watershed signed an agreement to improve the situation in the Aral Sea basin. Without funding, however, no unified water strategy has been forthcoming. Environmental concerns still take a back seat to short-term economic gains.

Donald J. Thompson

SUGGESTED READINGS: A complete and well-illustrated history of the changes in the Aral Sea and the environmental consequences is given in "The Aral: A Soviet Sea Lies Dying," by William S. Ellis, *National Geographic* (February, 1990). A brief discussion of the problems related to the Aral Sea are contained in Paul Ehrlich and Anne Ehrlich, *Healing the Planet* (1991). Don Hinrichson, "Requiem for a Dying Sea," *People and the Planet* 4 (1995), provides an excellent discussion of the health issues related to the shrinking Aral Sea.

SEE ALSO: Environmental illnesses; Irrigation; Runoff: agricultural; Soil, salinization of; Water pollution.

Arctic National Wildlife Refuge

DATE: established 1962
CATEGORY: Preservation and wilderness issues

The Arctic National Wildlife Refuge is an intact ecosystem covering 19.5 million acres in northeastern Alaska.

Called "America's Serengeti," the Arctic National Wildlife Refuge is roughly the same size as the state of Maine and encompasses many different topographical elements: mountains, broad valleys, and boreal forest. The northern portion consists of tundra and an icebound coastal plain that provides a vital onshore polar bear denning area and the calving grounds for more than 150,000 caribou. Other animals living in the refuge include bears (black, grizzly, and polar), musk oxen, Dall sheep, moose, wolves, arctic foxes, and approximately 180 species of birds, including golden eagles and arctic peregrine falcons.

Established in 1962, the Arctic National Wildlife Refuge is unique because it essentially remains a wilderness. There are no campgrounds, hiking trails, or souvenir stands; the only sustained human contact comes from two small settlements of indigenous people. At the southern edge of the refuge, the Gwich'in Indians rely on the caribou migrations for survival, while the Inupiat to the north survive on whaling and fishing. What moisture there is comes from the seasonal thawing of the permafrost; the low annual rate of precipitation technically makes it a desert.

Alaska has a long history of environmental exploitation, and the Arctic National Wildlife Refuge is not immune from the controversy. A portion of the refuge known as Area 1002 has become a major battleground between conservationists and the petroleum industry. At issue is whether unspoiled wilderness is more valuable than oil. Though Area 1002 comprises only 1.5 million acres of the refuge's 19.5 million acres, 75 percent of the area is fragile coastal plain. Using the sites in Prudhoe Bay as an example, conservationists argue that the ecosystem is too fragile to support oil drilling. Heavy vehicle traffic cuts into the tundra's mat of vegetation, allowing sunlight to thaw the top layer of permafrost, which in turn creates deep ruts and gullies. Since permafrost limits the biologically active layer of ground to only a few feet, any disruption could wreak havoc within the fragile ecosystem.

The oil companies counter by claiming that the technology for locating oil deposits and drilling into them has improved to the point where the environmental impact of drilling installations is negligible. These arguments carry a lot of weight because revenue from the oil industry finances 85 percent of Alaska's state government and provides an annual dividend to every resident. Consequently, the state government and congressional representatives tend to be prodevelopment in general and pro-oil in particular. In 1998 the U.S. government authorized drilling in more than 87 percent of Alaska's National Petroleum Reserve, located west of the Prudhoe Bay and Kuparuk oil fields. This move was made, in part, to deflect development plans in the Arctic National Wildlife Refuge.

P. S. Ramsey

See also: Alaska Highway; Alaska National Interest Lands Conservation Act; Oil drilling; Trans-Alaskan Pipeline; Wildlife refuges.

Argo Merchant oil spill

Date: December 15, 1976
Category: Water and water pollution

On December 15, 1976, the tanker Argo Merchant *grounded off the coast of Rhode Island, spilling its cargo of heavy fuel oil into the sea.*

The *Argo Merchant* was built as the *Arcturus* in Hamburg, Germany, in 1953 and renamed the *Permina Samudia III* in 1968. It was then renamed the *Vari* in 1970 and finally the *Argo Merchant* in 1973. Regardless of the name or the owners, the ship had a long history of accidents. The *Argo Merchant* was not an extremely large vessel. It was 195 meters (641 feet) in length and 25.6 meters (84 feet) wide, and had a draft of 10.7 meters (35 feet). It was owned by the Thebes Shipping Company of Greece, chartered to Texaco, and carried a crew of Greek officers and Filipino crewmembers under the Liberian flag.

The vessel departed Puerto La Cruz in Venezuela bound for Salem, Massachusetts, with a cargo of 7.6 million gallons of heavy fuel oil in its thirty cargo tanks. The ship's problems began the evening before the grounding. The *Argo Merchant*'s gyrocompass had broken, and the officers were unable to get the ship's position.

At 6:00 a.m. on December 15, 1976, the *Argo Merchant* grounded on the southern end of Fish Rap Shoal in 5.5 meters (18 feet) of water. The ship was 48 kilometers (30 miles) north of the Nantucket Lightship, 24 kilometers (15 miles) outside the normal shipping lanes, and 39 kilometers (24 miles) off its charted track line. When the crew of the *Argo Merchant* called the United States Coast Guard to report the grounding, they had no idea of their position. The position they gave was 48 kilometers (30 miles) from where the tanker was located.

The Coast Guard, however, responded quickly and soon located the grounded vessel. They tried both floating it and towing it, but neither method worked. The sea continued to batter the grounded ship, which began to bend and twist on the reef until oil started leaking from the cargo tanks. Within one week the vessel had broken in half and was leaking large amounts of oil. As in the case of the *Torrey Canyon*, attempts were made to ignite the oil, but no attempts were made to destroy the oil remaining in the ship's tanks.

By Christmas the slick extended 160 kilometers (100 miles) from the *Argo Merchant*. Although the weather was rainy and windy, which made oil transfer and salvage operations difficult, the wind blew predominately from the northwest, and this drove the oil offshore. Thus, very little oil came ashore in the rich fishing and shellfish areas along the coast of New England. However, the northwesterly wind drove the slick across the Georges Bank area, which is one of the richest fishing areas in the Atlantic Ocean.

The short-term effects of the wind-blown oil were marginal, and only a limited number of birds and marine mammals were killed. The long-term effects of oil in the water column and the oil's impact on bottom-dwelling creatures have not been documented.

Robert J. Stewart

SEE ALSO: *Amoco Cadiz* oil spill; *Braer* oil spill; *Exxon Valdez* oil spill; Oil spills; *Sea Empress* oil spill; Tobago oil spill; *Torrey Canyon* oil spill.

Asbestos

CATEGORY: Pollutants and toxins

Asbestos is an industrial term for certain silicate minerals that occur in the form of long, thin fibers. The adverse health effects of breathing high concentrations of asbestos over many years have been known since the early 1970's. The Clean Air Act classified asbestos as a carcinogenic material, and in 1990 the U.S. Environmental Protection Agency (EPA) established a broad ban on the manufacture, processing, importation, and distribution of asbestos products.

Asbestos-form minerals are natural substances that are common in many types of igneous and metamorphic rocks found over large areas of the planet. Erosion continually releases these fibers into the environment; most people typically inhale thousands of fibers each day or more than 100 million over a lifetime. Asbestos fibers also enter the body through drinking water. Drinking water supplies in the United States typically contain almost 1 million fibers per quart, but water in some areas may have as many as 100 million or more fibers per quart.

Many silicate minerals occur in fibrous form, but only six have been commercially produced as asbestos. In order of decreasing commercial importance, these are chrysotile (white asbestos), crocidolite (blue asbestos), amosite (brown asbestos), anthophyllite, tremolite, and actinolite. All the minerals except chrysotile are members of the amphibole group of minerals, which have a chainlike arrangement of the atoms. In contrast, chrysotile, as a member of the serpentine family, has atoms arranged in a sheetlike fashion.

Although the individual properties of these minerals differ greatly from one another, they share several characteristics that make them useful and cost-effective. These include great resistance to heat, flame, and acid attack; high tensile strength and flexibility; low electrical conductivity; resistance to friction; and a fibrous form, which allows them to be used for the manufacture of protective clothing. Therefore, asbestos was widely used until the 1970's in a great variety of building and industrial products. Such common materials as vinyl floor tiles, appliance insulation, patching and joint compounds, automobile brake pads, hair dryers, and ironing board covers all might have contained asbestos. Most such products now contain one or more of several substitutes for asbestos instead of asbestos itself. However, many of the substitutes may not be hazard-free, a fact that is starting to be recognized by legislators. For example, in 1993 the World Health Organization (WHO) stated that all substitute fibers must be tested to determine their carcinogenicity. Germany now classifies glass, rock, and mineral wools as probable carcinogens.

In the early 1970's asbestos was classified as carcinogenic by the United States government, prompting the removal of asbestos products from both public and private buildings. (Jim West)

The U.S. Department of Health and Human Services classifies asbestos as a carcinogen. Studies leading to this determination were mostly based on asbestos workers who were exposed to extremely high levels of fibers for many years. These studies concluded that the asbestos workers have increased chances of developing two types of cancer: mesothelioma (a cancer of the thin membrane surrounding the lungs) and cancer of the lung tissue itself. These workers were also at increased risk of developing asbestosis, an accumulation of scarlike tissue in the lungs that can cause great difficulty in breathing and permanent disability. None of these diseases develops immediately; instead, they have a long latency period, typically fifteen to forty years. Despite the common misconception, exposure to asbestos does not cause muscle soreness, headaches, or any other immediate symptoms. The effects typically are not noticed for many years.

It is generally agreed that the risk of developing disease depends on the number of fibers in a person's body, how long the fibers have been in the body, and whether one is a smoker, since smoking greatly increases the risk of developing disease. There is no agreement on the risks associated with low-level, nonoccupational exposure. The EPA has concluded that there is no safe level of exposure to asbestos fibers, but the Occupational Safety and Health Administration (OSHA) allows up to one thousand fibers per cubic meter during an eight-hour work day.

Another area of controversy stems from scientific studies showing that all forms of asbestos are not equally dangerous. Evidence has shown that the amphibole forms of asbestos, and particularly crocidolite, are hazardous, but the serpentine mineral chrysotile—accounting for 95 percent of all asbestos used in the past and 99 percent of current production—is not. For example, one case study involved a school that was located next to a 150,000-ton rock dump containing chrysotile. Thousands of children played on the rocks over a one-hundred-year period, but not a single case of asbestos-related disease developed in any of the children. The difference seems to be in how the human body responds to amphibole compared to chrysotile. The immune system can eliminate chrysotile fibers much more readily than amphibole, and there is also evidence that chrysotile in the lungs dissolves and is excreted. This remains a controversial area, and the U.S. government still treats all forms of asbestos the same. This is not true of some European governments.

The risk of developing any type of disease from exposure to normal levels of asbestos fibers in outdoor air or the air in closed buildings is extremely low. The calculations of Melvin Benarde in *Asbestos: The Hazardous Fiber* (1990) show that the risk of dying from nonoccupational exposure to asbestos is one-third the risk of being killed by lighting. The Health Effects Institute made similar calculations in 1991 and found that the risk of dying from asbestos is less than 1 percent the risk of dying from exposure to secondary tobacco smoke.

Gene D. Robinson

SUGGESTED READINGS: A good general discussion of asbestos is contained in *Industrial Minerals Geology and World Deposits* (1990), by Peter W. Harben and Robert L. Bates. H. Catherine W. Ross, Malcolm Ross, and Clifford Frondel provide a balanced look at the health effects of asbestos in *Asbestos and Other Fibrous Minerals* (1988). An example of a study conducted to determine the health effects of chrysotile is J. C. McDonald et al., "The Health of Chrysotile Asbestos on Mine and Mill Workers of Quebec," *Archives of Environmental Health* 28 (1974).

SEE ALSO: Asbestosis; Environmental illnesses; Hazardous and toxic substance regulation.

Asbestosis

CATEGORY: Human health and the environment

Asbestosis is a disease of the lungs caused by repeated exposure to asbestos.

Asbestos is a generic term applied to such minerals as amosite, anthophyllite, chrysotile, and crocidolite. These silicates, first used extensively in the 1940's, have remarkable qualities that initially made them desirable as both thermal and electrical insulators. After the insulating properties of asbestos were recognized, such substances were commonly used in the building trades, notably to insulate pipes and boilers. Some cements and floor tiles contained substantial quantities of asbestos, which was used as a fireproof filler. Forms of asbestos were also used to make blankets designed to smother fires and in manufacturing safety garments for firefighters. The substance is still used in the manufacture of automotive brake and clutch linings, where its insulating qualities are particularly valued.

Asbestosis, a disease of the lungs, occurs in people who have been exposed to this silicate, but the onset of symptoms may occur years after exposure. Even those whose exposure to asbestos has been moderate may contract the disease, an early symptom of which is perpetual shortness of breath, especially following strenuous physical activity. This symptom is often accompanied by a persistent, hacking cough. Asbestosis is a progressive disease that characteristically begins in the lower lung and spreads to the middle and upper lungs, with disabling, sometimes fatal, results as the air spaces in the lungs narrow. Little aggressive treatment exists for the disease. Asbestosis victims are urged not to smoke and are usually treated with oxygen to improve their breathing.

Direct exposure to asbestos, especially when it occurs over extended time periods, often results in asbestosis. It is found among those who have worked in asbestos mines or as pipe fitters, boilermakers, automotive mechanics working with brake and clutch linings, and demolition workers who raze buildings in which asbestos was used for insulation.

Indirect exposure can also result in asbestosis. The disease has been reported among people who have regularly laundered work clothes that have been directly exposed to asbestos or those who live or work in buildings where asbestos has been used as an insulator. Widespread community exposure has been noted in situations where steel girders in large buildings were sprayed with asbestos as a fire precaution. Cases occurred in a

London neighborhood near an asbestos plant among those who had no direct contact with the plant. South Africa had outbreaks among the general population near its asbestos mines.

After 1975 the use of asbestos declined substantially as it was replaced by other mineral fibers. People currently working with asbestos, particularly demolition workers, are specially trained to avoid exposure that could lead to the onset of the disease. Environmental laws in most jurisdictions prohibit the use of asbestos as an insulating material in new buildings. Many laws now protect workers from exposure. Those who were exposed earlier and currently suffer from the disease are usually eligible for workers' compensation.

R. Baird Shuman

See also: Asbestos; Environmental health; Environmental illnesses; Hazardous and toxic substance regulation; Sick building syndrome.

Ashio, Japan, copper mine

Date: began operating during the seventeenth century; closed in 1972
Category: Human health and the environment

Development of the Ashio copper mine propelled Japan's industrial revolution and set the stage for the nation's first conflict over environmental quality. Hazardous runoff from Ashio emphasized the tension between agriculture and industry; it also dramatized the human and environmental costs of industrial pollution.

The Ashio mine, located 110 kilometers (68 miles) north of Tokyo, first operated during the seventeenth century, but private ownership spurred development of the mine in 1877. By 1890 Ashio was the largest copper mine in Asia. Production expanded in 1950 because of the Korean War. The Ashio mine closed in 1972.

As copper production increased, the environmental impact on the surrounding area also increased. The Ashio mine, located at the headwaters of the Watarase River, caused environmental damage by depositing waste products in the river. By 1880 fish were beginning to die, and people who ate fish from the river became ill. Almost all marine life in the river had died by 1890. Deforestation compounded the pollution problem. The mine needed timber to shore up the shafts, for railroad ties, for mine buildings, and as fuel for steam engines. The mine obtained this timber by deforesting 104 square kilometers (40 square miles) of surrounding land, thus destroying the watershed at the head of the Watarase River. As a consequence, flooding became a serious problem in the Watarase Valley and the surrounding rice fields.

Although natural flooding had occurred before the development of the mine, such flooding had brought layers of rich silt that contributed to abundant crops. Later floods, however, produced vastly different results: Vegetation did not survive contact with the contaminated floodwaters. Floods became more frequent, more severe, and more damaging because they left poisons in the soil. Soil samples revealed concentrations of sulfuric acid, ammonia, magnesia, iron, arsenic, copper, and chlorine. These substances poisoned the rice fields, and new seeds would not grow. Earthworms, insects, birds, and animals succumbed to the contamination.

Although the Japanese government ordered pollution control measures, they were ineffective. In 1907 the government forced the evacuation and relocation of the inhabitants of a contaminated village. The collapse of a slag pile in 1958 introduced 2,000 cubic meters (70,630 cubic feet) of slag into the Watarase River, contaminating 14,820 acres of rice fields. The Japanese government set limits on the amount of copper that could be deposited in the river water and the soil, but the damage had already been done. In 1972 soil samples from 3 meters (9.8 feet) down still contained excessive amounts of copper as well as significant amounts of lead, zinc, and arsenic. The government ordered that rice grown in the area had to be destroyed.

Thousands of fishermen, rice farmers, and valley citizens suffered severe economic losses. Serious health problems in the area included high infant mortality rate, failure of new mothers to produce milk for their infants, sores on field workers, and high death rate. In 1973 the Environmental Agency's Pollution Adjustment Committee began to review farmers' claims. The mine was required to admit to being the source of the contamination, and the farmers were awarded US$5 million dollars in 1974. Attempts to reforest the area have failed.

Louise Magoon

SEE ALSO: Acid mine drainage; Fish kills; Heavy metals and heavy metal poisoning.

Aswan High Dam

DATE: completed 1970
CATEGORY: Preservation and wilderness issues

Since its completion in 1970, the Aswan High Dam on the Nile River in Egypt has supplied electricity to the region and has helped control seasonal flooding. These benefits, however, have been accompanied by several negative environmental effects, including loss of fertility on downriver floodplains, increased erosion, and increased incidence of earthquakes.

The Aswan High Dam was built with the aid of Soviet engineers on the Nile River approximately 960 kilometers (600 miles) south of Cairo, Egypt, between 1960 and 1970 at a cost of US$1 billion. The dam lies 7 kilometers (4.5 miles) south of Aswan City and several kilometers from a smaller dam constructed by British engineers between 1898 and 1902. The building of the high dam followed the signing of the Nile Water Agreement between Egypt and Sudan in November of 1959.

The high dam is a rock-fill structure with a core of impermeable clay. It measures 3,829 meters (12,562) feet long, 111 meters (364 feet) high, 980 meters (3,215 feet) wide at the base, and 40 meters (131 feet) wide at the crest. The volume of material contained in the structure, 1.6 million cubic meters (58 million cubic feet), would be enough to construct seventeen Great Pyramids. The flow of the river's waters through the dam is via six tunnels, each controlled by a 230-ton gate.

The reservoir impounded by the high dam is Lake Nasser, named for Egyptian president Gamal Abdel Nasser, who died the year the dam was finished. That portion of the reservoir that lies within Sudan, about 30 percent, is referred to as Lake Nubia. In total, the reservoir measures 499 kilometers (310 miles) in length and has a surface area of approximately 5,996 square kilometers (2,315 square miles) and 9,053 kilometers (5,625 miles) of shoreline. It averages 9.7 kilometers (6 miles) wide, with a maximum width of 16 kilometers (10 miles). Mean water depth is 70 meters (230 feet), maximum depth is 110 meters (360 feet), and the annual vertical fluctuation is 25 meters (82 feet). The reservoir contains enough water to irrigate more than 7 million acres.

The filling of Lake Nasser came with a high cost. Thousands of people were displaced, and natural habitats were significantly altered. Ancient temples and monuments that abound in the region were to be submerged beneath the rising waters of the reservoir. Some of these were saved by cutting them into large blocks that were reassembled at higher locations. High evaporation rates and water loss to infiltration into the underlying permeable Nubian sandstone caused the filling of the reservoir to take much longer than anticipated. A significant amount of water storage capacity has also been lost as sediment is carried in and occupies a part of the reservoir's volume.

The benefits derived from the high dam are principally hydroelectric power generation and the regulation of water flow along the lower Nile for flood control. Twelve turboelectric generators capable of producing ten billion kilowatt-hours provide 40 percent of Egypt's electrical power. The storage of water in Lake Nasser not only provides flood control but also allows for the irrigation of additional land and the ability to grow multiple crops over the course of a year. Since the filling of the reservoir, a fishing industry has also developed.

The yearly floods of the Nile are the result of late summer rains that fall in the plateau region to the south in Ethiopia. At peak floods, river volume may increase by as much as sixteen times. More than 100 million tons of soil are carried with the water each year. While the impoundment of water in Lake Nasser has probably saved the lower Nile Valley from disastrous floods and alleviated the effects of regional droughts, the loss of the yearly increments of silt on the floodplain, with the associated nutrients, has led to a decline in the floodplain's fertility. Without this natural fertilization, the Egyptians have had to rely on increasing use of artificial fertilizer. The floodwaters also provided a cleansing and draining action for the soil, preventing the accumulation of salts. Further, the floodwaters reduced the number of rats and disease-bearing snails. With the decline in flooding, incidences of disease have been on the increase.

The influx of sediment to the Nile Delta has, historically, replenished sediment lost to wave and current erosion at the delta's margins. Since the construction of the high dam, the front of the Delta is being eroded at a rate of 1.8 meters (6 feet) per year. As on the floodplain, the soil of the Delta, a region that has been farmed for more than 7,000 years, also shows evidence of declining fertility.

Since the high dam traps 98 percent of the Nile sediment, water passing through the dam has an enhanced ability to erode. Consequently, downstream erosion has become a significant problem, scouring the riverbed and undermining riverbanks and bridge piers. In some instances, the increased erosion has also affected the delicate balance of water irrigation systems.

The effects of trapped sediment are not confined to the floodplain and the Delta of the Nile. Prior to construction of the dam, the river brought sediment and nutrients into the normally nutrient-poor eastern Mediterranean Sea. This provided for blooms of phytoplankton that formed the base of a food pyramid that included sardines and other commercial varieties of fish. When construction of the dam began, the sardine fishing industry in the Mediterranean significantly declined. Beginning in the late 1980's, however, there has been a resurgence in sardine fishing, as the filling of Lake Nasser allowed for increased river discharge and nutrient enhancement.

One other environmental consequence of the construction of the high dam has been the occurrence of earthquakes in the region. These are related to stress that is placed on the earth's crust by the weight of the water impounded in Lake Nasser. A large shock of Richter magnitude 5.6, for example, occurred on November 14, 1981. This was followed by aftershocks for a period of seven months.

Donald J. Thompson

SUGGESTED READINGS: Excellent diagrams of the structure of the Aswan High Dam are given in "Yankee Cruises the Storied Nile," by Irving and Electra Johnson, *National Geographic* (May, 1965). A more recent consideration of the effects of the high dam is included in "Journey up the Nile," by Robert Caputo, *National Geographic* (May, 1985). A brief discussion of the political background and environmental consequences of the building of the high dam is included in Norman Smith's *A History of Dams* (1972).

SEE ALSO: Dams and reservoirs; Flood control; Hydroelectricity; Sedimentation.

Audubon, John James

BORN: April 26, 1785; Saint-Domingue (now Haiti)
DIED: January 27, 1851; New York, New York
CATEGORY: Animals and endangered species

John James Audobon was an American naturalist and wildlife artist widely known for his accurate and lifelike illustrations of North American birds. The National Audobon Society, founded in 1886, was named for him.

John James Audubon was born Jean Jacques Audubon, the son of a French naval officer, on April 26, 1785. He grew up in Saint-Domingue (now Haiti) and Nantes, France. At the age of eighteen he was sent to live on a farm at Mill Grove, near Philadelphia, Pennsylvania, to escape induction into Napoleon Bonaparte's army. Audubon spent his days in Pennsylvania roaming the woods and observing, collecting, and sketching wildlife. He experimented with bird banding and developed techniques to mount bird specimens so they could be drawn in lifelike poses. During this time he began thinking of himself as an American and calling himself John James.

As Audubon matured, the hobby of his youth became an obsession. The time he spent outdoors teaching himself to live in the forest and understand its inhabitants cost Audubon dearly in financial terms and tried the patience of his wife Lucy. However, it gave him unparalleled insights into the lives of his subjects. A perfectionist who periodically tore up drawings he considered less than his best, Audubon developed a distinctive style, illustrating birds and animals in lively interaction with their surroundings.

Audubon established a frontier store and sawmill in Henderson, Kentucky, but the ventures failed because Audubon so often neglected his business to pursue his art. Forced to declare bankruptcy in 1819, Audubon first thought of giving up painting altogether, then decided to devote himself completely to it. He resolved to publish a portfolio of American birds, more complete and accurately illustrated than any previous work. Lucy taught school to support the family while he worked furiously to produce the necessary sketches and find a publisher. Audubon suffered mixed reviews from the American scientific and artistic establishments but found a receptive audience in Europe.

John James Audubon, who gained fame with his detailed drawings of birds and other animals interacting with their environments. (Library of Congress)

In 1826 Audubon made arrangements with an engraver to make folio-sized prints of his illustrations and publish them under the title *The Birds of America.* This large scale book presented each bird study as a life-size portrait and preserved the fine details of Audubon's watercolors. The hand-colored printing process was laborious and very expensive. To finance the project, Audubon sold subscriptions to a series of separate folios, each containing five engravings. It took him twelve years to complete the 435 plates. Although only 176 subscriptions were sold, Audubon covered his expenses and established an international reputation. The smaller and less expensive edition of the bird book, which included excerpts from his *Ornithological Biography* (1839), was a popular and financial success. Finally, Audubon was able to buy an estate on the Hudson River in New York, where he and Lucy mentored many young scholars, artists, and naturalists. One of these students, George Bird Grinnell, founded the National Audubon Society, a worldwide organization promoting bird study and conservation, in 1886. As Audubon's energy and eyesight waned, his sons helped him complete paintings for *The Viviparous Quadrupeds of North America* (1849-1854).

Audubon's unique paintings demonstrated ecological relationships among organisms by illustrating their food plants, nesting sites, competition, and predators. In his writings, Audubon set new standards for field observation and foresaw the threat of species extinction. Above all, by illustrating the beauty of birds and animals, he promoted the popular study of natural history, laying foundations for a national environmental consciousness in the United States.

Robert W. Kingsolver

SEE ALSO: Conservation; Endangered species.

Automobile emissions

CATEGORY: Atmosphere and air pollution

Fossil fuel emissions from transportation vehicles contain carbon monoxide and other compounds that undergo secondary reactions under certain environmental conditions. Such emissions negatively impact the environment and human health.

Automobile emissions create ongoing and potentially dangerous environmental problems when gases and particulates are released into the atmosphere at a rate that exceeds the capacity of the atmosphere to dissipate or dispose of them. Smog, a term coined in 1905 in England to describe the combination of smoke and fog, comes from exhaust fumes from transportation vehicles and other sources of air pollution that contain carbon monoxide and a variety of complex hydrocarbons, nitrogen oxides, and other compounds.

This collection of chemicals changes in composition when exposed to the heat of sunlight, thus producing the brown smog seen in urban areas of the United States, such as Los Angeles, California, and Denver, Colorado. Carbon monoxide levels peak twice during the day, corresponding to morning and evening rush-hour traffic, with Los Angeles urban areas averaging 37 parts per million in general and 54 parts per million in heavy traffic, and peaking at 120 parts per million near stop signals. This elevated level extends up to 20 meters (65 feet) from the roadway. Policemen in Tokyo, Japan, often wear supplemental oxygen units when directing traffic at busy intersections. Temperature inversions cause a greenhouse effect, which reverses the ecosystem's normal atmospheric temperature gradient, heats the harmful chemicals, and enhances their negative effects via delayed photochemical reactions. Secondary pollutants formed by photochemical reactions under certain atmospheric conditions include ozone, formaldehyde, and peroxyacylnitrate, with ozone being the most highly reactive and dangerous to the environment and human health.

Researchers have discovered that nitrogen dioxide from exhaust fumes will create ozone one hundred times more rapidly when combined with hydrocarbons that are naturally produced by trees than with hydrocarbons that are produced by human sources. The Environmental Protection Agency (EPA) originally estimated that Atlanta, Georgia, could meet federal air-quality ozone standards by reducing exhaust levels by 30 percent, but more later data that took into account the contribution by trees indicated that a 70 to 100 percent reduction would be necessary to meet current standards.

Automobile emissions have been shown to exert their negative effects a considerable distance from the source depending upon atmospheric changes in wind and temperature. Suburbs surrounding rural areas often exhibit higher levels of pollution than the downtown areas where the emissions were originally produced. Fallout of tetraethyl lead from urban automobile exhausts has been observed in oceans and on the Greenland ice sheet.

Studies in cities such as London, England, have shown that major improvements in air quality can be achieved in fewer than ten years in urban areas with favorable climatic conditions by requiring cleaner-burning fuels and more combustion-efficient engines. In an attempt to reduce automobile emissions, the U.S. Congress passed legislation requiring that the exhaust gases of new vehicles pass through catalytic converters, which transform more of the carbon monoxide and hydrocarbons into carbon dioxide and water. However, these converters only

A pollution-control official in the Philippines inspects a taxi cab. Restrictions on automobile emissions, as well as such improved technologies as cleaner fuels and more efficient engines, have helped improve air quality in many urban areas around the world. (Reuters/Romeo Ranoco/Archive Photos)

minimally reduce carbon dioxide or nitrogen oxide levels. Automobiles with catalytic converters can meet emissions standards if they are appropriately tuned and burn only unleaded gasoline.

Environmental problems associated with automobile emissions include deleterious effects on many forms of agriculture and natural forests, as well as damage to metal, building materials such as stone and concrete, rubber, paint, textiles, and plastics. Automobile emissions can reduce visibility and cause lung and eye irritation and coughing, chest pain, shallow breathing, and headaches. Automobile-produced air pollution is also a contributing factor to allergies, asthma, emphysema, bronchitis, lung cancer, heart disease, and negative psychological states. Carbon monoxide quickly combines with blood hemoglobin and impairs oxygen delivery to the tissues, particularly in children and senior citizens, causing heart and lung problems to such an extent that some large urban areas regularly broadcast air-quality reports over television and radio.

The increased rate and depth of breathing during physical exertion exposes more polluted air to delicate lung tissues. Research indicates that exercise near a busy freeway may be more harmful than beneficial to the body. The evening rush-hour start of the 1984 Los Angeles Olympic men's marathon during a stage 2 California Health Advisory alert drew criticism that organizers were more interested in commercial revenues than the safety of the athletes and spectators. Later events were postponed during heavy air-pollution episodes.

Professional organizations such as the American Lung Association conservatively estimate that air pollution costs Americans nearly $100 billion each year in health care and related costs. The 1970, 1977, and 1990 amendments to the Clean Air Act set emissions standards for automobiles and ambient air-quality standards for urban areas for six pollutants—carbon monoxide, sulfur oxides, nitrogen oxides, particulates, ozone, and hydrocarbons—to better protect human health and the environment.

Daniel G. Graetzer

SUGGESTED READINGS: Robert Jennings Heinsohn and Robert Lynn Kabel, *Sources and Control of Air Pollution* (1999), and Kenneth Wark, Cecil F. Warner, and Wayne T. Davis, *Air Pollution: Its Origin and Control* (1998), provide excellent reviews of urban air pollution from automobile emissions, air purification methods, and the effects of the U.S. Clean Air Act amendments of 1990. Christopher J. Bailey's *Congress and Air Pollution: Environmental Policies in the USA* (1998) reviews the environmental politics behind the legislative battles and laws regarding automobile emissions reductions. Maureen Sevigny's *Taxing Automobile Emissions for Pollution Control* (1998) assesses the environmental impact of California's motor vehicle policies with respect to exhaust gases. Fred Schafer's and Richard Van Basshuysen's *Reduced Emissions and Fuel Consumption in Automobile Engines* (1995) describes pollution-control devices on motor vehicles.

SEE ALSO: Air pollution; Air pollution policy; Catalytic converters; Clean Air Act and amendments; Smog.

Bacillus thuringiensis

Category: Biotechnology and genetic engineering

Bacillus thuringiensis (B.t.) is one of a number of natural microbial pesticides used for biological control of insects.

Chemical control of plant parasites using pesticides, while often effective, has generally been found to create its own problems in areas of environmental pollution. The use of microbial pesticides has often proven as effective at pest management, without the problems of polluting soil and water. In addition, since many of these organisms are species specific, targeting only select insects, neither plants nor more desirable insects (such as ladybugs) are at risk.

B.t. is one of several bacteria species that have been used as biopesticides for control of insects. The bacterium produces a proteinaceous parasporal crystal within the cell when it sporulates, which is toxic only for a specific insect host. The crystalline protein is harmless in its natural state, becoming toxic only when it is cleaved by specific proteases. Once the bacterium has been ingested by the target insect, the alkaline secretions within the midgut of the insect activate the toxic character of the crystalline body. The toxin binds the intestinal cells, resulting in formation of pores and leakage of nutrients from the gut. The insect dies within three to five days. The ability of the bacterium to sporulate also provides a means to disseminate the organism as a spray. Some commercial treatments utilize the crystalline body itself, which is sprayed onto the plants. Though commercial application of *B.t.* first began in the 1960's, its widespread use only began decades later with the decline in use of chemical pesticides for protecting food crops and other types of plants.

Various strains of *B.t.* have been developed for use against numerous types of insects. The *israelensis* variety is used commercially for control of mosquitos and blackfly larvae. The variety *kurstaki* has been used effectively against both the gypsy moth and cabbage loopers. More recently developed strains of *B.t.* have been shown to effectively control insects that feed on fruit crops such as oranges and grapes. Varieties of *B.t.* are used commercially against nearly 150 different insect species, including cabbage worms, caterpillars, and other insects that feed on vegetable crops. The targeting of specific insects while using certain strains of *B.t.* also reduces the chance that desirable populations of organisms may be adversely affected.

Attempts have also been made through genetic engineering to introduce the gene that encodes the crystalline body directly into the plant genome. Since the protein is toxic only to insects, such an approach may result in naturally insect resistant plants without further need for spraying. The procedure has proven effective in controlled experiments using tobacco plants, and more widespread testing in other plants will also occur. The controversial nature of introducing such potentially toxic genes may preclude their use in food crops for some time, but such genetic experiments may prove useful in natural protection of ornamental trees and other plants.

Richard Adler

See also: Biopesticides; Genetically altered bacteria; Integrated pest management.

Bacterial resistance and super bacteria

Category: Human health and the environment

Inappropriate antibiotic use has caused bacteria to develop resistance against the most common antibiotics. Previously controlled infectious path-

ogens with multiple-drug resistance are becoming an increasing threat to public health safety.

Bacteria are the most adaptable living organisms on earth and are found in virtually all environments—from the lowest ocean depths to the highest mountains. Bacteria resist extremes of heat, cold, acidity, alkalinity, heavy metals, and radiation that would kill most other organisms. *Deinococcus radiodurans*, for example, grow within nuclear power reactors, and *Thiobacillus thiooxidans* can grow in toxic acid mine drainage. Super bacteria, however, generally refers to bacteria that have either intrinsic (naturally occurring) or acquired resistance to multiple antibiotics. For example, two soil organisms—*Pseudomonas aeruginosa* and *Burkholderia* (*Pseudomonas*) *cepacia*—are intrinsically resistant to many antibiotics. Because many of the bacteria that acquire resistance are pathogens that were previously controlled by antibiotics, development of antibiotic resistance is now regarded as a serious public health crisis, particularly for those individuals who have compromised immune systems.

History of Antibiotic Use

The history of antimicrobial compounds reaches into the early twentieth century when German chemist Paul Ehrlich received worldwide fame for discovering Salvarsan, the first relatively specific prophylactic agent against the microorganisms that caused syphilis. Salvarsan had serious undesirable side effects since it contained arsenic as an active ingredient. In addition, despite advances in antiseptic surgery, secondary infections resulting from hospitalization were a leading cause of death in the early twentieth century. Consequently, when Scottish bacte-

Milestones in Antibiotic Use

Year	Event
1908	Photographs reveal the antimicrobial activity of "ray" fungi, which is regarded as a minor curiosity and ignored.
1910	Paul Ehrlich discovers Salvarsan, the "magic bullet" that proves that chemical agents can attack pathogens without attacking their hosts.
1928	Alexander Fleming's chance discovery of penicillin on a contaminated plate launches the era of antibiotics.
1940	Howard Florey and Ernst Chain isolate and purify penicillin for general clinical use, which allows the increased demand for antibiotics during World War II to be met.
1943	Streptomycin is discovered by Selman Waksman; tuberculosis, which is controlled by this antibiotic, dramatically declines.
1967	Penicillin-resistant *Streptococcus pneumonia* is isolated, and widespread antibiotic use begins to exert selective pressure for resistant bacteria.
1970's	The prophylactic use of antibiotics in animal science promotes animal growth but raises concerns about resistant bacteria.
1988	Resistance to Vancomycin, the last effective antimicrobial against multiple drug-resistant enterococci, increases deadly infections where enterococci are common.
1990's	Widespread antibiotic resistance is reported; health professionals fear that the war against infectious disease is in danger of being lost.

riologist Alexander Fleming reported his discovery of a soluble antimicrobial compound called penicillin, produced by the fungus *Penicillium*, it attracted worldwide attention.

Antibiotics such as penicillin are low-molecular-weight compounds excreted by bacteria and fungi. Antibiotic-producing microorganisms most often belong to a group of soil bacteria called actinomycetes. *Streptomyces* are good examples of antibiotic-producing actinomycetes, and most of the commercially important antibiotics are isolated from *Streptomyces*. Antibiotics are produced by microorganisms late in growth, and it is not entirely clear what ecological role the antibiotics play in natural environments.

Fleming's discovery was not of much clinical importance until two English scientist, Howard Florey and Ernst Chain, took Fleming's fungus and produced purified penicillin just in time for World War II. The success of penicillin as a therapeutic agent with almost miraculous effects on infection prompted other microbiologists to look for naturally occurring antimicrobial compounds. In 1943 Selman Waksman, an American biochemist born in Ukraine, discovered the antibiotic streptomycin, the first truly effective agent to control *Mycobacterium tuberculosis*, the bacteria that causes tuberculosis. Widespread antibiotic use began shortly after World War II and was regarded as one of the great medical advances in the fight against infectious disease. By the late 1950's and early 1960's, pharmaceutical companies had extensive research and development programs devoted to isolating and producing new antibiotics.

Antibiotics were so effective, and their production ultimately so efficient, that they became inexpensive enough to be routinely prescribed for all types of infections, particularly to treat upper respiratory tract infections. When it was discovered that low levels of antibiotics also promoted increased growth in domesticated animals, antibiotics began to routinely appear as feed supplements.

Development of Antibiotic Resistance

The widespread use and, ultimately, misuse of antibiotics inevitably caused antibiotic-resistant bacteria to appear as microorganisms adapted to this new selective pressure. There are now many strains of pathogenic organisms for which antibiotics have little or no effect.

Streptococcal infections are the leading bacterial cause of morbidity and mortality in the United States. In the mid-1970's *Streptococcus pneumonia* was uniformly susceptible to penicillin. However, penicillin-resistant streptococci strains were being isolated as early as 1967. A study in Denver, Colorado, showed that penicillin-resistant *S. pneumonia* strains increased from 1 percent of the isolates in 1980 to 13 percent of the isolates in 1995. One-half of the resistant strains were also resistant to another antibiotic, cephalosporin. It was apparent from the Denver study that there was a high correlation between antibiotic resistance and whether a family member attended day care. Children attending day care are frequently exposed to preventative antibiotics, and these may have assisted in selecting for resistant bacteria. By 1996 penicillin-resistant *S. pneumonia* represented between 33 and 58 percent of the clinical isolates around the world.

Mycobacterium tuberculosis causes the disease tuberculosis, once the leading cause of death in young adults in industrialized countries. Tuberculosis was once so common and feared that it was known as the White Plague. Before 1990 multidrug-resistant tuberculosis was uncommon. However, by the mid-1990's there were increasing outbreaks in hospitals and prisons, in which the death rate ranged from 50 to 80 percent. Likewise, multiple drug resistance in *Streptococcus pyogenes*, the so-called flesh-eating streptococci, was once rare. There are now erythromycin- and clindimycin-resistant strains. In Italy, for example, antibiotic resistance in *S. pyogenes* increased from about 5 percent to more than 40 percent in some areas between 1993 and 1995.

Many old pathogens have become major clinical problems because of increased antibiotic resistance. *Salmonella* serotypes, for example, including those causing typhoid fever, have been discovered with resistance to at least five antibiotics, including ampicillin, chloramphenicol, streptomycin, sulfanilamide, and tetracycline. The number of these resistant isolates in England rose from 1.5 percent in 1989 to more than 34 percent in 1995. Gonorrhea, caused by

The Growth of Antibiotic Resistance

ANTIBIOTIC	ENTEROCOCCAL SPECIES	% OF RESISTANT BACTERIA		
		1995	1996	1997
Ampicillin	*Enterococcus faecium*	69.0	77.0	83.0
	Enterococcus faecalis	0.9	1.6	1.8
Vancomycin	*Enterococcus faecium*	28.0	50.0	52.0
	Enterococcus faecalis	1.3	2.3	1.9

Measurements taken over a three-year period indicate a general rise in enterococcal resistance to two common antibiotics.

Source: Centers for Disease Control.

Neisseria gonorrhoeae, is the most common sexually transmitted disease. Physicians began using a class of broad-spectrum cephalosporin antibiotics called fluoroquinolones because *N. gonorrhoeae* had become resistant to penicillin, tetracycline, and streptomycin. There is now real concern in the medical community that exclusive use of fluoroquinolones, the only antibiotics to which *N. gonorrhoeae* are routinely susceptible, may lead to rapidly developing resistance; in fact, this already appears to be occurring. Even newly discovered pathogens such as *Helicobacter pylori,* which is associated with peptic ulcers, are rapidly developing resistance to the antibiotics used to treat them.

The development of resistance to some antibiotics appears to be linked to antimicrobial use in farm animals. Shortly after antibiotics appeared, it was discovered that subtherapeutic levels could promote growth in animals and treat acute infections in such settings as aquaculture. One such antimicrobial drug, avoparcin, is a glycopeptide (a compound containing sugars and proteins) that is used as a feed additive. Vancomycin-resistant enterococci such as *Enterococcus faecium* were first isolated in 1988 and appeared to be linked to drug use in animals. Antibiotic resistance in enterococci has been more prevalent in farm animals exposed to antimicrobial drugs, and prolonged exposure to oral glycoproteins in tests led to vancomycin-resistant enterococci in 64 percent of the subjects.

HOW ANTIBIOTIC RESISTANCE OCCURS

Antibiotic-resistant bacteria have no competitive advantage over other cells, which is one of the unanswered questions about their production in nature. Antibiotic resistance occurs because the antibiotics exert a selective pressure on the bacterial pathogens. The selective pressure eliminates all but a few bacteria that can persist through evasion or mutation. One reason antibiotic treatments may persist for several weeks is to ensure that bacteria that have evaded the initial exposure are killed. The rare bacteria that are resistant can persist and grow regardless of continued antibiotic exposure. Terminating antibiotic treatment early, once symptoms disappear, has the unfortunate effect of stimulating antibiotic resistance without completely eliminating the original cause of infection.

Mutations that promote resistance occur with different frequencies. For example, spontaneous resistance of *Mycobacterium tuberculosis* to cycloserine and viomycin may occur in 1 in 1,000 cells; resistance to kanamycin may occur in only 1 in 1 million cells; and resistance to rifampicin may occur in only 1 in 100 million cells. Consequently, 1 billion bacterial cells will contain several individuals resistant to at least one antibiotic. Using multiple antibiotics further reduces the likelihood that an individual cell will be resistant to all antibiotics. However, it can cause multiple antibiotic resistance to develop in bacteria that already have resistance to some of the antibiotics.

Bacterial pathogens may not need to spontaneously mutate to acquire antibiotic resistance. There are several mechanisms by which bacteria can acquire the genes for antibiotic resistance from microorganisms that are already antibiotic resistant. These mechanisms include conjugation (the exchange of genetic information through direct cell-to-cell contact), transduction (the exchange of genetic information from one cell to another by means of a virus), transformation (acquiring genetic information by taking up deoxyribonucleic acid, or DNA, directly from the environment), and transfer of plasmids, small circular pieces of DNA that frequently carry genes for antibiotic resistance. Exchange of genes for antibiotic resistance on plasmids is one of the most common means of developing or acquiring antibiotic-resistant bacteria in hospitals because of the heavy antibiotic use in these settings.

The genes for antibiotic resistance take many forms. They may make the bacteria impermeable to the antibiotic. They may subtly alter the target of the antibiotic within the cell so that it is no longer affected. The genes may code for production of an enzyme in the bacteria that specifically destroys the antibiotic. For example, fluoroquinolone antibiotics inhibit DNA replication in pathogens by binding to the enzyme required for replication. Resistant bacteria have mutations in the amino acid sequences of this enzyme that prevent the antibiotic from binding to this region. Some resistant pathogens produce an enzyme called penicillinase, which degrades the antibiotic penicillin before it can prevent cell wall formation.

New Strategies

The increased use of antibiotics has led to increases in morbidity, mortality caused by previously controlled infectious diseases, and health costs. Some of the recommendations to deal with this public health problem include changing antibiotic prescription patterns, changing patient attitudes about the necessity for antibiotics, increasing the worldwide surveillance of drug-resistant bacteria, improving techniques for susceptibility testing, and investing in the research and development of new antimicrobial agents.

Gene therapy is regarded as one promising solution to antibiotic resistance. In gene therapy, the genes expressing part of the pathogen's cell are injected into a patient and stimulate a heightened immune response. Some old technologies are also being revisited. There is increasing interest in using serum treatments, in which antibodies raised against a pathogen are injected into a patient to cause an immediate immune response. Previous serum treatment techniques have yielded mixed results. However, with the advent of monoclonal antibodies and the techniques for producing them, serum treatments can now be made much more specific and the antibodies delivered in much higher concentrations.

There have been numerous reports from Russia about virus treatment for pathogenic infections. Viruses attack all living organisms, including bacteria, but are extremely specific, so that they will not infect other types of cells. In essence, virus treatments are a form of biocontrol. In virus treatments, the patient is injected with viruses raised against specific pathogens. Once injected, the viruses begin specifically attacking the pathogenic bacteria. Although this technology has not been widely used, it is the subject of growing research in the United States.

Mark Coyne

Suggested readings: The best source for information on antibiotic resistance is *Emerging Infectious Disease*, published by the Centers for Disease Control (CDC), a part of the U.S. Department of Health and Human Services. The history of antibiotics is succinctly explained by P. E. Baldry in *Battle Against Bacteria* (1965). The story of Ehrlich's discovery of Salvarsan and Fleming's discovery of penicillin are enthusiastically portrayed in *The Microbe Hunters* (1953), by Paul DeKruif. Steven Witt outlines the adaptability of microbes in *Biotechnology, Microbes, and the Environment* (1990). This is an easy book to read in which the mechanisms of microbial genetic transformation are clearly illustrated. For a guide to understanding the infectious diseases that antibiotics are supposed to stop, read *A Field Guide to Germs* (1995), by Wayne Biddle. It is an entertaining, irreverent, but informative book on disease.

SEE ALSO: Biotechnology and genetic engineering; Genetically altered bacteria.

Balance of nature

CATEGORY: Ecology and ecosystems

The ecological concept of the balance of nature—a view that proposes that nature, in its undisturbed state, is constant—has never been legitimized in science as either a hypothesis or a theory. However, it persists as a designation for a healthy environment.

Greek natural philosophers in the fifth and sixth centuries B.C.E. attempted to naturalistically explain how nature works rather than depending upon myths. The atomistic theory of Leucippus and Democritus taught that matter can be transformed but is never created or destroyed. The Pythagoreans heard musical harmony in the universe. Hippocratic medicine taught that the balance of humors within the body produce health and an imbalance produces disease, and Greek physicians believed in the healing power of nature. Within this worldview, ecological balance would have been a compelling expectation.

Herodotus, the father of history, wrote not only about the human histories of Greece, Persia, and Egypt but also about their geography and natural history. He was influenced by ideas in natural philosophy, and he was concerned with concrete examples that might illustrate generalities. In organizing his information on the lions, snakes, and hares of Arabia, he asked why predatory species did not eat up all of their prey. His answer was that a superintending "Providence" had created the different species with different capacities for reproduction. Predatory species, such as lions and snakes, produce fewer offspring than species that they eat, such as hares. From Egypt he obtained a report about a mutually beneficial relationship between Nile crocodiles and plovers: Crocodiles allow plovers to sit on their teeth and eat the leeches that infest their mouths.

Plato lived after the natural philosophers and Herodotus, but he lacked their trust in sensory data, and therefore he explained nature with naturalistic myths. In a dialogue called *Protagoras* (c. 390 B.C.E.; English translation, 1804), Plato asked Herodotus's question about why some species do not eat up the others, but he asked it more abstractly and not for its own sake. The point of this creation myth was to explain why humans do not have specialized traits such as wings or claws. The gods assigned to Epimetheus the task of creating each species with traits that would enable it to survive. He had given out all the specialized traits before he got around to creating humans, and his brother Prometheus had to save humanity by giving it reason and fire. The secondary point of the myth supplements what Herodotus had concluded about differences in reproduction with a conclusion about differential traits that ensure survival.

In the writings of Aristotle and his colleagues in Athens, full-fledged science emerged. These scholars realized the importance of both collecting data and ordering it in order to draw conclusions. However, these scholars focused upon physiology and anatomy and neglected to look for what we call ecological explanations of how nature works. For example, they explained that the greater number of offspring in hares than in lions was because of their size. Since hares are smaller, it is easier for more of them to grow within a female than the larger lion cubs to grow in their mother. The balance of nature concept was not distinct enough to require either defending or refuting.

Scientific Revolution

The Romans excelled in engineering, not science. They were mostly content with abbreviated Latin versions of Greek science. Roman writings are still worth mentioning, however, because of their influence on later European naturalists. Aelianus compiled a popular natural history book that, among other things, explained that jackdaws are friends to the farmer because jackdaws eat the eggs and young of locusts that would eat the farmer's produce. The philosopher Cicero wrote an influential book, *De natura deorum* (44 B.C.E.; *On the Nature of the Gods* 1683), in which he saw the work of Providence in endowing plants with the capacity to feed humans and animals and still be able to have seeds left over to ensure their own reproduction. Another philosopher, Plotinus, pondered the evils of suffering when predators kill animals for food. He decided that the existence of predation allowed a greater diversity of life to exist than would be possible if all animals ate plants. These miscellaneous observations were insufficient for a theory, but they kept alive the notion of a balance of nature.

During the scientific revolution, fresh observations and conclusions appeared. Most significantly, John Graunt, a merchant, analyzed London's baptismal and death records in 1662 and discovered the balance in the sex ratio and the regularity of most causes of death (excluding epidemics). England's chief justice, Sir Matthew Hale, was interested in Graunt's discoveries, but he nevertheless decided that the human population, in contrast to animal populations, must have steadily increased throughout history. He surveyed the known causes of animal mortality and in 1677 published the earliest explicit account of the balance of nature.

English scientist Robert Hooke studied fossils and in 1665 concluded that they represented the remains of plants and animals, some of which were probably extinct. However, a clergyman-naturalist, John Ray, replied that the extinction of species would contradict the wisdom of the ages, by which he seems to have meant the balance of nature. Ray also studied the hydrologic cycle, which is a kind of environmental balance of water. Antoni van Leeuwenhoek, one of the first investigators to make biological studies with a microscope, discovered that parasites are more prevalent than anyone had suspected and that they are often detrimental or even fatal to their hosts. Before that, it was commonly assumed that the relationship between host and parasite was mutually beneficial.

Richard Bradley, a botanist and popularizer of natural history, pointed out in 1718 that each species of plant has its own kind of insect and that there are even different insects that eat the leaves and bark of a tree. His book *A Philosophical Account of the Works of Nature* (1721) explored aspects of the balance of nature more thoroughly than had been done before. Ray's and Bradley's books may have inspired the comment in Alexander Pope's *Essay on Man* (1733) that the all species are so closely interdependent that the extinction of one would lead to the destruction of all living nature.

Toward a Science of Ecology

Swedish naturalist Carl Linnaeus was an important protoecologist. His essay *Oeconomia Naturae* (1749; *The Economy of Nature*, 1749) attempted to organize the aspect of natural history dealing with the balance of nature, but he realized that one must study not only ways that plants and animals interact but also their habitat. He knew that while balance had to exist, there occurred over time a succession of plants, beginning with a bare field and ending with a forest. In *Politia Naturae* (1760; *Governing Nature*, 1760) he discussed the checks on populations that prevent some species from becoming so numerous that they eliminate others. He noticed the competition among different species of plants in a meadow and concluded that feeding insects kept them in check. French naturalist Comte de Buffon developed a dynamical perspective on the balance of nature from his studies on rodents and their predators. Rodents can increase in numbers to plague proportions, but then predators and climate reduce their numbers. Buffon also suspected that humans had exterminated some large mammals, such as mammoths and mastodons.

However, a later Frenchman, Jean-Baptiste Lamarck, published his book on evolution called *Philosophie zoologique* (1809; *Zoological Philosophy*, 1914), which cast doubt on extinction by arguing that fossils only represent early forms of living species: Mammoths and mastodons evolved into African and Indian elephants. In developing this idea, he minimized the importance of competition in nature. An English opponent, the geologist Charles Lyell, argued in 1833 that species do become extinct, primarily because of competition among species. Charles Darwin was inspired by his own investigations during a long voyage around the world and by his reading of the works of Linnaeus and Lyell. Darwin's revolutionary book, *On the Origin of Species* (1859), argued an intermediate position between Lamarck and Lyell: Species do evolve into different species, but in the process, some species do indeed become extinct.

Darwin's theory of evolution might have brought an end to the balance of nature concept, but it did not. Instead, American zoologist Stephen A. Forbes developed an evolutionary concept of the balance of nature in his essay, "The Lake as a Microcosm" (1887). Although the reproductive rate of aquatic species is enormous and the struggle for existence among them is severe, "the little community secluded here is as prosperous as if its state were one of profound and perpetual peace." He emphasized the stabilizing effects of natural selection.

Ecology

The science of ecology became formally organized between the 1890's and the 1910's. One of its important organizing concepts was that of "biotic communities." An American plant ecologist, Frederic E. Clements, wrote a large monograph titled *Plant Succession* (1916), in which he drew a morphological and developmental analogy between organisms and plant communities. Both the individual and the community have a life history during which each changes its anatomy and physiology. This supraorganismic concept was an extreme version of the balance of nature that seemed plausible as long as one believed that a biotic community was a real entity rather than a convenient approximation of what one sees in a pond, a meadow, or a forest. However, the studies of Henry A. Gleason in 1917 and later indicated that plant species merely compete with one another in similar environments; he concluded that Clements's superorganism was poetry, not science.

While the balance of nature concept was giving way to ecological hypotheses and theories, Rachel Carson decided that she could not argue her case in *Silent Spring* (1962) without it. She admitted, "The balance of nature is not a *status quo*; it is fluid, ever shifting, in a constant state of adjustment." Nevertheless, to her the concept represented a healthy environment, which humans could upset. Her usage of the phrase has persisted within the environmental movement.

In 1972 English medical chemist James E. Lovelock developed a new balance-of-nature idea, which he calls Gaia, named for a Greek earth goddess. His reasoning owed virtually nothing to previous balance of nature notions that focused upon the interactions of plants and animals. His concept emphasized the chemical cycles that flow from the earth to the waters, atmosphere, and living organisms. He soon had the assistance of a zoologist named Lynn Margulis. Their studies convinced them that biogeochemical cycles are not random, but exhibit homeostasis, just as some animals exhibit homeostasis in body heat and blood concentrations of various substances. They believe that living beings, rather than inanimate forces, mainly control the earth's environment. In 1988 three scientific organizations sponsored a conference of 150 scientists from all over the world to evaluate their ideas. Although science more or less understands how homeostasis works when a brain within an animal controls it, no one has succeeded in satisfactorily explaining how homeostasis can work in a world "system" that lacks a brain. The Gaia hypothesis is as untestable as were earlier balance of nature concepts.

Frank N. Egerton

SUGGESTED READINGS: The history of the balance of nature concept is told in two articles by Frank N. Egerton, "Changing Concepts of the Balance of Nature," *Quarterly Review of Biology* (June, 1973), and "The History and Present Entanglements of Some General Ecological Perspectives," in *Humans as Components of Ecosystems* (1993), edited by Mark J. McDonnell and S. T. A. Pickett. Two biologists, Lorus J. Milne and Margery Milne, have written a popular account, *The Balance of Nature* (1960). Within the volume *Scientists on Gaia* (1991), edited by Stephen H. Schneider and Penelope J. Boston, see James W. Kirchner, "The Gaia Hypotheses: Are They Testable? Are They Useful?"

SEE ALSO: Carson, Rachel; Darwin, Charles; Ecology; Ecosystems; Food chains; Gaia hypothesis; Lovelock, James.

Berry, Wendell

Born: August 5, 1934; Henry County, Kentucky
Category: Agriculture and food

Wendell Berry's integrated professions of farmer, writer, and critic of industrial development have placed him among the major figures of the twentieth century in both conservation and literature.

Born to a tobacco farm family during the Great Depression, Wendell Berry grew up in a simple environment of small farms in Henry County, Kentucky, that were oriented to crop diversification, organic fertilizing, and use of draft animals. Berry's family had deep roots in the community, as did their neighbors. Farmers in Henry County were largely self-sufficient, depending little on resources beyond their region.

Berry's rural upbringing affected every facet of his adult life. After receiving bachelor's and master's degrees in English from the University of Kentucky at Lexington, he was awarded a prestigious Wallace Stegner Fellowship in creative writing at Stanford University in 1958. This opportunity moved Berry into a circle of scholars and writers, notably Stegner, who at that time was a prominent novelist and conservationist. In 1960 Berry returned to Henry County. With the exception of a brief time in France and Italy on a Guggenheim Fellowship and a teaching position at New York University, Berry remained in his native area as a farmer, as a writer of poetry, fiction, and nonfiction, and as an English professor at the University of Kentucky.

Berry is considered by many to be one of the most important nature poets of his generation. He gained widespread recognition for his early volumes *The Broken Ground* (1964) and *Openings* (1968), both of which center on rural themes. Berry's writing on environmental issues focuses mainly on agriculture and the simple life. His most successful novel, *The Memory of Old Jack* (1974), uses flashbacks of a ninety-two-year-old farmer reliving his simple, agrarian life. Old Jack reflects the longing of Berry to return to his pre-World War II life of rural self-sufficiency. Among Berry's best known nonfiction works are *The Long-Legged House* (1969), *The Hidden Wound* (1970), *A Continuous Harmony: Essays Cultural and Agricultural* (1972), *The Unsettling of America* (1977), and *The Gift of Good Land* (1981).

Berry's writings emphasize the connectedness of human beings with the rest of nature. He is critical of the destruction of the land by mechanized monoculture farming, use of pesticides and fertilizers, clear-cutting of forests, and strip mining. He decries the movement away from the family farm to corporate farming, noting that the corporate sector has killed rural America.

Berry views farming as an art and prioritizes ecology over economics, which values efficiency and specialization as means of maximizing income in the short term. For him, technology has dehumanized agriculture by replacing the self-fulfilling labor of farmers and their families. In addition, people have been driven from their land into the cities, generating social and environmental problems in urban areas.

Wendell Berry's poems and novels promote a sustainable, agrarian lifestyle in which humans peacefully coexist with nature. (Dan Carraco)

Berry believes that a return to sustainable agriculture is an ecological imperative for maintaining a high quality of life. In his writings he praises the Amish for their stewardship of the land and their farming on a scale appropriate to the needs of their communities. He likewise admonishes Christians to heed the biblical message that "the earth is the Lord's and the fullness thereof."

Ruth Bamberger

SEE ALSO: Environmental ethics; Sustainable agriculture.

Bhopal disaster

DATE: December 2-3, 1984
CATEGORY: Human health and the environment

The escape of methyl isocyanate from a pesticide production plant in Bhopal, India, resulted in the world's worst chemical disaster. Afterward, the issues involved in the introduction of technology to less-developed countries were thoroughly reviewed and debated.

Union Carbide India, which was jointly owned by Union Carbide and the Indian public, ran a pesticide production plant in Bhopal, a city in central India with a population of approximately one million people. The plant was surrounded by a heavily populated shanty town. Methyl isocyanate (MIC), a very reactive and toxic chemical used in the production of pesticides, was stored at the plant.

The Bhopal disaster began on the evening of December 2, 1984. The plant was closed to reduce inventory and to perform routine maintenance. Around 11:30 P.M., workers realized that a leak of MIC had occurred when their eyes started to tear and burn. They reported the leak to their supervisor, but it was time for tea break. The supervisor told the workers that they would deal with the leak after the break. By the time they returned, the pressure and temperature in the MIC storage tank had risen to dangerous levels. A rupture disk and a pressure release valve burst, and a cloud of MIC was released over Bhopal.

Some people died in their beds, while others awoke gasping for breath with their eyes and throats burning. The gas cloud quickly spread over an area of 65 square kilometer (25 square miles), engulfing panicked residents who were trying to flee. In a matter of minutes people began to collapse and die. Although the leak started about 12:45 A.M, the plant's public alarm was not sounded until 2:15 A.M. The police were only informed of the leak after it had been stopped and after the alarm had been sounded. All the details surrounding the disaster have not been released because of litigation. Enough is know to establish that culpability spread from high levels of management at Union Carbide to mistakes made by plant workers. The chemical cloud that spread over much of Bhopal claimed an estimated 2,500 lives within hours and another 3,000 to 4,000 over the next ten years. A total of 170,000 people experienced some kind of treatable injury, mainly to the eyes or respiratory system, and 11,500 were hospitalized.

When MIC comes into contact with water, it causes a spontaneous reaction that releases heat. In the presence of a variety of catalysts, including iron ions, three molecules of MIC will join together to form a trimer. This reaction also releases heat. It is thought that both of these reactions occurred in the MIC storage tank. The heat released during the chemical reactions raised the MIC's temperature, which caused the rates of the MIC reactions to increase, releasing even more heat. The low boiling point of MIC caused it to vaporize into a gas by the heat released by the chemical reactions. The gas expanded, increasing the pressure inside the storage tank until it burst a rupture disk on the line leading to the safety valve. When the safety valve was forced open, the gases from the tank began to escape.

Four months after the disaster, Union Carbide released a report detailing failings that led to the MIC escape. Warren Anderson, the chairman of Union Carbide, acknowledged that operating conditions at the plant were so poor that it should not have been in operation. The report recounted that sometime before midnight on December 2, 1984, a considerable quantity of water entered the tank and started the heat-

A clinic in India distributes medicine to women exposed to methyl isocyanate, which leaked from the nearby Union Carbide plant in 1984. A total of 170,000 people were adversely affected by the accident. (Reuters/Marty Wolfe/Archive Photos)

releasing reactions. Carbide claimed that the water was intentionally introduced by a disgruntled employee. Plant workers said it occurred accidentally when water being used to clean pipes leaked into the tank. The tank was equipped with a refrigeration system designed to keep the contents at a low temperature, but it had been shut down for more than five months. At the higher storage temperature, the MIC reacted more rapidly with the water.

An alarm should have sounded when the temperature of the tank started to rise. The alarm failed to sound because it had not been reset for the higher storage temperature resulting from the lack of refrigeration. Carbide hypothesized that the water and high temperature in the tank caused rapid corrosion of its stainless steel walls. This led to iron contamination that would have catalyzed the heat-releasing trimerization reaction. The temperature in the tank probably rose to at least 200 degrees Celsius (392 degrees Fahrenheit). A sodium hydroxide vent scrubber designed to destroy leaking MIC was started manually by a control room operator, but apparently the sodium hydroxide failed to circulate. The gas rushed to the flare tower, where any MIC escaping the scrubber was supposed to be burned. This was the last safety control, but it was also out of service. A basic design fault of the plant was that the safety systems were only designed to deal with minor leaks, so even if they had been operative, they would have been overwhelmed by the volume of gas released.

A suit against Union Carbide by the Indian government was adjudicated by the Indian Supreme Court. The case was settled in February, 1989, with an award of US$470 million to the victims. As a result of the Bhopal disaster, India's regulation of hazardous industries was significantly strengthened. One of the most important actions was the passage of the Environmental Protection Act of 1986. However, environmental activists have continued to criticize the government for ignoring environmental laws and regulations to promote economic development.

Francis P. Mac Kay

SUGGESTED READINGS: David Weir's *The Bhopal Syndrome* (1987) is an environmental activist's readable account of the accident and its aftermath in India and around the world. The essays in *Learning from Disaster* (1994), edited by Sheila Jasanoff, examine the legal, environmental, and business consequences of Bhopal. The aftereffects in India and the United States are concisely summarized in Wil Lepkowski's "Ten Years Later: Bhopal," *Chemical and Engineering News* (1994).

SEE ALSO: Environmental legislation and lobbying; Hazardous and toxic substance regulation; Pesticides and herbicides.

Bikini Atoll bombing

DATE: March 1, 1954
CATEGORY: Nuclear power and radiation

On March 1, 1954, the United States tested a hydrogen bomb at the Bikini Atoll in the Marshall Islands. The explosion was more powerful than had been anticipated, and an unexpected shift in the wind resulted in radioactive fallout landing in populated regions of the Marshall Islands, on ships participating in the test, and on a Japanese fishing boat.

The hydrogen bomb that the United States tested at the Bikini Atoll was code-named *Bravo.* It had an explosive force of approximately fifteen megatons. However, because of its new design, it used only about the same amount of the rare isotope uranium 235 as the Hiroshima bomb, which had an explosive yield one thousand times weaker. Although the residents of Bikini had been relocated before the test, the radiation spread to two inhabited islands, Rongelap and Utrick, located about 160 kilometers (100 miles) and 480 kilometers (300 miles) from Bikini, respectively. U.S. military personnel on Rongerik Atoll were also exposed to radiation, as was the crew of a U.S. Navy gasoline tanker en route from the Marshall Islands to Hawaii.

The crew of a Japanese tuna fishing boat, the *Lucky Dragon,* received the most serious radiation exposure. At the time of the explosion, the boat was located about 160 kilometers (100 miles) from Bikini, apparently outside the "danger zone" established by United States authorities. For about three hours the crew observed white, sandy dust falling onto the deck of their boat. Most of the twenty-three crewmembers of the *Lucky Dragon* later indicated they had suffered from skin inflammation and nausea over the next few days. On March 14, when the boat returned to port in Japan, the crew was hospitalized and treated for radiation exposure. Medical tests indicated that the crewmembers had suffered bone marrow damage, resulting in anemia, as well as temporary sterility. The exposed areas of their skin were also damaged.

The radiotelegraph operator of the *Lucky Dragon* died in September, 1954. The cause of his death has been disputed. There were claims that he died as a direct result of radiation exposure, but U.S. officials said his death was caused by hepatitis contracted from the blood transfusions he received to treat the radiation exposure. The other crewmembers of the *Lucky Dragon* all recovered.

By the time the crew's radiation exposure was discovered, tuna from boats fishing in the region of the bomb test had been shipped to the markets. Tests indicated that some of this tuna was radioactive. By the time officials realized this, some radioactive tuna had already been sold. About one hundred people in Japan consumed this contaminated fish before it was removed from the marketplace. No adverse health effects were reported. Nonetheless, when word of the contamination spread, many consumers simply stopped buying fish. Tuna canneries on the West Coast of the United States were alerted to the possibility of contamination, and radiation monitoring was instituted for a short time.

In January, 1955, the United States government donated $2 million to the government of Japan to compensate for the injuries to the crew of the *Lucky Dragon* and damage to the Japanese fishing industry.

George J. Flynn

SEE ALSO: Nuclear testing; Nuclear weapons; Radioactive pollution and fallout.

Bioassays

CATEGORY: Ecology and ecosystems

A bioassay is the use of biological organisms to either detect the presence of a given chemical substance or determine the biological activity of a known substance in a particular environment.

In many instances, a scientist may suspect that a certain chemical is present in a given environment but may not have access to a specific piece of equipment designed to measure the presence of the chemical. In some cases, an experimental protocol for the detection of the chemical may not exist. In either of these cases, the scientist may be able to detect the presence of the chemical by using a biological organism that responds in a specific manner when exposed to that particular chemical agent. At other times, a scientist may know that a certain chemical is present but not know how a particular organism will respond when exposed to the agent. In this case, the scientist will expose the test organism to the chemical and measure a particular physiological response.

Bioassays are utilized in many different areas of the biological sciences, including environmental studies, but some bioassay methods work better than others. A good bioassay will meet two basic criteria. First, it should be specific for a given physiological response. For example, if a given chemical is responsible for inhibiting the feeding response of a particular insect, then the bioassay for that chemical should measure only the inhibition of feeding of that insect and not some other physiological response to the chemical. Second, a good bioassay will measure the same response in the laboratory that is observed in the field. Again, if a particular chemical inhibits the feeding response in the field, then the laboratory bioassay for that chemical should also inhibit feeding. There is a continuing need to develop accurate bioassay methods as well to improve existing techniques.

Many different bioassays are used in environmental studies. One of the most common is the measure of the LD 50—the concentration or dose of the chemical that will result in the deaths of one-half of a population of organisms—of a new pesticide on species of pest and nonpest organisms. To conduct this bioassay, the test species is exposed to a wide range of different concentrations of the chemical. The concentration of the pesticide that kills one-half of the test organisms represents the LD 50.

Another common environmental bioassay is the measure of resistance of plants to a particular insect pest. In order to reduce the dependence on chemical insecticides, plant breeders are continually trying to develop insect-resistant plants, either through traditional breeding programs or by using biotechnology to transfer resistance genes to susceptible crop strains. Bioassays are used to measure the degree of success of these attempts. In these bioassays, the same number of susceptible and resistant plants are subjected to infestation by equal numbers of the insect pest for which the breeder is trying to develop resistance. The two groups of plants are observed, and the degree of resistance, if any, is recorded.

D. R. Gossett

SEE ALSO: Agricultural chemicals; Biopesticides; Biotechnology and genetic engineering; Pesticides and herbicides; Sustainable agriculture.

Biodiversity

CATEGORY: Ecology and ecosystems

In 1993 the Wildlife Society defined biodiversity as "the richness, abundance, and variability of plant and animal species and communities and the ecological process that link them with one another and with soil, air, and water." Included in this concept is the recognition that life on earth exists in great variety and at various levels of organization.

Many kinds of specialists—including organismic biologists, population and evolutionary biologists, geneticists, and ecologists—investigate biological processes that are encompassed by the concept of biodiversity. Conservation biologists are concerned with the totality of biodiversity, including the process of speciation that forms new species, the measurement of biodiversity, and factors involved in the extinction process. However, the primary thrust of their efforts is the development of strategies to preserve biodiversity. The biodiversity paradigm connects classical taxonomic and morphological studies of organisms with modern techniques employed by those working at the molecular level.

It is generally accepted that biodiversity can be approached at three levels of organization, commonly identified as species diversity, ecosystem diversity, and genetic diversity. Some also recognize biological phenomena diversity.

Species Diversity

No one knows how many species inhabit the earth. Estimates range from five million to several times that number. Each species consists of individuals that are somewhat similar and capable of interbreeding with other members of their species but are not usually able to interbreed with individuals of other species. The species that occupy a particular ecosystem are a subset of the species as a whole. Ecosystems are generally considered to be local units of nature; ponds, forests, and prairies are common examples.

Conservation biologists measure the species diversity of a given ecosystem by first conducting a careful, quantitative inventory. From such data, scientists may determine the "richness" of the ecosystem, which is simply a reflection of the number of species present. Thus, an island with three hundred species would be 50 percent richer than another with only two hundred species. Some ecosystems, especially tropical rain forests and coral reefs, are much richer than others. Among the least rich are tundra regions and deserts.

A second aspect of species diversity is "evenness," defined as the degree to which each of the various elements are present in similar percentages of the total species. As an example, consider two forests, each of which has a total of twenty species of trees. Suppose that the first forest has a few tree species represented by rather high percentages and the remainder by low percentages. The second forest, with its species more evenly distributed, would rate higher on a evenness scale.

Species diversity, therefore, is a value that combines both species richness and species evenness measures. Values obtained from a diversity index are used in comparing species diversity among ecosystems of both the same and different types. They also have implications for the preservation of ecosystems; other things being equal, it would be preferable to preserve ecosystems with a high diversity index, thus protecting a larger number of species.

Considerable effort has been expended to predict species diversity as determined by the nature of the area involved. For example, island biogeography theory suggests that islands that are larger, nearer to other islands or continents, and have a more heterogeneous landscape would be expected to have a higher species diversity than those possessing alternate traits. Such predictions apply not only to literal islands but also to other discontinuous ecosystems; examples would be alpine tundra of isolated mountaintops, or ponds several miles apart.

The application of island biogeography theory to designing nature preserves was proposed by Jared Diamond in 1975. His suggestion began the "single larger or several smaller," or SLOSS, area controversy. Although island biogeography theory would, in many instances, suggest selecting one large area for a nature preserve, it is often the case that several smaller areas, if carefully selected, could preserve more species.

The species diversity of a particular ecosystem is subject to change over time. Pollution, deforestation, and other types of habitat degradation invariably reduce diversity. Conversely, during the extended process of ecological succession that follows disturbances, species diversity typically increases until a permanent, climax ecosystem with a large index of diversity results. It is generally assumed by ecologists that more diverse ecosystems are more stable than are those with less diversity. Certainly, the more species present, the greater the opportunity for various interactions, both with other species and with the environment. Examples of interspecific reactions include mutualism, predation, and parasitism. Such interactions apparently help to integrate a community into a whole, thus increasing its stability.

ECOSYSTEM DIVERSITY

Ecology can be defined as the study of ecosystems. From a conservation standpoint, ecosystems are important because they sustain their particular assemblage of living species. Conservation biologists also consider ecosystems to have an intrinsic value beyond the species they harbor. Therefore, it would be ideal if representative global ecosystems could be preserved.

However, this is far from realization. Just deciding where to draw the line between interfacing ecosystems can be a problem. For example, the water level of a stream running through a forest is subject to seasonal fluctuation, causing a transitional zone characterized by the biota from both adjoining ecosystems. Such ubiquitous zones negate the view that ecosystems are discrete units with easily recognized boundaries.

The protection of diverse ecosystems is of utmost importance to the maintenance of biodiversity. However, ecosystems throughout the world are threatened by global warming, air and water pollution, acid deposition, ozone depletion, and other destructive forces. At the local level, deforestation, thermal pollution, urbanization, and poor agricultural practices are among the problems affecting ecosystems and therefore reducing biodiversity. Both global and local environmental problems are amplified by rapidly increasing world population pressures.

In the process of determining which ecosystems are most in need of protection, it has be-

Construction of the Trans-Amazonian Highway in Brazil fragmented the fragile ecology of the Amazon River Basin. Such development threatens the diversity of plant and animal life in tropical rain forests and other ecosystems. (Archive Photos)

come apparent to many scientists that a system for naming and classifying ecosystems is highly desirable, if not imperative. Efforts are being made to establish a system similar to the hierarchical system applied to species that was developed by Swedish botanist Carl Linnaeus during the eighteenth century. However, a classification system for ecosystems is far from complete. Freshwater, marine, and terrestrial ecosystems are recognized as main categories, with each further divided into particular types. Though tentative, this has made possible the identification and preservation of a wide range of representative, threatened ecosystems.

In 1995 conservation biologist Reed F. Noss of Oregon State University and his colleagues identified more than 126 types of ecosystems in the United States that are threatened or critically endangered. The following list illustrates their diversity: southern Appalachian spruce-fir forests; eastern grasslands, savannas, and barrens; California native grasslands; Hawaiian dry forests; caves and Karst systems; old-growth forests of the Pacific Northwest; and southern forested wetlands.

Not all ecosystems can be saved. Establishing priorities involves many considerations, some of which are economic and political. Ideally, choices would be made on merit: rarity, size, number of endangered species they include, and other objective, scientific criteria.

Genetic and Biological Phenomena Diversity

Most of the variation among individuals of the same species is caused by the different genotypes (combinations of genes) that they possess. Such genetic diversity is readily apparent in cultivated or domesticated species such as cats, dogs, and corn, but also exists, though usually to a lesser degree, in wild species. Genetic diversity can be measured only by exacting molecular laboratory procedures. The tests detect the amount of variation in the deoxyribonucleic acid (DNA) or isoenzymes (chemically distinct enzymes) possessed by various individuals of the species in question.

A significant degree of genetic diversity within a population or species confers a great advantage. This diversity is the raw material that allows evolutionary processes to occur. When a local population becomes too small, it is subject to a serious decline in vigor from increased inbreeding. This leads, in turn, to a downward, self-perpetuating spiral in genetic diversity and further reduction in population size. Extinction may be imminent. In the grand scheme of nature, this is a catastrophic event; never again will that particular genome (set of genes) exist anywhere on the earth. Extinction is the process by which global biodiversity is reduced.

Biological phenomena diversity refers to the numerous unique biological events that occur in natural areas throughout the world.

Examples include the congregation of thousands of monarch butterflies on tree limbs at Point Pelee in Ontario, Canada, as they await favorable conditions before continuing their migration, or the return of hundreds of loggerhead sea turtles each April to Padre Island in the Gulf of Mexico in order to lay their eggs.

Although biologists have been concerned with protecting plant and animal species for decades, only recently has conservation biology emerged an a identifiable discipline. Conceived in a perceived crisis of biological extinctions, conservation biology differs from related disciplines, such as ecology, because of its advocative nature and its insistence on maintaining biodiversity as intrinsically good. Conservation biology is a value-laden science, and some critics consider it akin to a religion with an accepted dogma.

The prospect of preserving global ecosystems and the life processes they make possible, all necessary for maintaining global diversity, is not promising. Western culture does not give environmental concerns a high priority. For those who do, there is more often a concern over issues relating to immediate health effects than concern over the loss of biodiversity. Only when education in basic biology and ecology at all levels is extended to include an awareness of the importance of biodiversity will there develop the necessary impetus to save ecosystems and all their inhabitants, including humans.

Thomas E. Hemmerly

Suggested readings: Textbooks written for college courses in conservation biology are largely devoted to biodiversity and its preservation; examples include *Fundamentals of Conservation Biology* (1996), by Malcolm L. Hunter, Jr., and *Essentials of Conservation Biology* (1998), by Richard B. Primack. The books by Pulitzer-winning author and "dean of biodiversity studies" E. O. Wilson are without equal: *The Diversity of Life* (1992), *Biophilia* (1984), and *The Naturalist* (1994). *Extinction: The Causes and Consequences of the Disappearance of Species* (1981), by Paul Ehrlich and Anne Ehrlich, is a classic text by well-known environmentalists. *Conservation Biology: The Theory and Practice of Nature Conservation, Preservation, and Management* (1992), edited by Peggy L. Fielder and Subodh K. Jain, includes several good readings on biodiversity.

See also: Ecology; Ecosystems; Endangered species; Extinctions and species loss; Global biodiversity assessment; Rain forests and rain forest destruction.

Biofertilizers

Category: Biotechnology and genetic engineering

Biofertilizers provide a means by which biological systems can be utilized to supply plant nutrients such as nitrogen to agricultural crops. The use of biofertilizers could reduce the dependency on chemical fertilizers, which are often detrimental to the environment.

Plants require an adequate supply of the thirteen mineral nutrients necessary for normal growth and reproduction. These nutrients, which must be supplied by the soil, include both macronutrients (those nutrients required in large quantities) and micronutrients (those nutrients required in smaller quantities). As plants grow and develop, they remove these essential mineral nutrients from the soil. Since normal crop production usually requires the removal of plants or plant parts, the nutrients are continuously removed from the soil. Therefore, the long-term agricultural utilization of any soil requires periodic fertilization to replace lost nutrients.

Nitrogen is the plant nutrient that is most often depleted in agricultural soils, and most crops respond to the addition of nitrogen fertilizer by increasing their growth and yield; therefore, more nitrogen fertilizer is applied to cropland than any other fertilizer. In the past, nitrogen fertilizers have been limited to either manures, which have low levels of nitrogen, or chemical fertilizers, which usually have high levels of nitrogen. However, the excess nitrogen in chemical fertilizers often runs off into nearby waterways, causing a variety of environmental problems.

Biofertilizers offer a potential alternative: They supply sufficient amounts of nitrogen for maximum yields yet have a positive impact on the environment. Biofertilizers generally consist of either naturally occurring or genetically modified microorganisms that improve the physical condition of soil, aid plant growth, or increase crop yield. Biofertilizers provide an environmentally friendly way to increase plant health and yields with reduced input costs, new products and additional revenues for the agricultural biotechnology industry, and cheaper products for consumers.

While biofertilizers could potentially be used to supply a number of different nutrients, most of the interest is focused on enhancing nitrogen fertilization. The relatively small amounts of nitrogen found in soil come from a variety of sources. Some nitrogen is present in all organic matter in soil; as this organic matter is degraded by microorganisms, it can be used by plants. A second source of nitrogen is nitrogen fixation, the chemical or biological process of taking nitrogen from the atmosphere and converting it to a form that can be utilized by plants. Bacteria such as *Rhizobia* can live symbiotically with certain plants such as legumes, which house nitrogen-fixing bacteria in their roots. The *Rhizobia* and plant root tissue form root nodules, which house the nitrogen-fixing bacteria; once inside the nodules, the bacteria use energy supplied by the plant to convert atmospheric nitrogen to ammonia, which nourishes the plant. Natural nitrogen can also be supplied by free-living microorganisms, which can fix nitrogen without forming a symbiotic relationship with plants. The primary objective of biofertilizers is to enhance any one or all of these processes.

One of the major goals for the genetic engineering of biofertilizers is to transfer the ability to form nodules and establish effective symbiosis to nonlegume plants. The formation of nodules in which the *Rhizobia* live requires plant cells to synthesize many new proteins, and many of the genes required for the expression of these proteins are not found in the root cells of plants outside the legume family. Many research programs have been devoted to efforts to transfer such genes to nonlegume plants so that they can interact symbiotically with nitrogen-fixing bacteria. If this is accomplished, *Rhizobia* could be used as a biofertilizer for a variety of plants.

There is also much interest in using the free-living, soil-borne organisms that fix atmospheric nitrogen as biofertilizers. These organisms live in the rhizosphere (the region of soil in immediate contact with plant roots) or thrive on the surface of the soil. Since the exudates from these microorganisms contain nitrogen that can be utilized by plants, increasing their abundance in the soil could reduce the dependency on chemical fertilizers. Numerous research efforts have been designed to identify and enhance the abundance of nitrogen-fixing bacteria in the rhizosphere. Soil microorganisms primarily depend on soluble root exudates and decomposed organic matter to supply the energy necessary for fixing nitrogen. Hence, there is also an interest in enhancing the biodegradation of organic matter in the soil. This research has primarily centered on inoculating the soil with cellulose-degrading fungi and nitrogen-fixing bacteria or applying organic matter such as straw that has been treated with a combination of the fungi and bacteria to the soil.

D. R. Gossett

SUGGESTED READINGS: *Plant Physiology* (1985), by F. B. Salisbury and C. W. Ross, contains excellent chapters on plant nutrition and nitrogen metabolism. *Soil-Plant Relationships* (1988), by C. A. Black, provides an in-depth discussion of the relationship between soil microbes and soil fertility. *Plants, Genes, and Agriculture* (1994), by M. J. Crispeels and D. E. Sadava, offers a discussion on the potential genetic modifications necessary to increase the use of microorganism to supply nitrogen fertilizer. *Soil Biotechnology: Microbiological Factors in Crop Production* (1983), by J. M. Lynch, contains some excellent information on the potential for genetically engineering microorganism to improve crop production.

SEE ALSO: Biopesticides; Biotechnology and genetic engineering; Genetically altered bacteria; Sustainable agriculture.

Biomagnification

CATEGORY: Ecology and ecosystems

Biomagnification is the accumulation of toxic contaminants in the environment as they move up through the food chain. As members of each level of the food chain are progressively eaten by those organisms found in higher levels of the chain, the concentration of toxic chemicals within the tissues of the higher organisms increases.

Not all chemicals, potentially toxic or not, are equally likely to undergo biomagnification. However, molecules susceptible to biomagnification have certain characteristics in common. They are resistant to natural microbial degradation and therefore persist in the environment. They are also lipophilic, tending to accumulate in the fatty tissue of organisms. In addition, the chemical must be biologically active in order to have an effect on the organism in which it is found. Such compounds are likely to be absorbed from food or water in the environment and stored within the membranes or fatty tissues.

The process usually begins with the spraying of pesticides for the purpose of controlling insect populations. Industrial contamination, including the release of heavy metals, can be an additional cause of such pollution. Biomagnification results when these chemicals contaminate the water supply and are absorbed into the lipid membranes of microbial organisms. This process, often referred to as bioaccumulation, results in the initial concentration of the chemical in an organism in a form that is not naturally excreted with normal waste material. Levels of the chemical may reach anywhere from one to three times that found in the surrounding environment. Since the nature of the chemical is such that it is neither degraded nor excreted, it remains within the organism.

The widespread use of toxic chemicals susceptible to biomagnification before the 1960's, such as dichloro-diphenyl-trichloroethane (DDT), led to dramatic declines in the populations of bald eagles and other predatory birds. (Jim West)

As organisms on the bottom of the food chain are eaten and digested by members of the next level in the chain, the concentration of the accumulated material significantly increases; at each subsequent level, the concentration may reach one order of magnitude (a tenfold increase) higher. Consequently, the levels of the pollutant at the top of the environmental food chain, such fish or carnivorous birds, and potentially even humans, may be as much as one million times more concentrated than the original, presumably safe, levels in the environment. For example, studies of dichloro-diphenyl-trichloroethane (DDT) levels in the 1960's found that zooplankton at the bottom of the food chain had accumulated nearly one thousand times the level of the pollutant in the surrounding water. Ingestion of the plankton by fish resulted in concentration by another factor of several hundred. By the time the fish were eaten by predatory birds, the level of DDT was concentrated by a factor of over two hundred thousand.

DDT is characteristic of most pollutants subject to potential biomagnification. It is relatively stable in the environment, persisting for decades. It is soluble in lipids and readily incorporated into the membranes of organisms. While DDT represents the classic example of biomagnification of a toxic chemical, it is by no means the only representative of potential environmental pollutants. Other pesticides with similar characteristics include pesticides such as aldrin, chlordane, parathion, and toxaphene. In addition, cyanide, polychlorinated biphenyls (PCBs), and heavy metals—such as selenium, mercury, copper, lead, and zinc—have also been found to concentrate within the food chain.

Some heavy metals are inherently toxic or may undergo microbial modification to increase their toxic potential. For example, mercury does not naturally accumulate in membranes and was therefore not originally viewed as a significant danger to the environment. However, some microorganisms are capable of adding a methyl group to the metal and producing methyl mercury, a highly toxic material that does accumulate in fatty tissue and membranes.

Since pesticides are, by their nature, biologically active compounds, which reflects their ability to control insects, they are of particular concern if subject to biomagnification. DDT remains the classical example of how bioaccumulation and biomagnification may have an effect on the environment. Initially introduced as a pesticide for control of insects and insect-borne disease, DDT was not thought to be particularly toxic. However, biomagnification of the chemical was found to result in the deaths of birds and other wildlife. In addition, DDT contamination was found to result in formation of thin egg shells that greatly reduced the birth rate among birds. Before the use of DDT was banned in the 1960's, the population levels of predatory birds such as eagles and falcons had fallen to a fraction of the levels found prior to use of the insecticide. Though it was unclear whether there was any direct effect on the human population in the United States, the discovery of elevated levels of DDT in human tissue contributed to the decision to ban the use of the chemical.

Several procedures have been adopted since the 1960's to prevent the biomagnification of toxic materials. In addition to outright bans, pesticides are often modified to prevent their accumulation in the environment. Most synthetic pesticides contain chemical structures that are easily degraded by microorganisms found in the environment. Ideally, the pesticide should survive no longer than a single growing season before being rendered harmless by the environmental flora. Often such chemical changes require only simple modification of the basic structure.

Richard Adler

SUGGESTED READINGS: The classic study of environmental pollution and the effects of biomagnification remains Rachel Carson's *Silent Spring* (1962). A similar subject is addressed by Theo Colborn, *Our Stolen Future* (1997). More academic approaches to the subject of biomagnification are provided in *Essentials of Conservation Biology* (1993), by Richard Primack, and *Microbial Ecology* (1993), by Ronald Atlas and Richard Bartha.

SEE ALSO: Dichloro-diphenyl-trichloroethane (DDT); Heavy metals and heavy metal poisoning; Mercury and mercury poisoning; Pesticides and herbicides; Polychlorinated biphenyls.

Biomass conversion

CATEGORY: Energy

Biomass conversion involves burning organic plant matter to produce energy. Biomass energy is derived from the earth's plant life, including grass, straw, wood, waste paper, agricultural plant residue, manure, and gases and liquids obtained from these sources. Most biomass is burned for heat rather than for the generation of electricity.

In 1996 the United Nations estimated that wood and animal dung used as fuel for cooking and heating accounted for approximately 14 percent of the world's total energy usage. Wood is the primary heating source for about 50 percent of the world's population, including most people in Third World countries, where costs of fossil fuels are high in relation to income. Methane gas derived from animal dung and used as a fuel has helped many farm families in developing countries improve their standard of living. Many landfills collect methane that evolves from the biodegradation of organic material and sell it as fuel.

All energy sources have some drawbacks and environmental impacts. Burning biomass releases the greenhouse gas carbon dioxide into the atmosphere and contributes to particulate air pollution. In many developing countries, especially in semiarid regions, live trees are being harvested faster than they can be regrown, leading to widespread deforestation. As trees are harvested, the roots that hold soil are lost, sharply increasing soil erosion. Killing live vegetation for fuel usage not only increase erosion but also slows the rate of removal of carbon dioxide from the atmosphere.

Much wood in the United States is used in fireplaces by individual homeowners to generate heat. However, heating a house using a fireplace is not very efficient, and the smoke and particulate pollution can reach such high levels that wood-burning stoves have been banned in some cities and restricted in others in order to reduce winter air pollution.

More than 50 percent of the household trash

in the United States and Canada consists of paper, which could be collected and used as fuel. By burning this material in controlled incinerators, the volume of solid waste is reduced, and fossil fuel resources are conserved. However, if pollution-control devices are not adequately maintained, the toxic materials in household waste generate particulate air pollution. As a result, the construction of these incinerators declined in the late 1990's.

Not all biomass energy is obtained directly from the burning of plant matter. Charcoal is a form of biomass energy in which wood is not burned directly in the stove or fireplace, but is first transformed into a more convenient and longer-lasting form of fuel. Where charcoal and wood are used for cooking and heating, the smoke generated can create indoor air pollution and pose a hazard to health. In regions where people rely heavily on charcoal as a primary fuel for cooking, deforestation is again a problem.

Alvin K. Benson

See also: Air pollution; Alternative energy sources; Deforestation; Greenhouse effect; Particulate matter.

Biopesticides

Category: Biotechnology and genetic engineering

Biopesticides are biological agents, such as viruses, bacteria, fungi, and mites, that are used to control insect and weed pests in an environmentally and ecologically friendly manner.

Pests are any unwanted animal, plant, or microorganism. When the environment has no natural resistance to a pest and when no natural antagonists are present, pests can run rampant. For example, the fungus *Endothia parasitica*, which entered New York in 1904, caused the nearly complete destruction of the American chestnut tree because there was no natural control present.

Biopesticides represent the biological, rather than the chemical, control of pests. Many plants and animals are protected from pests by passive means. For example, plant rotation is a traditional method of insect and disease protection that is achieved by removing the host plant long enough to reduce pathogen and pest populations. Biopesticides have several significant advantages over commercial pesticides. They appear to be ecologically safer than commercial pesticides because they do not accumulate in the food chain. Some biopesticides provide persistent control since more than a single mutation is required to adapt to them and because they can become an integral part of a pest's life cycle. In addition, biopesticides have slight effects on ecological balances because they do not affect nontarget species. Finally, biopesticides are compatible with other control agents. The major drawbacks to using biopesticides are the time required for them to kill their targets and the inefficiency with which they work.

Viruses, bacteria, fungi, protozoa, mites, and flowers have all been used as biopesticides. Viruses have been developed against insect pests such as *Lepidoptera*, *Hymenoptera*, and *Dipterans*. These viruses cause hyperparasitism. Gypsy moths and tent caterpillars, for example, periodically suffer from epidemic virus infestations.

Many saprophytic microorganisms that occur on plant roots and leaves can protect plants against microbial pests. *Bacillus cereus* has been used as an inoculum on soybean seeds to prevent infection by the fungal pathogen *Cercospora*. Some microorganisms used as biopesticides produce antibiotics, but the major mechanism for protection is probably by competitively excluding a pest from sites on which the pest must grow. For example, *Agrobacterium radiobacter* antagonizes *Agrobacterium tumefaciens*, which causes the disease crown gall. Two bacteria—*Bacillus* and *Streptomyces*—added as biopesticides to soil help control the damping off disease of cucumber, peas, and lettuce caused by *Rhizoctonia solani*. *Bacillus subtilis* added to plant tissue also controls stem rot and wilt rot caused by the fungus *Fusarium*. *Mycobacteria* produce cellulose-degrading enzymes, and their addition to young seedlings helps control fungal infection by *Pythium*, *Rhizoctonia*, and *Fusarium*. *Bacillus* and

Pseudomonas are bacteria that produce enzymes that dissolve fungal cell walls.

The best examples of microbial insecticides are *Bacillus thuringiensis* (*B.t.*) toxins, which were first used in 1901. They have had widespread commercial production and use since the 1960's and have been successfully tested on 140 insect species, including mosquitoes. *B.t.* produces insecticidal endotoxins during sporulation and also produces exotoxins contained in crystalline parasporal protein bodies. These protein crystals are insoluble in water but readily dissolve in an insect's gut. Once dissolved, the proteolytic enzymes paralyze the gut. *Bacillus* spores that have also been consumed germinate and kill the insect. *Bacillus popilliae* is a related bacteria that produces an insecticidal spore that has been used to control Japanese beetles, a pest of corn.

Saprophytic fungi can compete with pathogenic fungi. There are several examples of fungi used as biopesticides, such as *Gliocladium virens*, *Trichoderma hamatum*, *Trichoderma harzianum*, *Trichoderma viride*, and *Talaromyces flavus*. For example, *Trichoderma* competes with the pathogens *Verticillium* and *Fusarium*. *Peniophora gigantea* antagonizes the pine pathogen *Heterobasidion annosum* by three mechanisms: It prevents the pathogen from colonizing stumps and traveling down into the root zone, it prevents the pathogen from traveling between infected and uninfected trees along interconnected roots, and it prevents the pathogen from growing up to stump surfaces and sporulating.

Nematodes are pests that interfere with commercial button mushroom (*Agaricus bisporus*) production. Several types of nematode-trapping fungi can be used as biopesticides to trap, kill, and digest the nematode pests. The fungi produce structures such as constricting and nonconstricting rings, sticky appendages, and spores, which attach to the nematodes. The most common nematode-trapping fungi are *Arthrobotrys oligospora*, *Arthrobotrys conoides*, *Dactylaria candida*, and *Meria coniospora*.

Protozoa have occasionally been used as biopesticide agents, but their use has suffered because of such difficulties as slow growth and complex culture conditions associated with their commercial production. Predaceous mites are used as a biopesticide to protect cotton from other insect pests such as the boll weevil.

Dalmatian and Persian insect powders contain pyrethrins, which are a toxic insecticidal compounds produced in *Chrysanthemum* flowers. Synthetic versions of these naturally occurring compounds are found in products used to control head lice. Molecular genetics have also been used to insert the gene for the *B.t.* toxin into cotton and corn. *B.t.* cotton and *B.t.* corn both express the gene in their roots, which provides

Comparison of the Properties of *Bacillus Thuringiensis* and *Bacillus Popilliae* as Microbial Biocontrol Agents

	Bacillus thuringiensis	*Bacillus popilliae*
Pest controlled	Lepidoptera (many)	Coleoptera (few)
Pathogenicity	low	high
Response time	immediate	slow
Formulation	spores and toxin crystals	spores
Production	in vitro	in vivo
Persistence	low	high
Resistance in pests	developing	reported

Source: Data adapted from J. W. Deacon, *Microbial Control of Plant Pests and Diseases* (1983).

them with protection from root worms. Ecologists and environmentalists have expressed concern that constantly exposing pests to the toxin will cause insect resistance to rapidly develop and reduce the effectiveness of traditionally applied *B.t.*

Mark Coyne

Suggested readings: B. W. Churchill's informative article on industrial methods of producing biocontrol agents appears in *Biological Control of Weeds with Plant Pathogens* (1982), edited by R. Charudattan and H. Walker. An even better reference, which gives a broad perspective of biocontrol agents, is a short monograph by J. W. Deacon called *Microbial Control of Plant Pests and Diseases* (1983).

See also: *Bacillus thuringiensis*; Biotechnology and genetic engineering; Genetically engineered bacteria; Integrated pest management; Organic gardening and farming; Pesticides and herbicides; Sustainable agriculture.

Bioremediation

Category: Ecology and ecosystems

Bioremediation is a waste management technology that employs naturally occurring plants, microorganisms, or enzymes and genetically engineered organisms to clean contaminated environments by degrading toxic organic and inorganic compounds into environmentally harmless products.

Bioremediation uses biological agents, such plants, microorganisms, and enzymes, to degrade or decompose toxic environmental compounds into less toxic forms. It is a beneficial and inexpensive waste management strategy that is environmentally friendly compared to other remediation technologies. The products of waste decomposition are usually simple inorganic nutrients or gases.

Bioremediation works because, as a general rule, all naturally occurring compounds in the environment are ultimately degraded by biological activity. Toxic and industrial wastes, and even some chemically synthesized compounds that do not naturally occur, can also be decomposed because parts of their structures resemble naturally occurring compounds that are sources of carbon and energy for biological systems. Wastes are either metabolized, in which case they are used as a source of carbon and energy, or cometabolized, in which case they are simply modified so that they lose their toxicity or are bound to organic material in the environment and rendered unavailable.

Bioremediation can occur in situ (at the contaminated site) or ex situ, in which case contaminated soil or water is removed to a treatment facility where bioremediation takes place under controlled environmental conditions. Bioremediation can use organisms that naturally occur at a site, or it can be stimulated by adding organisms, sometimes genetically engineered organisms, to the contaminated site in a process called "seeding." The first organism ever patented was a genetically engineered bacterium that had been designed to degrade the components of oil.

There are numerous approaches to bioremediation. One of the simplest is to fertilize a contaminated site to optimal nutrient levels and allow naturally occurring biodegrading populations to increase and become active. Organic contaminants have been mixed with decomposed and partially decomposed organic material and composted as a bioremediation process. In a method analogous to the activated sludge process in wastewater treatment, contaminants are mixed in slurries and aerated to promote their decomposition. It is possible to obtain biosolids specially adapted for slurry systems because they have previously been exposed to similar organic wastes.

In situ restoration of contaminated groundwater is often accomplished by injecting nutrients and oxygen into the aquifers to promote the population and activity of indigenous microorganisms. Trichloroethylene (TCE), for example, is cometabolized by methane-oxidizing bacteria and can be bioremediated by injecting oxygen and methane into contaminated aquifers to stimulate the activity of these bacteria.

Nitrate-contaminated aquifers have been successfully treated by pumping readily available carbonlike methanol or ethanol into the aquifers to stimulate denitrifying bacteria, which subsequently convert the nitrate to harmless nitrogen gas.

Bioreactors have been used in which the contaminant is mixed with a solid carrier, or the organisms are immobilized to a solid surface and continuously exposed to the contaminant. This has been used with both bacteria and fungi. For example, *Phanerochaete chysosporium*, which produces an extracellular peroxidase and hydrogen peroxide (H_2O_2), has been used to cleave various organic contaminants such as dichlorodiphenyl-trichloroethane (DDT) in bioreactors.

Highly chlorinated organic contaminants such as TCE and polychlorinated biphenyls (PCBs) resist degradation aerobically, but the contaminants can be dechlorinated by anaerobic bacteria, which decreases their toxicity and makes them easier to decompose. High concentrations of PCBs in the Hudson River in New York have been dechlorinated to less toxic forms by anaerobic bacteria. Methanogens—anaerobic bacteria that produce methane—have been observed to dechlorinate TCE in anaerobic bioreactors.

One of the problems with some wastes is that they are mixed with radioactive materials that are highly toxic to living organisms. One solution to this problem has been to genetically engineer radiation-resistant bacteria so that they also have the ability to bioremediate. For example, *Deinococcus radiodurans*, a bacterium that can survive in nuclear reactors, has been genetically engineered to contain genes for the metabolism of toluene, which will enable it to be used in the bioremediation of radiation- and organic waste-contaminated sites.

Phytoremediation is a special type of bioremediation in which plants—grasses, shrubs, trees, and algae—are used to biodegrade or immobilize environmental contaminants, usually metals. Types of phytoremediation include phytoextraction, in which the contaminant is extracted from soil by plant roots; phytostabilization, in which the contaminant is immobilized in the vicinity of plant roots; phytostimulation, in which the plant root exudates stimulate rhizosphere microorganisms that bioremediate the contaminant; phytovolatilization, in which the plant helps to volatilize the contaminant; and phytotransformation, in which the plant root and its enzymes actively transform the contaminant. For example, horseradish peroxidase is a plant enzyme that is used to oxidize and polymerize organic contaminants. The polymerized contaminants become insoluble and relatively unavailable. Plants such as Indian mustard (*Brassica juncea*) and loco weed (*Astragalus*) are heavy metal accumulators and remove selenium and lead from soil. The above-ground plant parts are harvested to dispose of the metals. Algae are used to accumulate dissolved selenium in some treatments. Poplar trees have even been genetically engineered to contain a bacterial methyl reductase that lets them methylate and volatilize arsenic, mercury, and selenium absorbed by their roots.

A 1992 U.S. Environmental Protection Agency (EPA) survey indicated that of 132 well-documented bioremediation studies, seventy-five involved petroleum or related compounds, thirteen involved wood preservatives such as creosote, seven involved agricultural chemicals, five examined tars, four treated munitions such as trinitrotoluene (TNT), and the rest involved miscellaneous compounds. As this list suggests, bioremediation of oil spills has been the single best example of successful bioremediation in practice.

In March, 1989, the *Exxon Valdez* oil tanker spilled millions of gallons of crude oil in Prince William Sound, Alaska. On many beaches, the EPA authorized the use of simple bioremediation techniques, such as stimulating the growth of indigenous oil-degrading bacteria by adding common inorganic fertilizers. Beaches cleaned by this method did as well as beaches cleaned by mechanical methods. In another instance of successful bioremediation, selenium-contaminated soil in the Kesterson National Wildlife Refuge in California was partially decontaminated by supplying indigenous fungi with organic substrates such as casein and waste orange peels. This promoted as much as 60 percent selenium volatilization in fewer than two months.

Mark Coyne

Suggested readings: *Biodegradation and Bioremediation* (1994), by Martin Alexander, is an excellent introductory textbook on the topic of bioremediation. Another excellent reference from the American Society of Agronomy is *Bioremediation: Science and Applications* (1995), edited by H. D. Skipper and R. F. Turco. For specific examples of bioremediating heavy metals and metalloids such as selenium, read *Selenium in the Environment* (1994), edited by William Frankenberger and Sally Benson.

See also: Biotechnology and genetic engineering; Detoxification; *Diamond v. Chakrabarty*; Genetically altered bacteria; Genetically engineered organisms.

Biosphere concept

Category: Ecology and ecosystems

The biosphere is the 20-kilometer-thick zone extending from the floor of the oceans to the top of mountains within which all life on earth exists. The biosphere supports nearly one dozen biomes, regions of similar climatic conditions within which distinct biotic communities reside.

The biosphere, a term coined in the nineteenth century by Austrian geologist Eduard Suess, is thought to be more than 3.5 billion years old. Compounds of hydrogen, oxygen, carbon, nitrogen, potassium, and sulfur are cycled among the four major spheres—biosphere, lithosphere, hydrosphere, and atmosphere—to make the materials that are essential to the existence of life. The most critical of these compounds is water, and its movement between the spheres is called the water cycle. Dissolved water in the atmosphere condenses to form clouds, rain, and snow. The annual precipitation for a region is one of the major controlling factors in determining the terrestrial biome that can exist. The water cycle follows the precipitation through various paths leading to the formation of lakes and rivers. These flowing waters interact with the lithosphere to dissolve chemicals as they flow to the oceans, where about one-half of the biomes on earth occur. Evaporation of water from the ocean resupplies the vast majority of the moisture existing in the atmosphere. This cycle supplies continuous water needs for both the terrestrial and oceanic biomes.

The biosphere is also dependent upon the energy that is transferred from the various spheres. The incoming solar energy is the basis for all life. Light enters the life cycle as an essential ingredient in the photosynthesis reaction. Plants take in carbon dioxide, water, and light energy, which is converted into chemical energy in the form of sugar, with oxygen generated as a by-product. Most animal life reverses this process during the respiration reaction, where chemical energy is released to do work by the oxidation of sugar to produce carbon dioxide and water.

The incoming solar energy also has a dramatic interaction with the water cycle and the worldwide distribution of biomes. Because of the earth's curvature, the equatorial regions receive a greater amount of solar heat than the polar regions. Convective movements in the atmosphere—such as winds, high- and low-pressure systems, and weather fronts—and the hydrosphere—such as water currents—are generated during the redistribution of this heat. The weather patterns and general climates of earth are a response to these energy shifts. The seven types of climates are defined by the mean annual temperature and the mean annual precipitation, and there is a strong correspondence between the climate at a given location and the biome that will flourish.

Desert Biome

The major deserts of the world are located between 20 to 30 degrees latitude north and south of the equator. The annual precipitation in a desert biome is less than 25 centimeters (10 inches) per year. Deserts are located in northern and southwestern Africa, parts of the Middle East and Asia, Australia, the southwestern United States, and northern Mexico.

Deserts are characterized by life that is unique in its ability to capture and conserve water. Deserts show the greatest extreme in temperature fluctuations of all biomes: Daytime temperature can exceed 49 degrees Celsius (120

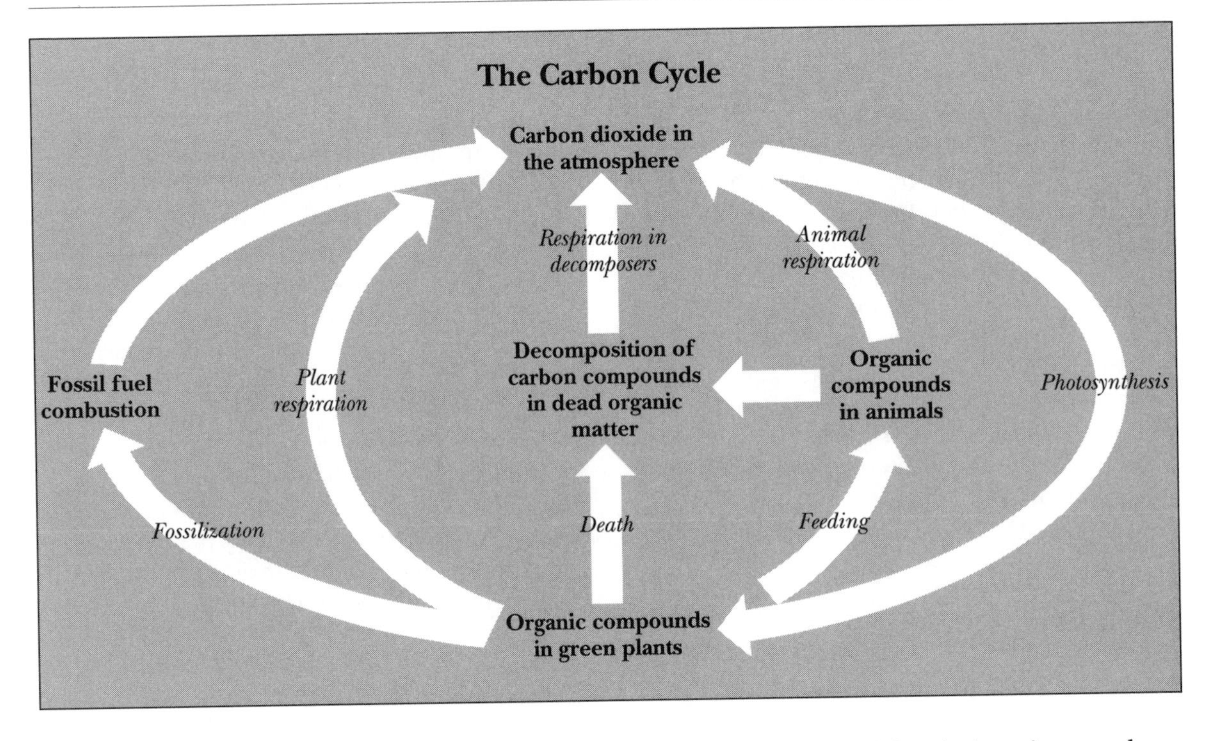

degrees Fahrenheit), and night temperatures can drop to 4.5 degrees Celsius (40 degrees Fahrenheit). Most of the animals that live in desert biomes are active at night and retreat to underground burrows or crevices during the day to escape the heat. The water cycle in deserts rarely provides surface water, so plant life usually finds its water with a wide distribution of shallow roots to capture the near-surface infiltration or a deep tap-root system that finds groundwater located below the surface of the dry stream beds. The plant life is characterized by scattered thorny bushes, shrubs, and occasional cacti. Animal life consists of an abundance of reptiles (mostly lizards and snakes), rodents, birds (many predatory types such as owls and hawks), and a wide variety of insects.

Although deserts and semideserts cover approximately one-third of the land surface on earth, they are growing in size because of human influences such as deforestation and overgrazing.

Grassland Biome

Grasslands are found in a wide belt of latitudes higher than those in which desert biomes exist. Large grassland regions occur in central North America, central Russia and Siberia, subequatorial Africa and South America, northern India, and Australia. This biome flourishes in moderately dry conditions, having an annual rainfall between 25 and 150 centimeters (10 and 60 inches). Precipitation and solar heating are unevenly divided throughout the year, providing a wet, warm growing season and a cool, dry dormant season.

The animal life in grassland regions is characterized by large grazing mammals, such as wild horses, bison, antelopes, giraffes, zebras, and rhinos, as well as smaller herbivores, such as rabbits, prairie dogs, mongooses, kangaroos, and wart hogs. This abundance of herbivores allows for a large development of secondary and tertiary consumers in the food chain, such as lions, leopards, cheetahs, wolves, and coyotes. Grasslands have rich soils that provide the fertile growing conditions for a wide variety of tall and short grasses. Within a single square meter of this healthy soil, several hundred thousand living organisms can be found, from microbes to insects, beetles, and worms. The profusion of these smaller life forms fosters an abundance of small birds.

Grasslands have been environmentally stressed by humans converting them to farmland because of their rich soils and to rangeland because of the grass supply. It is estimated that only 25 percent of the world's original grasslands remain undisturbed by human development. Worldwide overgrazing and mismanagement of rangeland have caused large tracts of fertile grassland to become desert or semidesert.

Tropical Rain Forest Biome

Rain forests receive a heavy rainfall almost daily, with an annual average of more than 240 centimeters (95 inches). The temperature is fairly constant from day to day and season to season, with an annual mean value of about 28 degrees Celsius (82 degree Fahrenheit). The combination of high rain and high temperature causes high humidity, allowing some plants to utilize the atmosphere for their water supply via "air roots." This biome is also unique because the chemical nutrients needed to sustain life are almost entirely contained within the lush vegetation of the biosphere itself and not in the upper layers of soil of the lithosphere. The soils are thin and poor in nutrients.

The tropical rain forests contain a wider diversity of plant and animal species per unit area than any other biome. It is estimated that nearly two-thirds of all the plants and insects found on earth are contained in tropical rain forests. This enormous biodiversity is accommodated in part because each form of plant or animal occupies a specialized niche based on its ability to thrive with different levels of sunlight, which correspond to different heights above the forest floor within the forest canopy. Numerous exotic insects, amphibians, reptiles, birds, and small mammals can coexist within a given canopy level. The plants growing in tropical rain forests currently provide ingredients found in 25 percent of the world's prescription and nonprescription drugs. It is estimated that at least three thousand tropical plants contain cancer-fighting chemicals.

Tropical rain forests are being destroyed at a fast rate for farmland, timber operations, mining, and grazing. It has been estimated most of the animal and plant diversity of the forests will be lost by the year 2050. Further, if the high rate of destruction continues, the release of carbon dioxide into the atmosphere from burning of biomass and fossil fuels may no longer be offset by plant consumption, and the balance of the carbon cycle will shift toward a higher concentration of carbon dioxide in the atmosphere.

Temperate Forest Biome

Temperate forests exist in areas where temperatures change dramatically during the four distinct seasons. Temperatures fall below freezing during the winter, with warmer, more humid conditions during the summer. Rainfall averages between 75 and 200 centimeters (30 and 80 inches) per year. This biome is often divided into broad-leaved deciduous (leaf-shedding) trees and conifer (cone-bearing) trees. Deciduous forests develop in regions with higher precipitation values, whereas the needle-like evergreen leaves of the conifers have scales and thick, waxy coatings that allow them to flourish at the lower end of the precipitation range.

Deciduous forests develop more solid canopies with widely branching trees such as elm, oak, maple, ash, beach, and other hardwood varieties. The forest floor often contains an abundance of ferns, shrubs, and mosses. Coniferous

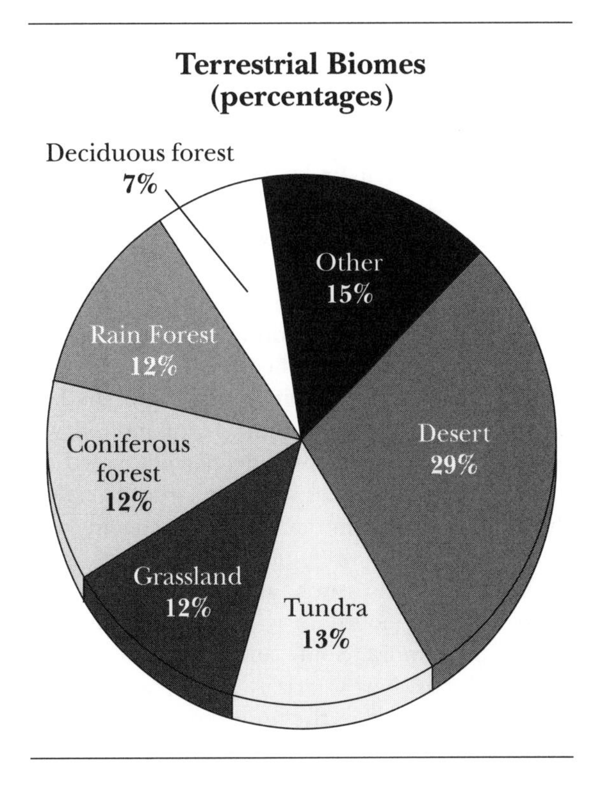

forests are usually dominated by pine, spruce, fir, cedar, and hemlock trees. Some deciduous trees, such as aspen and birch, often occur with the conifers. The coniferous forest floor is so acidic from decomposing evergreen needles that often only lichens and mosses can grow.

Temperate forests are home to diverse animal life. Common mammals of deciduous forests include squirrels, chipmunks, porcupines, raccoons, opossums, deer, foxes, black bears, and mice. Snakes, toads, frogs, and salamanders exist alongside a large bird population of thrushes, warblers, woodpeckers, owls, and hawks. The larger animals of the coniferous forests include moose, elk, wolves, lynx, grouse, jays, and migratory birds.

Forests have been traditionally viewed as a limitless timber resource by the lumber industry. The vast forests of Europe were cleared one thousand years ago. This liquidation of temperate forest continues in Siberia, where 25 percent of the world's timber reserves exist.

Tundra Biome

Tundras occur in areas near the Arctic ice cap and extends southward across the far northern parts of North America, Europe, and Asia. During the majority of the year, these largely treeless plains are covered with ice and snow and are battered by bitterly cold winds. Tundras are covered with thick mats of mosses, lichens, and sedges (grass-like plants). Because the winters are long and dark, tundra vegetation grows during the three months of summer when there is almost constant sunlight.

Bogs, marshes, and ponds are common on the summer landscape because permafrost, a thick layer of ice that remains beneath the soil all year long, prevents drainage of melted waters. These wet areas provide perfect breeding grounds for mosquitoes, deerflies, and blackflies during the brief summer. These insects in turn serve as a source of food for migrating birds. Larger mammals, such as caribou, reindeer, musk ox, and mountain sheep, migrate in and out of the tundra. Some animals, such as lemmings, Arctic hares, grizzly bears, and snowy owls, can be found in the tundra during all times of the year.

The tundra is the earth's most fragile terrestrial biome. Vegetation disturbed by human activity can take decades to replenish itself. Roads and pipelines must be constructed on bedrock or layers of added gravel in order to avoid melting the upper layers of the permafrost.

Oceanic Biomes

Oceans cover 70 percent of the earth's surface and contain 97.6 percent of the water of the hydrosphere. They play a primary role in regulating the earth's distribution of heat, and they are central to the water cycle. Oceans are instrumental in the survival of all life on earth. In addition, oceans house more than 250,00 species of marine plants and animals that occur as six common biomes.

Coral reefs are coastal biomes that develop on continental shelves in regions of clear, tropical waters. Collectively, coastal biomes comprise only 10 percent of the world's ocean area but contain 90 percent of all ocean species. It is the region where most commercial fishing is done. The vast open ocean contains only 10 percent of all oceanic species. Vegetation is mostly limited to free-floating plankton. Exotic bottom fauna exist on deep hydrothermal vents. Animal life includes whales, dolphins, tuna, sharks, flying fish, and squids.

A number of transitional biomes at the ocean-land interface also exist. The intertidal biome can be composed of sandy beaches or more rocky zones that are covered by water only during periods of high tide. A variety of crustaceans and mollusks are found on wet sandy beaches, whereas rock tidal pools contain kelp, Irish moss, and rockweed, all of which compete for space with snails, barnacles, sea urchins, and star fish.

Toby Stewart and Dion Stewart

SUGGESTED READINGS: *The Biosphere* (1970), edited by *Scientific American*, contains a thorough discussion of biosphere processes. *Biodiversity* (1990), edited by E. O. Wilson, contains articles from noted biologists on topics about biodiversity and problems facing biodiversity in biomes. *The Next One Hundred Years* (1990), by Jonathon Weiner, discusses threats to the earth's biosphere. *Conserving the World's Biodiversity* (1990), by Jeffrey A. McNeely, covers strategies being used to conserve the biosphere around the world. *Extinctions: The Causes and Consequences of the Disappearance of Species* (1981), by Paul Ehrlich and Anne Ehrlich, is a well-written book on threats to biodiversity.

See also: Balance of nature; Biodiversity; Ecosystems; Global Biodiversity Assessment.

Biosphere reserves

Category: Preservation and wilderness issues

Through discussions that started in 1970, the United Nations Educational, Scientific, and Cultural Organization (UNESCO) initiated the Man and the Biosphere Programme to recognize sites where the preservation of natural resources is integrated with research and the sustainable management of those resources.

The first biosphere reserve was designated in 1976, and by the end of 1997 there were 352 such sites in eighty-seven countries. In 1995 UNESCO convened a conference in Spain and developed the Seville Strategy, which was designed to strengthen the international network and encourage the use of the sites for research, monitoring, education, and training. In the early years of the program, preservation was stressed. The adoption of the Seville Strategy by UNESCO emphasized the role of people in the use of their natural resources.

All biosphere reserves contain a legally protected core area where there has been minimal disturbance by people. Only uses that are compatible with the preservation of biological diversity are permitted in the protected core. Model biosphere reserves also include a managed use or buffer zone that surrounds the protected core. Research and environmental education are examples of activities suitable for the buffer zone. Surrounding the buffer zone is a zone of cooperation or transition zone. The boundaries

Isle Royal National Park in Michigan has been designated a biosphere reserve by UNESCO's Man and the Biosphere Programme. (Jim West)

of the transition zone are loosely defined and often include local towns and communities. Economic activities of people, such as farming, logging, mining, and recreation occur within the transition zone and are not restricted by the biosphere reserve.

Only the boundaries of the core area are legally defined. The designation of a biosphere reserve does not alter the legal ownership of the land or water that is included within its zones. UNESCO does not have jurisdiction over any nation's biosphere reserves. In many cases there is a mosaic of land ownership, including federal, state, local, and private ownership. Even the core area may be privately owned, as long as it is managed for its preservation.

In the United States, the core areas of some biosphere reserves are within national parks, such as Olympic or Yellowstone National Parks, whereas other biosphere reserves are composed of a cluster of core areas, such as the ten units within the California Coastal Range Biosphere Reserve. The management and administration of a biosphere reserve often involves a number of interested citizens, government agencies, and owners.

The worldwide network of biosphere reserves represents the only international network of protected areas that also emphasizes sustainable development and wise use of the natural resources. Hence, they are sites where the objective of integrating conservation and development can be examined, demonstrated, and tested. Research at these sites serves to solve practical problems in resource management.

William R. Teska

See also: Biodiversity; Biosphere concept.

Biosphere II

Date: began operation September 26, 1991
Category: Ecology and ecosystems

Biosphere II was an attempt to reproduce the atmosphere of the earth (Biosphere I) inside a sealed structure and show that it could support humans.

Biosphere II is a glass and steel structure located near Oracle, Arizona, that looks similar to a series of connected greenhouses. The structure was sealed with four men and four women inside on September 26, 1991; they were released exactly two years later. A second team was sent in for ten months during 1994, then the whole enterprise went into receivership and was turned over to Columbia University in 1996 for management until 2001.

Biosphere II was intended to be a prototype for a Mars settlement, and the experiments were designed to show that it was possible to maintain a self-sustaining human colony in a hostile environment. The crew members had to grow their own food, recycle the water they drank, and produce the oxygen they breathed. They were responsible for the health of the plant and animal species inside the sealed structure, and they had to maintain the complex apparatus that kept Biosphere II functioning.

The biomes of earth were represented by a rain forest, desert, savannah, ocean, and salt marsh, plus an area of intensive agriculture. It proved to be impossible to keep the biomes separate: The desert, for example, was too wet. Also, the structure shut out 40 percent of the sunlight, and the ceiling was sometimes so hot that the tops of the trees burned. Nineteen out of twenty-five vertebrate species died, while others (such as ants and cockroaches) swarmed out of control.

Among the difficulties encountered during the first two-year experiment was a lack of oxygen: The integrity of the sealed environment had to be broken when it became apparent that the local soil and the concrete consumed more oxygen than predicted. The biospherians produced about 80 percent of the food they needed but suffered crop failures because of unanticipated depredations by mites and other insects. One crew member had to be taken out of Biosphere II for emergency surgery on a wounded finger. Upon her return, she took some supplies with her. Such incidents caused critics to claim that the giant terrarium was not as self-sufficient as had been claimed, but the project directors maintained that they had never expected perfection in this first model and that finding flaws in the design was precisely the purpose of the experiment.

Crewmembers also reported inevitable personality conflicts between the biospherians who had been part of the construction and the newcomers. While the professionalism of the biospherians saw them through the conflict, it nevertheless had the potential for seriously disrupting the mission. By the end of the stay, three of the biospherians were having therapy sessions by telephone.

Biosphere II attracted nationwide attention. Because of the structure's thousands of windows, spectators were able to view everything going on inside. All this public exposure made Biosphere II an excellent tool for generating interest in ecological concerns as well as in the human potential in space.

Robert B. Bechtel

See also: Biosphere concept; Ecosystems.

Biotechnology and genetic engineering

Category: Biotechnology and genetic engineering

Biotechnology is the use of living organisms, or substances obtained from these organisms, to produce products or processes of value to humankind. Genetic engineering, also known as recombinant deoxyribonucleic acid (DNA) technology, refers to the manipulation of DNA and the transfer of genes or gene components from one species to another. This technology has tremendous potential in agriculture, human health, and solving major environmental problems, but it also has the potential to introduce a new range of agents that may be detrimental to the environment.

The term biotechnology is relatively new, but the practice of biotechnology is as old as civilization. Civilization did not evolve until humans learned to produce food crops and domestic livestock through the controlled breeding of selected plants and animals. The pace of modifying organisms accelerated during the twentieth century. Through carefully controlled breeding programs, plant architecture and fruit characteristics of crops have been modified to facilitate mechanical harvesting. Plants have been developed to produce specific drugs or spices, and microorganisms have been selected to produce antibiotics such as penicillin or other useful medicinal or food products.

The ability to utilize artificial media to propagate plants has led to the development of a technology called tissue culture; more than one thousand plant species have been propagated by tissue culture techniques. In some of these instances, the plant tissue is treated with the proper plant hormones to produce masses of undifferentiated cells called callus tissue, which can also be separated into single cells to establish a cell suspension culture. Specific drugs or other chemicals can be produced with callus tissue and cell suspensions, or this tissue can be used to regenerated entire plants.

Numerous advances have also occurred in animal biotechnology. Artificial insemination, the process in which semen is collected from the male animal and deposited into the female reproductive tract by artificial techniques rather than natural mating, has been in use for more than one hundred years. Males in species such as cattle can sire hundreds of thousands of offspring with this technique, while only fifty or fewer could be sired through natural means.

Embryo transfer is a technique used to increase the number of offspring that can be produced by a superior female. In this procedure, an embryo in its early stage of development is removed from its own mother's reproductive tract and transferred to another female's reproductive tract. Superovulation is the process in which females that are to serve as embryo donors are injected with hormones to stimulate increased egg production. Embryo splitting is the mechanical division of an embryo into identical twins, quadruplets, sextuplets, and so on. Both superovulation and embryo splitting have increased the feasibility of routine embryo transfers. It is now possible for the sperm from a superior male to be used to fertilize several ovules, each of which can then be split into sev-

Ian Wilmut, who cloned an adult sheep named Dolly, works in his laboratory. The Scottish scientist believes that it would be "inhumane" to experiment with cloning humans. (AP/Wide World Photos)

eral offspring, from a superior female. The resulting embryos can then be transferred to the reproductive tracts of inferior surrogate females.

RECOMBINANT DNA TECHNOLOGY

Biotechnological advances have provided the ability to tap into the world gene pool. This technology has such great potential that its full magnitude is just beginning to be appreciated. Theoretically, it is possible to transfer one or even more genes from any organism in the world into any other organism. Since genes ultimately control how an organism functions, gene transfer can have a dramatic impact on agricultural resources and human health.

Research has provided the means by which genes can be identified and manipulated at the molecular and cellular levels. This identification and manipulation is primarily dependent upon recombinant DNA technology. In concept, recombinant DNA methodology is fairly easy to comprehend, but in practice it is rather complex. The genes in all living cells are very similar in that they are all composed of the same chemical, the nucleic acid called DNA. The DNA of all cells, whether from bacteria, plants, lower animals, or humans, is very similar, and when DNA from a foreign species is transferred into a different cell, it functions exactly as the native DNA functions; that is, it codes for protein.

The simplest protocol for this transfer involves the use of a vector, usually a piece of circular DNA called a plasmid, which is removed from a microorganism such as bacteria and cut open by an enzyme called a restriction endonuclease or restriction enzyme. A section of DNA from the donor cell that contains a previously identified gene of interest is cut out from the donor cell DNA by the same restriction endonuclease. The section of donor cell DNA with the gene of interest is then combined with the open plasmid DNA, and the plasmid closes with the new gene as part of its structure. The recombinant plasmid (DNA from two sources) is placed back into the bacteria, where it will replicate and code for protein just as it did in the donor cell. The bacteria can be cultured and the gene product (protein) harvested, or the bacteria can be used as a vector to transfer the gene to another species, where it will also be expressed. This transfer of genes, and therefore inherited traits, between different species has revolutionized biotechnology and provides the potential for genetic changes in plants and animals that have not yet been envisioned.

Biotechnology and Agriculture

Biotechnology can have a tremendous impact on agriculture. Traditional breeding programs may be too slow to keep pace with the rapidly increasing growth of the human population. Biotechnology provides a means of developing these higher yielding crops in one-third of the time it takes to develop them though traditional plant breeding programs because the genes for the desired characteristics can be inserted directly into the plant without having to go through several generations to establish the trait. Also, there is often a need or desire to diversify agriculture production in a given area, but soil or climate conditions may severely limit the amount of diversification that can take place. Biotechnology can provide the tools to help solve this problem; crops with high cash value can be developed to grow in areas that heretofore would not have supported such crops. In addition, biotechnology can be used to increase the cash value of a crop by developing plants that can produce new and novel products such as antibiotics, drugs, hormones, and other pharmaceuticals.

There is also going to be increased pressure for crop production to be more friendly to the environment, and biotechnology will play a major role in the development of a long-term, sustainable, environmentally friendly agricultural system. Biotechnology is already being used to develop crops with improved resistance to pests. For example, a gene from the bacterium *Bacillus thuringiensis* (*B.t.*) codes for an insecticidal protein that kills insects but is harmless to other organisms. When this gene is transferred from the bacterium to a plant, insect larvae are killed if they eat from the leaves or roots of the plant. A number of *B.t.* plants have been developed, including cotton and potatoes. The availability of crop varieties with improved pest resistance will reduce the reliance on pesticides. Nevertheless, agricultural pollutants will still be present, and the need to remediate these polluting agents will continue to exist. Biotechnology will play an important role in the development of bioremediation systems for agriculture as well as other industrial pollutants.

Biotechnology will also play an important role in livestock production. Bovine somatotropin, a hormone that stimulates growth in cattle, is already being harvested from recombinant bacteria and fed to dairy cattle to enhance milk production. Future experimentation might lead to increased productivity and at the same time a reduction in the cost of production. Just as with plants, disease-resistant animals could be genetically engineered, and animals could also be genetically engineered to produce novel and interesting products such as pharmaceuticals. The cloning of Dolly the sheep in Scotland in 1997 opened a whole new avenue in the use of biotechnology for livestock production. The use of cloning technology, in conjunction with surrogate mothers, will provide the means to produce a whole herd of genetically superior animals in a short period of time.

Biotechnology and Medicine

DNA technology will also have a direct impact on human health and can be used to produce a variety of gene products that could be utilized in the clinical treatment of diseases. Several human hormones produced by this methodology are already in use. The hormone insulin used to treat insulin-dependent diabetics was the first major success in using a product of recombinant technology. Recombinant DNA-produced insulin, marketed under the name Humalin, has been used to successfully treat thousands of diabetic patients since 1982. Protropin, the trade name for human growth hormone (HGH), has been released by the U.S. Food and Drug Administration (FDA) to treat a disease called hyposomatotropism. Without treatment with HGH, people suffering from this disease do not produce enough growth hormone and will not reach normal height.

Somatostatin, another pituitary hormone, has also been produced by recombinant DNA techniques. This hormone controls the release of insulin and HGH. Small proteins called interferons normally produced by cells to combat viral infections have been produced using recombinant DNA methodology. Some vaccines against viral diseases can also be produced by using this technology. Recombivax HB, the first of these vaccines, has been successfully used to vaccinate against hepatitis B, an incurable and sometimes fatal liver disease.

The potential for the future application of gene therapy has also been enhanced by advances in biotechnology. Among the forms of gene therapy currently being considered are gene surgery, where a mutant gene that may or may not be replaced by its normal counterpart is excised from the DNA; gene repair, where the defective DNA is repaired within the cell to restore the genetic code; and gene insertion, where a normal gene complement is inserted in cells that carry a defective gene. Gene surgery and gene repair techniques are currently too complex to be used on humans. Gene insertion can potentially be done in germ-line cells such as the egg or sperm, the fertilized ovum or zygote, the fetus, or the somatic cells (a nonreproductive cell) of children or adults. Although zygote therapy holds the most promise because completely normal individuals could potentially be produced, gene insertion in zygotes is still a very complex issue and will not be practiced in the near future. Gene insertion into somatic cells will most likely be the type of gene therapy to be attempted in the near future. The gene will be inserted into the cells of a tissue that is most influenced by a defective gene with the hope that the inserted gene will code for sufficient gene product to alleviate the symptoms of the disease.

Environmental Issues

In spite of biotechnology's potential benefits to human health, agriculture, and the environment, there are also potential drawbacks to its use. Since the first recombinant DNA experiments in 1973, there have been numerous social, ethical, and scientific questions raised about the possible detrimental effects of genetically engineered organisms on public health and the environment. The major environmental concerns are related to containment, or how to prevent genetically engineered organisms from escaping into the environment.

In the mid 1970's U.S. scientists invoked a self-imposed moratorium on genetic engineering experiments until the government could establish committees to develop safety guidelines that would apply to all recombinant DNA experimentation in the United States. This resulted in the formulation of guidelines that specified the degree of containment required for various types of genetic engineering experiments. Two types of containment, biological and physical, were prescribed. Physical containment refers to the methods required to prevent an engineered organism from escaping from the laboratory, while biological containment refers to the techniques utilized to ensure that an engineered organism cannot survive outside the laboratory. The guidelines associated with containment, particularly physical containment, are sometimes difficult to monitor and enforce.

Despite the rigors of the containment guidelines, there are those who fear that an engineered organism will eventually escape into the environment. Should this occur, the organism could cause as much, or even more, environmental damage than has occurred in the past when foreign species were introduced to new habitats. For example, the introduction of rabbits to Australia dramatically upset the ecological balance on that continent. Hence, field experiments with genetically engineered organisms must be strenuously controlled and monitored.4 Although there have been numerous safe field trials with genetically engineered organisms, such as *B.t.* plants, there is still widespread opposition to such practices. Although there appear to be few risks that cannot be ascertained within the laboratory associated with the release of genetically engineered higher plants, there is the fear that engineered genes could possibly be transferred by cross-pollination to other species of plants. While there is no evidence that such cross-pollinations occur in nature, such a transfer could, for example, produce a highly vigorous species of weed. In addition, such gene transfers could potentially result in a plant that produces a toxin that would be detrimental to other plants, animals, or humans.

Since viruses and bacteria are major components of numerous natural biochemical cycles and readily exchange genetic information in a variety of ways, it is even more difficult to envision all the ramifications associated with releasing these genetically altered organisms into the environment. Field testing of genetically engineered organisms will always involve some element of risk, and risk assessment is easier for some species, such as higher plants, than for other species, such as bacteria. There is no international agreement on the guidelines for field experiments with genetically engineered organisms. Some countries have imposed regulations for field experimentation, while others have not. There is clearly a need for rigid controls, and in order to minimize the risks, there is a need for integral cooperation among industry, governments, and regulatory organizations.

With advances in the cloning of plants and animals, there is concern about the loss of genetic variability. In nature, species survival is dependent on the genetic variability of the population. Genetic variability obtained through normal sexual reproduction provides a species with the ability to adapt to changes in the environment; since the environment is continually changing, the loss of genetic variability usually leads to extinction of the species. Cloning results in the production of large numbers of genetically identical individuals. Hence, the cloning of large numbers of animals or plants at the expense of those produced through sexual reproduction will lead to loss of genetic variability and could lead to their eventual extinction.

D. R. Gossett

SUGGESTED READINGS: *Understanding DNA and Gene Cloning* (1992), by Karl Drlica, provides an excellent discussion of the recombinant DNA techniques used in biotechnology. A very good overview of the potential role of biotechnology in human health can be found in *The New Human Genetics* (1989), by G. J. Stine. A good description of the potential use of biotechnology in the animal production industry is presented in *Scientific Farm Animal Production* (1992), by R. E. Taylor. An outstanding treatise on the use of biotechnology in crop production can be found in *Plants, Genes, and Agriculture* (1994), by M. J. Crispeels and D. E. Sadava. Neil Campbell provides an excellent discussion of the techniques used in recombinant DNA methodology in *Biology* (1996). *Environmental Issues in the 1990's* (1992), by Antionette M. Mannion and Sophia R. Bowlby, offers a good discussion of some of the environmental issues associated with biotechnology. "Biotechnology: An Overview: Its Application to Animals," in the *USDA Yearbook of Agriculture: Research for Tomorrow* (1986), is a government publication that gives the general reader an excellent summary of the uses of biotechnology in the animal industry.

SEE ALSO: *Bacillus thuringiensis*; Cloning; Dolly the sheep; Genetically altered bacteria; Genetically engineered foods; Genetically engineered organisms; Genetically engineered pharmaceuticals.

Birth defects, environmental

CATEGORY: Human health and the environment

Embryotoxic agents encountered through environmental exposure can cause a variety of birth defects, ranging from growth retardation to malformations and death. The severity of any particular birth defect results from a combination of factors, including the gestational age at which the fetus is exposed to the agent and the effective dose of the toxic substance.

About 150,000 children are born with birth defects in the United States each year; some of these defects are caused by environmental exposure to teratogens (substances that cause developmental malformations). The National Birth Defect Registry, assembled through a questionnaire designed by the Association of Birth Defect Children (ABDC), details maternal and paternal exposure to environmental agents, including chemicals, radiation, pesticides, lead, and mercury. In March, 1998, the Birth Defects Prevention Act was passed by the U.S. Congress. This act continued the National Birth Defects Prevention Study, which was begun in 1996 to provide data on birth defects, including possible environmental causes.

Prenatal exposure to ionizing radiation is known to cause birth defects. Exposure to large, fluctuating electromagnetic fields (EMFs) has also been implicated in causing environmental birth defects. Speculation that high-voltage electric power transmission lines may cause environmental birth defects or childhood cancer has led to a number of epidemiological studies using small mammals. Results of American studies, which have taken years to complete, have not established a link between high-voltage lines and birth defects. However, European epidemiological studies using different methodologies have not ruled out a link between EMFs and birth defects.

Chemical agents known to cause environmental birth defects in animals and humans include compounds of heavy metals (especially mercury, lead, and thallium), urethane, dioxinlike chemicals, various steroids, sex hormones (including xenoestrogens and antiandrogenic pesticides), and trypan blue, an agent once used to treat mange. Chemical agents disturb intracellular chemistry; embryonic changes usually precede placental changes. Deliberate testing of some agents has shown increases in the frequency of common malformations. For example, dioxin and similar chemicals have been observed to increase cleft palate and hydronephrosis in mice. Fetal resorption may cloud tests, in which case animals must have the pregnancy terminated by cesarean section. The animals and fetuses are then autopsied and carefully examined to determine if fetal resorption sites are detectable. A minimum of two species of animals, studied at a minimum of three-dose levels, are essential for teratogenic testing. Strong teratogenic agents usually produce similar malformations in different species. Agents that produce demonstrable birth defects in animals, especially in higher primates such as baboons, must be presumed capable of producing birth defects in a human fetus.

Because of the high cost and lengthy time involved in animal testing, the U.S. National Institute for Environmental Health (NIEH) is validating a model known as Frog Embryo Teratogenesis Assay-Xenopus (FETAX) to screen chemicals for their potential to cause birth defects.

About 3 percent of all newborn humans have some birth defect or congenital abnormality. Of these defects and abnormalities, 20 percent are attributed to gene mutations, 5 to 10 percent to chromosomal abnormalities, and 5 to 10 percent to known nonenvironmental teratogenic factors such the drug thalidomide or to a maternal factor. No more than 10 percent of defects and abnormalities are believed to arise from exposure to teratogenic factors in the environment. This estimate does not include chemicals that produce alterations in central nervous system (CNS) development during critical life stages or deviations in postpubescent behavior. The ABDC has found a pattern of disabilities in the children of 1,800 Vietnam War veterans. It has also found a pattern of craniofacial birth defects (Goldenhar syndrome) in children of Gulf War veterans.

Epidemics of environmental birth defects in human populations are occasionally seen as a result of exposure to a release of a teratogenic agent, especially when the food chain is contaminated. When consumed by pregnant women, organic mercury is a potent teratogen. The best-known epidemic of organic mercury poisoning occurred in Minamata Bay, Japan, where industrial effluents containing high levels of mercury were pumped into the ocean for many years. This mercury was assimilated into seafood consumed by the local population, which poisoned thousands and resulted in many cases of congenital Minamata disease.

One study found an increased risk of neural tube defects, malformations of the cardiac septa, and anomalies of great arteries and veins among children borne to mothers who lived within a 3-kilometer (2-mile) radius of a landfill. An epidemiological study of California Superfund hazardous waste sites found that women who lived within 0.4 kilometers (0.25 miles) of a Superfund site during the first three months of pregnancy were four times more likely to have a baby with conotruncal heart defects (3 of 201 births) and twice as likely to have a child with neural tube defects (8 of 507 births) than women who lived at greater distances.

Anita Baker-Blocker

SUGGESTED READINGS: Ted A. Loomis, *Essentials of Toxicology* (1974), includes a basic introduction to teratology. J. B. Bishop, K. L. Witt, and R. A. Sloane, "Genetic Toxicities of Human Teratogens," in *Mutation Research* 396 (December 12, 1997), provides a review of human teratogens. Birth defects near California Superfund sites are examined in "Maternal Residential Proximity to Hazardous Waste Sites and Risk for Selected Congenital Malformations," by L. A. Croen, et al., in *Epidemiology* 8 (1997). *Bitter Sea: The Human Cost of Minamata Disease* (1992), by Akio Mishima, translated into English by Richard L. Gage and Susan B. Murata, presents a nontechnical discussion (with photographs) of congenital birth defects resulting from maternal consumption of seafoods contaminated with mercury.

SEE ALSO: Environmental illnesses; Minamata Bay mercury poisoning; Superfund.

Black lung

CATEGORY: Human health and the environment

Black lung is a chronic respiratory disease caused by long-term inhalation of coal and mineral dusts in closed coal-mining environments.

Since the Industrial Revolution, coal has provided energy for society. With escalating demand, coal mining developed into a prevalent industry in many regions, such as Wales in Great Britain and the Appalachian Mountains of the United States. By the nineteenth century, economies of such regions were dependent upon coal through mining or burning.

Coal burning produced gaseous and particulate (soot) pollution. Cities were characterized by black, soot-coated buildings and coal residue. Coal dust explosions in mining and storage became common. Slower to appear were global warming (the greenhouse effect), acid rain, and coal workers' pneumoconiosis, also called black lung, caused by chronic inhalation of coal dust. This particle-induced fibrosis-emphysema produces lesions in respiratory bronchioles, which interfere with oxygen absorption and transport in the lungs. Its symptoms are consistent with similar lung diseases: shortness of breath and oxygen starvation, heart disease, immune system irregularities, weakness and poor health, and lung cancers. It resembles diseases caused by fine, airborne, respirable particles of asbestos (asbestosis); cotton, wood, and other plant-based dusts (farmers' lung); and silicas such as quartz, glass, and sand (silicosis).

The greatest incidence of black lung occurs in underground mining. Drilling, pulverizing, loading, and transporting coal generate large dust concentrations, which are breathed into lungs to produce, over years, black lung disease. Numerous safeguards now minimize this risk. First, air is ventilated through mine passages by fans and baffles, while sophisticated routing produces multiregion flows that keep the air at face level fresh. Second, mines utilize wetting and scrubber systems to keep coal wet or water mists to remove dust from the air.

More recent developments include ventilation and filtering systems worn on the body—for example, helmets that provide continuous streams of filtered air across workers' faces. Research is ongoing in the development of drill bits and other components that minimize dust. Replacement of steel bits with those coated with tungsten carbide, polycrystalline diamond, or other hard ceramic or metallic films is receiving attention, as is optimization of thread geometry and bit speed.

Since the Federal Coal Mine Health and Safety Act of 1969 and the Mine Act amendments of 1977, federal and state agencies (for example, the Mine Safety and Health Administration, or MSHA) have mandated efforts by mine operators to abolish black lung. Operators are required to sample mine air for deviations from permissible exposure limits (PELs) of coal and silica dusts, methane, carbon monoxide, sulfur dioxide, and hydrogen sulfide; to report these deviations; and to take appropriate measures to correct them.

Operators are required to medically screen employees over time. Periodic chest X rays, lung capacity measurements, cardiovascular checkups, and general blood, urine, and endocrine analyses warn of early signs of black lung. Periodic questionnaires provide worker health data and, hence, insight into overall risks in a given operation. Emphasis has been placed on education regarding black lung and the regulatory infrastructure that now addresses it. Numerous government agencies provide information and services related to the disease, its prevention, its control, and compensation or support for those affected.

The PEL for coal dust is 2 milligrams per cubic meter (mg/m^3) for unaffected workers and 1 mg/m^3 for workers with any signs of black lung. In 1995 the National Institute for Occupational Safety and Health (NIOSH) recommended a reduction for all workers to 1 mg/m^3. The Occupational Safety and Health Administration (OSHA) has a slightly different value of 2.4 mg/m^3 for coal dust with less than 5 percent silica. In 1968-1969 the average dust concentration in underground coal mines was 6 mg/m^3, but since the Coal Act and the Mine Act, these averages have dropped to below 2. It is estimated that the risk of eventually developing symptoms of black lung are 32 out of 1000 in workers thirty to forty years old and 8 out of 1000 for workers fifty-five to sixty-five years old.

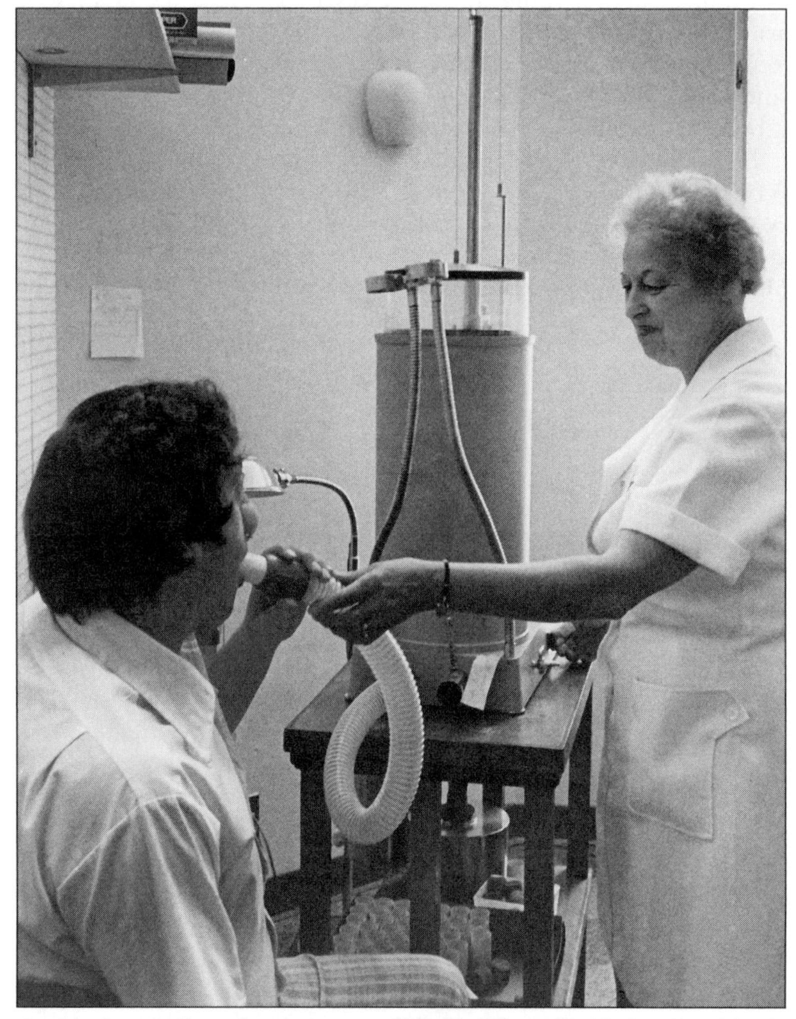

A coal miner in Pennsylvania is screened for black lung. Legislation passed in the United States during the 1960's and 1970's requires mine operators to provide periodic medical tests for their employees. (Jim West)

The silica content of coal dust impacts the epidemiology and PELs. Fewer than 1 mg/m^3 of silica can cause silicosis, and the current PEL for silicon is 0.1 mg/m^3, with reduction to 0.05 mg/m^3 being considered. Silica and mineral dusts are generated in coal deposits interspersed with bedrock and during initial drilling through rock to reach the coal.

Other hazards of coal mining include trace toxic heavy metals, asphyxiation or explosions caused by methane, other poisonous gases such as carbon monoxide and hydrogen sulfide, high underground temperatures, unhealthy fungi, and accidents. Environmental problems include destruction of native habitats, coal residue and associated aesthetic problems, noise pollution, emission of toxic gases, and runoff of water that contains toxic substances.

Significant progress was made over the last three to four decades of the twentieth century in fighting black lung and associated problems in coal mines. Government intervention, technological developments, and education should further improve the health, safety, and environmental status of the coal mining industry.

Robert D. Engelken

SUGGESTED READINGS: For more information on black lung, consult the *Coal Handbook* (1981), edited by Robert A. Mayers, and *Environmental Effects of Mining* (1996), by Earle Ripley, Robert Redmann, and Adele Crowder, for general overviews of coal mining and associated health and safety concerns. *Mechanisms in Respiratory Toxicology* (1982), edited by Hanspeter Witschi and Paul Nettesheim, discusses the medical aspects of black lung. The U.S. Department of Health and Human Services Centers for Disease Control and Prevention's *Work-Related Lung Disease Surveillance Report 1996* (1996) presents extensive data on black lung occurrence and epidemiology. Extensive information is also available on the Internet in the "Report of the Advisory Committee on the Elimination of Pneumoconiosis Among Coal Mine Workers" through the Department of Labor's Mine Safety and Health Administration's (MSHA) Web site (http://www.msha.gov).

SEE ALSO: Environmental health; Environmental illnesses.

Black Wednesday

DATE: September 8, 1943
CATEGORY: Atmosphere and air pollution

The first severe episode of photochemical smog occurred in Los Angeles, California, on September 8, 1943. The Los Angeles Times dubbed the day Black Wednesday. Causative factors proved to be both natural and human-made in origin.

The Los Angeles basin is prototypical of a smog-producing area, with mountains rising on the east and north and persistent high-pressure weather systems during summer and early fall. Subsidence in upper levels of the high-pressure atmosphere results in air being compressed and heated. Westerly winds blowing over a cold ocean current carry cool, moist air in beneath the subsiding air, forming a temperature inversion at an altitude lower than the mountain peaks. With cool, moist air beneath hot, dry air, rising currents can ascend only to the inversion level. They then spread laterally and descend when they run into the natural barrier of the mountains. This situation places a lid over rising air, which causes pollutants to remain at low elevations for several days.

The human-made origins of Black Wednesday included increased industrialization and population growth in the area as a result of World War II. These factors led to a rise in the number of automobiles and increased automotive effluent, as well as a higher yield of effluent from industrial smokestacks. In addition, a shortage of natural rubber led to the manufacture of synthetic rubber. A plant in Boyle Heights produced a synthetic called butadiene, which appeared to be the main source of pollution. Other minor sources contributing to photochemical smog were outdoor burning, backyard incinerators, and smudge pots.

The impact of photochemical smog includes a variety of repercussions. The consequences on humans can vary. Carbon monoxide (CO) from automotive exhaust can result in a change in physiology: Exposure to low levels of CO can produce impaired functions, whereas exposure to high concentrations may end in death. There seems to be a positive correlation between polyaromatic hydrocarbons and lung, skin, and scrotal cancers. The main complaints from humans, however, are burning eyes and irritated throats. Peroxyacetyl nitrates (PAN) are the primary eye irritants, although acrolein and formaldehyde also contribute to the problem. Finally, reduced visibility is a product of aerosols contained in smog.

Vegetation is distressed by several components of photochemical smog. Ozone assaults leaf palisades, resulting in destruction of chlorophyll, lowered photosynthesis and respiration rates, and development of dark spots on leaf surfaces. Plant exposure to sulfur dioxide causes gas to combine with water to form sulphite ions, which cause leaves to darken, grow flaccid, and become dry. This condition eventually induces death of tissue. The effect of photochemical smog on animals was not studied or recorded during the Black Wednesday event. However, the consequence of animal exposure to smog is considered to be similar to that of humans.

Photochemical smog also alters inanimate objects. Rapid cracking and eventual deterioration of stretched rubber can result from exposure to smog. Likewise, breakdowns of natural and synthetic fabrics and fading of dyes can be traced to smog elements.

Ralph D. Cross

See also: Air pollution; Automobile emissions; London smog disaster; Smog.

Bookchin, Murray

Born: January 14, 1921; New York, New York
Category: The urban environment

Creator of the concept of social ecology, Murray Bookchin is an anarchist thinker and ecological activist. In the 1960's he, along with Rachel Carson, suggested that the prosperity of the post-World War II United States had been bought at the price of serious harm to the environment. Some critics believe his predictions were as insightful as Carson's.

Like German socialist philosopher Karl Marx, Bookchin argues that the human race cannot survive in a civilization based on life in the modern city, bureaucratic decision-making structures, and industrialized labor. However, he builds upon Marx's insights to further argue that both socialism and capitalism are heedless of modern industry's impact on the environment. Thus, neither socialism nor capitalism can be the basis for a sustainable society. He posits a close interconnection between human domination over nature and domination over one another, arguing that since the propensities to dominate both nature and other people sprang up together, they must be eliminated together.

Bookchin published his first two books under the pseudonym Lewis Herber: *Our Synthetic Environment* (1962) and *Crisis in Our Cities* (1965). He argued that it would not be an uprising of the proletariat but rather an uprising of an anti-authoritarian younger generation that would resolve the environmental crisis by revolting and dissolving all social hierarchies. While this has not come to pass—Bookchin's faith in the younger generation is criticized by some as naïve—his insight concerning the connection between social problems and the earth's environment—that the existence of hierarchy, in addition to the misuse of technology, has brought humankind to the brink of disaster—is seen as a valuable contribution to environmental thought.

Throughout his career, Bookchin has developed and refined his argument that posits a link between the "destructive logic behind a hierarchical social structure" and environmental crisis. He strongly criticizes environmentalists who are satisfied to save endangered species or ban harmful chemicals yet support underlying social structures that will only produce new toxins and similar problems. He calls for a return to the way of life of past societies that he sees as organic and harmonious, such as the Plains Indians in North America, to life as it was before the Industrial Revolution. This notion is also deemed naïve by some critics.

Bookchin is professor emeritus of social ecology at Ramapo College in New Jersey and director emeritus of the Institute for Social Ecology at Goddard College in Rochester, Vermont. He has published numerous journal and magazine articles. He has also published twelve books, under both the pseudonym Lewis Herber and his own name. Among these are *Remaking Society: Pathways to a Green Future* (1990), *The Philosophy of Social Ecology* (1990), *The Rise of Urbanization and the Decline of Citizenship* (1987), *The Modern Crisis* (1986), and *Toward an Ecological Society* (1980).

Anne Statham

See also: Environmental ethics; Social ecology.

Borlaug, Norman

Born: March 25, 1914; Cresco, Iowa
Category: Agriculture and food

American plant pathologist and environmental activist Norman Borlaug became known as the father of the Green Revolution for his efforts to develop high-yield crops to increase food production throughout the world.

Norman Ernest Borlaug credited his childhood experiences on his family's farm with providing him with a practical approach to agriculture. In the 1930's Borlaug studied forest management and plant pathology at the University of Minnesota, earning a doctorate by 1942. In 1943 the Rockefeller Foundation, which U.S. secretary of agriculture Henry Wallace had convinced to fund agricultural aid for Mexican farmers, hired Borlaug to breed disease-immune crops that could be grown in varied climates. He perfected a strain of high-yield dwarf spring wheat.

While Borlaug was experimenting with plant breeding, the post-World War II global population rapidly increased. Some environmentalists, such as Paul Ehrlich, predicted that not enough food could be produced and mass starvation would occur. Borlaug believed that his wheat could prevent such a disaster. In 1963 the Rockefeller Foundation and the Mexican government established the International Maize and Wheat Center (CIMMYT), naming Borlaug director of the Wheat Improvement Program. He traveled to India and other developing nations to share his high-yield agricultural techniques. Critics thought that Borlaug should plant indigenous crops rather than Western grains, but he emphasized that wheat provided necessary calories and nutrients.

The Indian and Pakistani governments resisted Borlaug's efforts until famine became extreme. Delayed seed shipments and the outbreak of war between India and Pakistan also hindered Borlaug's work. He persisted, however, and yields increased approximately 70 percent during the first season. By 1968 Pakistan was agriculturally self-sufficient and increased its yields from 3.4 million tons of wheat in 1963 to 18 million by 1997. India boosted its yields from 11 million tons to 60 million tons, even briefly exporting wheat.

Borlaug's expansion of food production, which saved hundreds of millions of people from starvation, was called the Green Revolution. He won the Nobel Peace Prize in 1970 for his humanitarian efforts to secure the basic human right of freedom from hunger. However, this honor did not ensure continued support of his work. When he expressed interest in using his techniques to assist African agriculture, Borlaug's CIMMYT patrons stopped funds because of an environmental backlash that protested high-yield agriculture, claiming that its use of inorganic fertilizers and irrigation damaged the environment. Borlaug explained that high-yield agriculture preserved habitats from slash-and-burn techniques to create farmland. He criticized theorists that he thought did not comprehend the reality of what was economically and

Norman Borlaug won the Nobel Peace Prize in 1970 for developing agricultural practices that increased food supplies in Third World countries. (Nobel Foundation)

politically possible in developing countries. With colleagues Haldore Hanson and R. Glenn Anderson, he wrote *Wheat in the Third World* (1982).

Former U.S. president Jimmy Carter and Japanese industrialist Ryoichi Sasakawa financed Sasakawa-Global 2000, which sponsored Borlaug's projects in African nations. Continuing development of high-yield crop strains at CIMMYT in the 1990's, Borlaug wrote articles, lectured, and testified to the U.S. Congress about the opposition posed by environmental lobbyists. After receiving government support in Ethiopia and Poland, he planned similar agricultural programs in the former Soviet Union and Latin America. Warning that arable land was finite, Borlaug stressed that uncontrolled population growth could result in starvation in the twenty-first century and demanded the storage of food reserves.

Elizabeth D. Schafer

SEE ALSO: Agricultural revolution; Green Revolution; High-yield wheat; Population growth.

Boulder Dam

DATE: completed 1935
CATEGORY: Preservation and wilderness issues

Completed in 1935, Boulder Dam (officially renamed Hoover Dam in 1947) was built to regulate the flow of the lower Colorado River in order to prevent floods, provide consistent water levels necessary for irrigation, and produce hydroelectric power. Water collected behind the dam forms Lake Mead, which serves as a reservoir and a recreation area.

At the time of its construction, Boulder Dam was rightfully hailed as an engineering marvel, a concrete wall 221 meters (726 feet) high and 379 meters (1,244 feet) long at its crest. The dam significantly altered the environment of the southwestern United States, often in unanticipated ways.

Spanish explorers named the Colorado River for its reddish-brown color, a result of the sediment that the river carried. Boulder Dam captured the sand and silt, which settled to the bottom of Lake Mead. Experts knew that the silt would eventually fill the reservoir, but they estimated that this process would take several hundred years. In the meantime, they sought ways to reduce the amount of silt deposited behind the dam. Arguing that overgrazing by Navajo sheep herds caused soil erosion, which was the major source of silt in the Little Colorado and San Juan Rivers (two tributaries of the Colorado), government officials successfully forced the Navajo to accept a stock reduction program. Nonetheless, silt continued to build up behind the dam, and by 1949 the sediment was more than 82 meters (270 feet) deep in some areas. This changed the flow of the river upstream, slowing rapids in the Grand Canyon more than 160 kilometers (100 miles) north of the dam. Silt buildup in Lake Mead slowed after the completion of the Glen Canyon Dam north of the Grand Canyon in 1963.

Ironically, the reduction in the Colorado's flow downstream of the dam caused severe floods in Needles, California, during the 1940's. The water released from the newly finished dam picked up large deposits of sediment that stood at the base of the dam, carrying it downriver. Because the river's flow had been reduced, it could not carry the silt great distances, and sand quickly built up where the river widened near Needles. Plants soon filled the river bottom, providing yet another obstacle to water flow. The magnitude of the problem became apparent in 1941, when water releases from the dam increased in size and frequency. The sand, silt, and plant life diverted the rushing waters into the nearby town, forcing the evacuation of several families. During the 1940's the region became a veritable swamp. Flooded cesspools polluted the groundwater, creating a health hazard. In response, the federal government authorized levee construction and river dredging, expensive solutions to problems caused by a dam that was intended to prevent flooding.

One of the benefits to which the government pointed in publications praising the dam was the creation of new wildlife habitats along the shores of Lake Mead. A 1985 brochure noted that more

Boulder Dam as seen from the Nevada side. After its completion in 1935, the dam caused tremendous changes in both the natural environment and patterns of human settlement in the southwestern United States. (Archive Photos)

than 250 species of birds had been identified in the region, including migratory species that used the lake as a stopover. It also claimed that the lake provided water for native animals, including desert bighorn sheep. In addition, several species of game fish lived in the lake, a boon to the area's reputation as a recreational site and a plus for the region's economy.

The brochure neglected to mention the negative impacts of the dam on wildlife. For centuries the Colorado had deposited its sediment in a delta located at its mouth at the Gulf of California. This delta was a haven for wildlife, such as deer, birds, bobcats, and numerous other species. The reduced flow of the river and the reduction in sediment altered the delta, and wildlife began to disappear. Some observers likened the area to a desert. Changes also occurred upriver from the dam. The game fish in Lake Mead competed with native fish upriver, possibly contributing to the extinction of some species in the lower reaches of the Grand Canyon.

The construction of Boulder Dam may have also increased the possibility of earthquakes in the region. When Lake Mead began to fill during the late 1930's, scientists recorded several seismic shocks, a phenomenon that had not been noted in the area prior to the dam's construction. Arguing that several faults existed in the region, some scientists dismissed any notion of a general link between earthquakes and the weight of water in reservoirs. However, earthquakes at several major dams around the world during the 1960's led engineers to reconsider the connection between seismic activity and dams. Two major shocks in the Lake Mead area in 1972 confirmed suspicions that the reservoir contributed to earthquake activity.

Perhaps the two most important environmental impacts of Boulder Dam were the precedent it created and the population growth that it fostered in the American southwest. The success of the dam prompted the construction of several dams on the Colorado River, including the Glen

Canyon Dam, which significantly altered the ecosystem within the Grand Canyon. In addition, the electricity that the dam generated and the lure of Lake Mead increased human migration to the area, which in turn led to increased pollution and greater demand for water from the already beleaguered Colorado River.

Thomas Clarkin

SUGGESTED READINGS: Edward Goldsmith and Nicholas Hilyard's *The Social and Environmental Effects of Large Dams* (1984) discusses the question of Lake Mead and earthquakes. Richard White's *The Roots of Dependency Subsistence, Environment, and Social Change Among the Choctaws, Pawnees, and Navajos* (1983) touches upon the dam's impact on the Navajo. Although devoted to Glen Canyon Dam, Steven W. Carothers and Bryan T. Brown's *The Colorado River Through Grand Canyon* (1991) includes some material on Hoover Dam. Richard L. Berkman and W. Kip Viscusi's *Damming the West* (1973) examines the issue of sedimentation. The U.S. government publication *Hoover Dam* (1985) details the benefits of the dam without considering its negative impacts.

SEE ALSO: Dams and reservoirs; Glen Canyon Dam; Flood control; Grand Canyon; Hydroelectricity.

Braer oil spill

DATE: January 5, 1993
CATEGORY: Water and water pollution

The 1993 Braer *oil spill killed or sickened thousands of shellfish and led to severe economic setbacks for the local fishing industry.*

The oil tanker *Braer* ran aground at the southern tip of the Shetland Islands, located about 210 kilometers (130 miles) north of the Scottish mainland, in January 5, 1993, spilling more than one million barrels of crude oil into the Atlantic Ocean. Under the severe wind and wave conditions that prevailed around the Shetland Islands at the time, the spilled oil thoroughly mixed into the turbulent sea water and rapidly dispersed. Ten days after the spill, the concentrations in the vicinity of the wreck had fallen from several hundred parts per million (ppm) to 4 ppm, still about two thousand times the normal level.

Being less dense than water, most of the oil floated. The lightest, most volatile hydrocarbons started to evaporate, decreasing the volume of the spill but polluting the air. Subsequently, a slow decomposition process occurred, caused by sunlight and bacterial action. After several months, the oil mass was reduced by approximately 80 percent.

Shortly after the *Braer* oil spill, all fishing activities were prohibited in the surrounding areas. For most shellfish species, this ban remained in effect until the spring of 1995, when they were judged to be free of any significant levels of oil contamination. Shortly after fishing resumed, however, fisherman found that their catches of lobsters and queen scallops were very poor and that the proportion of young lobsters and scallops was abnormally low.

Supported by the Shetland Fishermen's Association, the North Atlantic Fisheries College carried out a number of laboratory trials in 1996 and 1997 to determine what effects crude oil may have on lobsters and scallops and to investigate whether the *Braer* oil spill could have had adverse effects on these stocks. During the experiments, researchers used Norwegian Gullfaks crude oil, the same type as the *Braer*'s cargo, to simulate the spill conditions. Results of the study showed that in the short term, adult scallops and lobsters could survive exposure to relatively high concentrations of oil, although lobster eggs and larvae suffered high mortality rates. Although the oil did not kill the adult lobsters, it did cause major behavioral abnormalities, including significant reductions in feeding, movement, responsiveness to stimuli, and aggression.

In May, 1997, the Scottish Office Fisheries Department released a document indicating that the *Braer* oil spill polluted a much wider area than previously thought. The report showed that levels of oil in prawns and mussels from a 1,036-square-kilometer (400-square-mile) zone of excluded fishing were still rising in 1996, and there were indications that the tides had spread the oil underwater all around the Shetland Island's 1,450-kilometer (900-mile) coastline. Fishermen reported that their fishing grounds were ruined, with nothing replacing what was being caught. By the late 1990's considerable work still needed to be done to rectify the damage done by the *Braer* oil spill, particularly on the large quantities of oil that had become incorporated into subtidal sediments.

Alvin K. Benson

SEE ALSO: *Amoco Cadiz* oil spill; *Argo Merchant* oil spill; *Exxon Valdez* oil spill; Oil spills; *Sea Empress* oil spill; Tobago oil spill; *Torrey Canyon* oil spill.

Brazilian radioactive powder release

Date: September-October, 1987
Category: Nuclear power and radiation

The release of radioactive powder in Goiania, Brazil, in 1987 led to the contamination of 244 people.

On September 13, 1987, two men found an old radiation-therapy machine in an abandoned medical clinic in Goiania, a Brazilian state capital with a population of one million people. When they took the machine apart they found a cylindrical lead container inside, which they sold to a local junk dealer. Unaware that the canister contained radioactive cesium 137, an isotope of cesium used in the treatment of cancer, the junk dealer opened the canister to investigate the curious blue light visible inside. He shared his discovery with friends, who were fascinated by the blue powder and the shimmer it made on their skin. The powder was rapidly distributed throughout the community as people gave samples to friends as gifts. Children played with it as if it were a new toy. Symptoms of radiation poisoning began days later.

Radiation destroys the reproductive mechanisms of cells, affecting to the greatest extent those cells that divide the most rapidly. These include bone marrow, gastrointestinal tract, skin, and hair cells. Effects include burns, hemorrhaging, and, because bone marrow cells are involved in the immune response of the body, decreased white blood cell count, with consequent susceptibility to infections.

A team of doctors was dispatched to the area, and helicopters equipped with radiation detection equipment determined the "hot spots" in the city. Tons of materials were found to have traces of the powder, including furniture, buses, money, and animals. The doctors found that 244 people had been contaminated with doses up to 600 rads (1 rad is roughly equivalent to seven chest X rays) and immediately hospitalized fifty-four. Patients were thoroughly washed to remove any excess cesium that may still have been on their skin and were fed with Prussian blue, which is known to complex with cesium and block its further absorption into the body. Doctors treated infections with antibiotics and hemorrhaging with blood-clotting factors. The Brazilian government enlisted the aid of the International Atomic Energy Agency (IAEA) and the Radiation Emergency Center Training Sites (REACTS), a World Health Organization unit that acts as the response unit for radiation incidents in the Western Hemisphere.

Some controversy arose over the use of an experimental treatment involving granulocyte-macrophage colony-stimulating factor (GM-CSF) on severely ill patients. GM-CSF is one of five hormones that increase the production of white blood cells in bone marrow. The treatment is used in cancer patients because it offsets the effects of radiation and chemotherapy, allowing larger doses to be used. The six patients treated with GM-CSF in Brazil represented the worst cases of contamination. Four of them died, while the other two seemed to respond well to the treatment. By the end of the year, although there were no further fatalities, twenty-eight people remained hospitalized with radiation sickness.

The Brazilian Nuclear Energy Commission came under strong criticism for the accident, which caused widespread panic among the citizens of Goiania and other cities in Brazil. There was confusion regarding which government agency had the responsibility to license, monitor, and ensure proper disposal of radioactive waste in the country. The Brazilian Ministry of Health and the Ministry of Labor were also implicated in negligence of their responsibilities with regard to monitoring radioactive materials.

Suzanne Jones and Massimo D. Bezoari

SEE ALSO: Nuclear accidents; Radioactive pollution and fallout.

Brent Spar occupation

TIME: April 30-May 23, 1995
CATEGORY: Water and water pollution

Greenpeace protesters occupied an abandoned oil platform that the Shell corporation had planned to sink, forcing the company to adopt alternate disposal methods.

On April 30, 1995, fourteen Greenpeace volunteers boarded the abandoned oil-storage platform *Brent Spar*, which was anchored off the western coast of Norway in the North Sea. The *Brent Spar*, a floating cylinder moored to the sea bottom, had been built in 1976, but it had been taken out of commission in 1991. Its owner, Royal Dutch Shell, intended to lower the platform into the Atlantic Ocean and then sink it more than six thousand feet of water.

Greenpeace had engaged in months of negotiations with Shell prior to the occupation. Greenpeace representatives urged Shell officials to develop a plan for dismantling the platform on land, citing studies that concluded that on-land disposal and recycling of oil rigs was environmentally preferable to sea dumping. Shell replied with studies of its own asserting that deep-sea disposal was the environmentally sounder option.

When it became clear that Shell planned to proceed with the ocean dumping, Greenpeace sent fourteen volunteers to the platform. It also stepped up its publicity campaign, drawing support from the European Union (EU) commissioner for the environment, Denmark's minister for environment and energy, and other important sources.

In May, the European Parliament passed a resolution opposing the dumping of the *Brent Spar*.

On May 22, Shell dispatched a team to the *Brent Spar* to remove the protesters, but bad weather delayed the evacuation. The volunteers ended their occupation the next day but vowed to continue the fight in the courts. However, an English judge refused to hear the Greenpeace case, stating that the English courts had no jurisdiction over the North Sea matter.

In June, German chancellor Helmut Kohl raised the question of the *Brent Spar* with British prime minister John Major. By then, Germany had joined the list of nations opposing the dumping of oil platforms in the sea.

Shell nevertheless proceeded with its plan. In late June, therefore, Greenpeace activists again boarded the platform by helicopter. Meanwhile, demonstrations against Shell took place in Great Britain, Denmark, the Netherlands, Germany, and Switzerland. In Germany, a Shell gas station was firebombed, and the company's sales fell more than 15 percent.

On June 20, Shell announced that it would not dump the *Brent Spar* in the sea. Nine days later, the member nations of the Oslo Paris Commission (OSPAR) voted 11-2 (Norway and the United Kingdom opposing) to impose a moratorium on sea disposal of oil platforms. The following month, Shell received permission from the Norwegian government to store the *Brent Spar* in an inlet on Norway's west coast while working out plans for its dismantling.

Christopher Kent

SEE ALSO: Greenpeace; oil spills.

Brower, David

BORN: July 1, 1912; Berkeley, California
CATEGORY: Preservation and wilderness issues

David Brower was one of the twentieth century's most influential and controversial environmental activists and writers. He was vigorously involved in battles concerning environmental issues for over fifty years.

David Ross Brower served as the first executive director of the Sierra Club from 1952 to 1969. He is credited by many with helping the San Francisco-based organization grow from two thousand to seventy-seven thousand members and developing it into a powerful, national organization. He led the club in aggressive campaigns against federal government projects to develop wild areas, most notably fights that successfully stopped the construction of the Echo Park Dam, which would have flooded part of Dinosaur National Monument in Utah in the 1950's, and two different dams across the Colorado River in the Grand Canyon in the 1960's and 1970's. His enterprising tactics included full-page advertisements in *The New York Times* and the *San Francisco Chronicle*, which resulted in the Internal Revenue Service (IRS) reclassifying the nonprofit Sierra Club as a lobbying organization and removing their tax-deductible status.

For over twenty-five years, Brower focused much of his passion and energy on the Glen Canyon Dam in northeastern Arizona. "Glen Canyon died, and I was partly responsible for its needless death," Brower wrote in *The Place No One Knew* (1963). In the mid-1950's the U.S. Bureau of Reclamation was planning to construct dams across the Colorado River in the Grand Canyon and Glen Canyon, Arizona, and across the Green and Yampa Rivers in Utah. Following the directives of the Sierra Club board of directors, Brower agreed to drop the club's opposition to the Colorado River dams if the Bureau of Reclamation discontinued plans to build the two dams in Utah. The bureau agreed to the deal and moved forward to build the Glen Canyon Dam. Before the dam construction was completed, Brower and the Sierra Club decided the compromise had been a mistake. They blamed their decision on a lack of familiarity with the spectacular beauty of Glen Canyon.

Brower and the Sierra Club successfully led an effort in the mid-1960's to prevent construction of the Bureau of Reclamation's proposed Marble Canyon Dam in the Grand Canyon and helped cripple the bureau's effort to build Bridge Canyon Dam farther downstream in the Grand Canyon. In 1996 the Sierra Club directors unanimously passed a motion by Brower to support draining Lake Powell, the reservoir behind the Glen Canyon Dam, and return the Colorado River flow to the most natural state possible. Brower did not advocate dismantling the dam but leaving it "as a tourist attraction, like the Pyramids, with passers-by wondering how humanity ever built it, and why."

By 1969 the majority of the Sierra Club's board of directors found Brower's tactics too reckless, both financially and politically. They removed him as executive director. He then formed the preservation-oriented Friends of the Earth and the League of Conservation Voters, both of which flourished under his leadership. He also facilitated the establishment of independent Friends of the Earth organizations in other countries. In 1982, after conflicts with members the Friends of the Earth's professional staff and its directors, Brower moved on to form another group, Earth Island Institute, whose stated mission was to globalize the environmental movement. He returned to the Sierra Club as a director in 1983 and was reelected in 1986 and 1995. In the fall of 1994 Brower helped develop the Ecological Council of Americas to improve cooperation among Western Hemisphere organizations that were attempting to better integrate environmental and economic needs.

In the 1990's Brower called on the federal government to replace the U.S. Bureau of Land Management with a new agency called the National Land Service. Its mission would be to protect and restore private and public land in the United States. He also strongly advocated the creation of a national biosphere reserve system.

Throughout his life, Brower pushed the edges of environmental thought of the day. He pioneered ideas and methods to preserve the environment and create a global approach to issues. For many years Brower advocated the establishment of international natural reserves in areas of rich biodiversity and ecosystems. The United Nations Educational, Scientific, and Cultural Organization (UNESCO) has established such a system of World Heritage Sites.

Brower also advocated a method called CPR to guard against the destruction of natural areas and biodiversity. "C" is for *conservation*, or the rational use of resources. "P" represents the *pres-*

ervation of threatened, endangered, and yet undiscovered species. "R" calls for *restoration* of the land already damaged by human activities. Many of his tactics and ideas seemed radical when he introduced them but became standard practice among mainstream environmentalists in later years. Russell Train, chairman of the Council on Environmental Quality during President Richard M. Nixon's administration, once said, "Thank God for Dave Brower; he makes it so easy for the rest of us to be reasonable."

Louise D. Hose

SUGGESTED READING: David Brower shares his views in *Let the Mountains Talk, Let the Rivers Run* (1995) and an earlier autobiography, *For Earth's Sake* (1990). *The Place No One Knew* (1963), also by David Brower, is a beautiful coffee-table book showing the canyons that were lost when water backed up behind the Glen Canyon Dam. John McPheer's *Encounters with the Archdruid* (1971) entertainingly documents informal debates between Brower and influential developers during three separate trips into wild areas.

SEE ALSO: Echo Park Dam proposal; Glen Canyon Dam; Preservation; Sierra Club.

Brown, Lester

BORN: March 28, 1934; Bridgeton, New Jersey
CATEGORY: Agriculture and food

Agricultural scientist and writer Lester Brown is the founder and president of the Worldwatch Institute, a Washington, D.C.-based environmental think tank. The mission of the institute is to analyze the state of the earth and to act, according to Brown, as "a global early warning system."

Lester Brown was raised on a small tomato farm in Bridgeton, New Jersey. He joined the 4-H Club and the Future Farmers of America at his local school. When he was fourteen, he and his brother purchased a used tractor and a small plot of land to grow tomatoes. Within a brief time, they became two of the most successful tomato farmers on the East Coast. Brown graduated from Rutgers University in 1955 with a degree in agricultural science and immediately put his education to practical use. He worked for six months in a small farming community in India, becoming intimately acquainted with hunger problems created by population growth and unsustainable agricultural practices.

In 1959 Brown earned a master's degree in agricultural economics and soon after joined the Department of Agriculture as an international agricultural analyst. After leaving that post in 1969, he helped organize the Overseas Development Council, a private group that analyzed issues relevant to relations between the United States and Third World countries.

Brown's educational and career background prepared him well for his work and leadership in the Worldwatch Institute. While living and working in the Third World, he became acutely aware of extensive poverty caused by economic systems dependent on cash crops for export to wealthy industrial countries and agricultural practices causing deforestation and desertification. Brown believes that food security could replace military security as the major concern of governments in the twenty-first century.

Shortly after Brown established the Worldwatch Institute in 1974, he and other staff members initiated the Worldwatch Papers, focusing on population growth and the resulting stress on natural resources, transportation trends, and the human and environmental impact of urbanization. In 1984 Brown established the institute's annual report, *State of the World*, a comprehensive overview of specific global environmental issues. This publication, now available in over twenty-five languages, is used by political leaders, educators, and citizens as a reference on environmental problems and ways to address them. In 1992 Brown inaugurated the publication of *Vital Signs: The Trends That Are Shaping Our Future.* This annual handbook features environmental, economic, and social statistical indicators on trends shaping the future.

To address world environmental problems, Brown favors international funding through a tax on currency exchanges, taxes for pollution emissions, and a greater involvement by the

United Nations. He advocates a shift away from national spending on unsustainable economic growth to investing in research and development that enhances environmental quality and protection of natural resources.

Brown is the author of several books independent of the Worldwatch Institute, including *In the Human Interest: A Strategy to Stabilize World Population* (1974) and *Building a Sustainable Society* (1981). In these and other writings, Brown warns people about the ecological dangers of exploiting the earth's resources and urges them to change their lifestyles. He likewise advises governments and scientists to cooperate in finding solutions to environmental problems.

Ruth Bamberger

SEE ALSO: Population growth; Sustainable agriculture; Sustainable development; Urbanization and urban sprawl.

Gro Harlem Brundtland initiated many innovative environmental programs while she served as Norway's prime minister. Shortly after her resignation in 1996, she was named director general of the World Health Organization. (Reuters/Ron Thomas/Archive Photos)

Brundtland, Gro Harlem

Born: April 20, 1939; Oslo, Norway
Category: Ecology and ecosystems

Gro Harlem Brundtland has been called the "Green Goddess" because of the innovative environmental programs she initiated during her career as prime minister of Norway.

Gro Harlem Brundtland grew up in a politically active family. She graduated from Oslo University Medical School in 1963 and completed a master's degree in public health at Harvard University two years later. Brundtland was a medical officer in Norwegian public health offices and participated in party politics.

In 1974 Brundtland was appointed minister of the environment. One year later she was named deputy leader of the Norwegian Labor Party. Elected to Parliament in 1977, Brundtland became prime minister and leader of the Norwegian Labor Party from February to October, 1981. She was the youngest person and the first woman to achieve that office. Brundtland was also the first environmental minister to become a country's leader, contemplating political and environmental problems together. A popular prime minister, Brundtland was reelected three times, serving from 1986 to 1989, 1990 to 1993, and 1993 to 1996.

In 1983 Brundtland was chosen to chair the World Commission on Environment and Development sponsored by the United Nations. The twenty-three-member commission worked for three years devising a global agenda for environmental protection, enhancement, and management. Brundtland traveled around the world to assess environmental conditions. The 1987 report *Our Common Future,* also known as the *Brundtland Report,* explained that environmental deterioration was caused by poverty. It supported the idea of sustainable development, which did not destroy resources to attain economic growth. Brundtland emphasized that citizens should be active in decision making and accountable for environmental quality. The report raised public awareness of the environment but was criticized by some for its overly optimistic and simplistic solutions.

Brundtland spoke at diplomatic gatherings to convince the world community to accept the report's proposals. She criticized economic development that depleted nonrenewable resources crucial for future generations and urged countries to stop such practices. Brundtland addressed the United Nations in 1988, stressing the need for international environmental cooperation by citing the example of how industrial pollution crossed national borders and caused acid rain in other countries. "The environment is where we all live," she said, "and development is what we all do in attempting to improve our lot within that abode. The two are inseparable." Brundtland urged coordinated political support of environmental issues because she believed that the world shared one economy and environment and thus politicians should adopt global solutions to environmental problems, such as formulating environmentally sound energy policies.

Brundtland wrote the introduction for *Story Earth: Native Voices on the Environment* (1993), a collection of comments by non-Western natives about environmental crises. Considered a champion of human rights and environmental quality in Norway and elsewhere, Brundtland received such awards as the Third World Prize. Despite her environmental rhetoric, Brundtland was criticized for her controversial support of Norwegian whaling in 1993. She was praised for providing opportunities for women in her cabinet and other government positions. Brundtland resigned as prime minister in October, 1996. In July, 1998, she was named director general of the World Health Organization. Brundtland focused on the worldwide prevention of disease, especially the improvement of children's health care.

Elizabeth D. Schafer

SEE ALSO: Environmental economics; Environmental policy and lobbying; Sustainable development.

Bureau of Land Management, U.S.

DATE: established 1946
CATEGORY: Land and land use

The Bureau of Land Management (BLM), a U.S. federal agency created in 1946, is responsible for managing the public lands of the United States.

The U.S. federal government's original policy was to encourage the disposal of public lands. The most widely known method for this disposal was through homesteading as a result of the Homestead Act, which was overseen by the General Land Office, a forerunner of the BLM created in 1812. The land that remained after settlement and the designation of national parks, national forests, and wildlife refuges was available for public use. Abusive use of these public lands became widespread, and by the 1930's there was need for correction. One of the most serious misuses of public land consisted of extensive overgrazing.

As a result of these abuses, the Grazing Service was created as part of the Department of Interior to manage some 80 million acres under the provisions of the Taylor Grazing Act in 1934. In 1946 the Grazing Service became the BLM. Part of the BLM's continuing responsibilities were based on the need to evaluate damage, classify public land for grazing purposes, and assess fees for grazing. Concern for environmental quality grew during the 1960's and 1970's, and this increased concern was extended to the public lands. As a result, the BLM was granted more authority under the provisions of the Federal Land Policy and Management Act of 1976, which encouraged the BLM to manage public lands in ways that were consistent with both multiple-use concepts and sustained yield principles. The multiple-use approach to land-use planning has a lengthy history in resource management, particularly in the forestry area. The potential for land to be used for a wide variety of purposes, such as timber production, grazing, and recreation, established a need for careful planning and also pointed out the advantages of managing public land in a way that would ensure that resources would be sustained and the environment would be protected. This need was articulated in the Multiple Use-Sustained Yield Act of 1960. These principles have, in turn, become a part of BLM policy.

Most lands managed by the BLM are in Alaska and the other states west of the Mississippi River. However, the management of onshore oil drilling, gas production, and mineral development on federal lands is also part of the BLM's responsibilities. As a result, the bureau maintains an office to deal with oil and mineral policies on public land east of the Mississippi River.

Jerry E. Green

SEE ALSO: Forest and range policy; Grazing; Land-use policy; Wise-use movement; Sagebrush Rebellion.

Burroughs, John

BORN: April 3, 1837; near Roxbury, New York
DIED: March 29, 1921; en route from California to New York
CATEGORY: Preservation and wilderness issues

John Burroughs was a best-selling nature writer during the late nineteenth and early twentieth centuries. In a writing career that lasted more than fifty years, Burroughs published twenty-nine nonfiction books, mostly about nature but also about travel and literary criticism, which sold more than 1.5 million copies.

John Burroughs grew up on a farm in the Catskill Mountains in New York, spending as much time as he could outdoors. Around the age of twenty, he decided that he would try to earn his living as a writer. After a brief teaching career, he spent ten years as a clerk in the United States Treasury Department in Washington, D.C. As a sideline, he published magazine essays about natural history and philosophy, always working to sharpen his writing skills. During these years he published his first book, *Notes on Walt Whitman as Poet and Person* (1867), the first biography of the great poet, who had also been a govern-

John Burroughs, on the right, poses in Alaska in 1899 with John Muir, one of the many admirers of Burroughs's nature books. (Library of Congress)

ment clerk and who was Burroughs's personal friend.

Burroughs's first volume of essays, *Wake Robin* (1871), was representative of the twenty-two collections that would follow: It featured close observations of natural history and commentary about simple country life, was made up mostly of essays (including such titles as "Birds' Nests" and "In the Hemlocks") that had been previously published in magazines, and won immediate acclaim. To Burroughs's first readers, the genre of nature writing was new and captivating, and Burroughs soon became its most popular practitioner. Sure now that he could live by his pen, Burroughs left his government job and moved back to New York, establishing a small fruit farm on the banks of the Hudson River in 1873. He continued to publish essays in some of the most popular magazines of his day and collected them into new books approximately every two years. His titles reveal something of the simple wonder that informs these books: *Fresh Fields* (1885), *Bird and Bough* (1906), and *Under the Apple-Trees* (1916).

Burroughs remained popular throughout his lifetime—a rare achievement for a writer. He formed lasting friendships with many of his admirers, including John Muir, Thomas Edison, and Henry Ford. Two more friendships led to books: *John James Audubon* (1902), an appreciation and biography, and *Camping with President Roosevelt* (1906).

After Burroughs's death in 1921, the John Burroughs Sanctuary was established to preserve his property and many of his books in West Park, New York. Seventy-five years after his death, Burroughs was still acknowledged as a pioneer and a master of the genre of nature writing, and more than a dozen of his books were still in print. The John Burroughs Association awarded a Burroughs medal annually to the best book of nature writing published that year. In 1997 Burroughs was named a charter member of the Ecology Hall of Fame in Santa Cruz, California.

Cynthia A. Bily

SEE ALSO: Audubon, John James; Muir, John; Roosevelt, Theodore.

C

Captive breeding

CATEGORY: Animals and endangered species

The selective breeding of an endangered or threatened species in captivity is sometimes the only way to save it from extinction. Captive-bred animals are often returned to their native habitat once the population has sufficiently recovered and the animals have been properly conditioned to survive in the wild.

As conditions in zoos steadily improved throughout the twentieth century and captive animals began breeding, scientists realized that the breeding of threatened and endangered species in captivity could save some species that would otherwise become extinct. Many zoos began the shift their priorities from entertainment to wildlife conservation in the late 1970's.

Initially, zoo animals were allowed to breed without consideration of their genetic or health status—the only guiding principle involved was "more is better." Zoo populations are far too small to sustain a healthy breeding population, which, depending on the species, might range from seventy-five to four hundred animals. The resulting inbreeding led to fewer offspring, an increased rate of birth defects, and susceptibility to disease. In response, the Species Survival Plan (SSP) was established to increase the number of animals in the breeding pool. A computerized mating system, the SSP maintains genetic records of all captive animals of a particular species, allowing zoos to exchange animals without fear of inbreeding. By 1997, there were eighty-two SSPs covering 134 species.

By the 1990's captive breeding had lost some of its appeal as the costs began to outweigh the benefits. In vitro fertilization of gorillas can cost up to $75,000, and the cost of the Condor Recovery Plan was an estimated $20 million between 1974 and the late 1990's. Critics of the captive breeding program—and zoos in general—argue that this money could be better spent preserving natural habitat. Opponents of the U.S. Endangered Species Act (1973) have also used the captive breeding pro-

A rare Fijian crested iguana and its offspring, which was bred at the Taronga Zoo in Australia. Captive breeding programs offer the only chance of survival for many endangered species. (AP/Wide World Photos)

gram as an excuse to eliminate habitat protection laws.

Another drawback of SSPs and captive breeding in general are surplus animals. Breeding endangered species may be a good public relations tactic, but the culling of surplus animals is definitely not. One large problem for captive breeding programs is the question of what to do with animals that are genetically inferior, are past breeding age, or have been hybridized in ways that would contaminate the gene pool. Contraceptive implants and separation are two common solutions, but some surplus animals end up in game parks and hunting preserves.

Despite its drawbacks, captive breeding remains the only option for some species, the most famous of which is the California condor. In an extraordinary measure of intervention, the last nine known condors in the wild were captured in 1987, bringing the total number in captivity to twenty-seven. By 1992 captive breeding had brought the population up to 134, and scientists gradually began returning birds to their natural habitat. Other species saved by captive breeding include the black-footed ferret, the Arabian oryx, Przewalski's horse, and Pere David's deer. Their continued survival in captivity buys time for conservation biologists to resolve the environmental issues threatening their original habitats.

P. S. Ramsey

See also: Endangered species; Extinctions and species loss; Whooping cranes; Zoos.

Carson, Rachel Louise

Born: May 27, 1907; Springdale, Pennsylvania
Died: April 14, 1964; Silver Spring, Maryland
Category: Ecology and ecosystems

Carson's 1962 bestseller Silent Spring *helped spark the modern environmental movement.*

Rachel Carson's 1962 book Silent Spring *alerted the general public to the dangers of widespread pesticide use.* (Library of Congress)

Even before the publication of *Silent Spring* made her a household name, Rachel Carson had a notable career. She was born on May 27, 1907, in Springdale, Pennsylvania, approximately eighteen miles from Pittsburgh; her mother, the daughter of a Presbyterian minister, instilled a love of nature in her three children. At the age of eighteen, Carson entered the Pennsylvania College for Women (later Chatham College), where she contributed many works to the school newspaper. Midway through her college career, Carson changed her major from English to biology, and she was accepted to graduate school at The Johns Hopkins University.

Following her graduation from Pennsylvania College for Women in the spring of 1929, Carson studied under a scholarship at the Marine Biological Laboratory at Woods Hole on Cape Cod. She completed her master's degree in marine zoology in 1932 and subsequently took a job with the Bureau of Fisheries, which later became the U.S. Fish and Wildlife Service. Carson continued her work for the bureau for sixteen years, rising to become editor in chief of the publications department.

In 1941, Carson's first book, *Under the Sea Wind*, a lyrical exploration of the sea and its life, was published to excellent reviews. Her 1951 *The Sea Around Us* proved even more successful, winning the 1951 National Book Award in 1951 and the 1952 John Burroughs Medal; it remained on the best-seller lists for more than a year. Carson was awarded a Guggenheim Fellowship, but she returned the money after receiving substantial royalties from the second book. The resulting financial independence allowed her to resign from her government post and devote herself to her writing. In 1955, she published *The Edge of the Sea*, another bestseller. It was the 1962 publication of *Silent Spring*, however, that transformed Carson from a successful nature writer to a controversial public figure. *Silent Spring* and related articles by Carson alerted the public to the fact that pesticides, particularly dichloro-diphenyl-trichloroethane (DDT), were decimating the bird populations of North America. Although chemical companies and other vested interests attacked the work savagely, *Silent Spring* attracted a broad readership and had a profound effect on public policy. Carson influenced environmental policy on two levels. First, *Silent Spring* and related works led to the virtual ban of the use of DDT. Public views of pesticides and toxic chemicals changed forever, and politicians responded with a generation of legislation regulating the use of pesticides and other chemicals. Despite the criticism it engendered, *Silent Spring* remains a defining critique of the indiscriminate use of chemicals, and the ecological dangers of modern technology have remained a scientific and public concern since its publication. At a broader level, Carson publicized the interdependence of humanity and nature. Her integration of scientific concepts into popular writings helped to educate the public on ecological principles and the beauty of natural systems. Carson's death in 1964 came at the height of her influence, but her vision of the need to appreciate, understand, and protect natural systems would remain forceful decades later.

Mark Henkels

SEE ALSO: Agricultural chemicals; DDT; extinctions and species loss; food chains; integrated pest management; pesticides and herbicides.

Carter, Jimmy

BORN: October 1, 1924; Plains, Georgia
CATEGORY: Preservation and wilderness issues

Jimmy Carter, governor of Georgia and thirty-ninth president of the United States, made many decisions that demonstrated an environmentalist agenda during his political career.

Jimmy Carter was born into a family that owned a general store and farm in the small community of Plains, Georgia. He was appointed to the United States Naval Academy, graduating in 1946, and served in the Navy until his father's death in 1953, when he assumed his father's business responsibilities.

Carter expanded his family's businesses and successfully ran for local political office. He was elected to the Georgia Senate in 1962 and 1964. He was elected governor of Georgia in 1970 and served in the office from 1971 to 1975. Governor Carter reorganized the state government, consolidating many functions and putting all environmental agencies under the Department of Natural Resources. He rejected a U.S. Army Corps of Engineers plan to dam the Flint River, the last free-flowing large river in western Georgia. It was apparently the first such action ever taken by a governor of a state. He helped to arrange for the greater part of Cumberland Island to be given to the state of Georgia; it would later become Cumberland Island National Seashore. He also began acquiring land along the Chattahoochee River in Atlanta for state parks; these later became parts of the Chattahoochee National Recreational Area.

Carter ran for the presidency in 1976, winning by a narrow margin over incumbent Gerald Ford. Early in his term he vetoed a public works bill that included nineteen water projects that President Carter considered economically unjustified and environmentally unsound. Under intense political pressure he later signed a compromise bill that included nine of the projects. He also issued executive orders that directed federal agencies to protect or restore wetlands and floodplains wherever possible as a matter of government policy.

President Jimmy Carter, center right, tours the Three Mile Island nuclear plant on April 1, 1979, four days after the plant suffered an accident. Carter demonstrated an intense concern for the well-being of the environment throughout his term as president. (AP/Wide World Photos)

During the Carter administration a large number of important environmental acts were passed and signed into law by the president. These included the Alaska National Interest Lands Conservation Act of 1980; the Marine Sanctuaries, Fish, and Wildlife Conservation Act; the Nuclear Waste Storage Act; and the Comprehensive Environmental Response, Compensation, and Liability Act, commonly referred to as the Superfund.

It is remarkable that all of this was accomplished in the midst of a very difficult term in office. President Carter's term began with a worldwide economic recession and instability in the international petroleum market, resulting in large national budget deficits, severe inflation, and high interest rates. The term ended with the overthrow of the shah of Iran and the seizure of the U.S. embassy in Iran by the revolutionary Islamic government.

In response to the petroleum market problem, President Carter requested creation of a cabinet-level Department of Energy. The mission of this department included research on reduction in dependence on fossil fuels and development of alternative energy resources. This research has had very desirable environmental effects.

Robert E. Carver

SEE ALSO: Energy policy; Environmental policy and lobbying; Nuclear regulatory policy.

Catalytic converters

CATEGORY: Atmosphere and air pollution

A catalytic converter is a device that converts carbon monoxide, hydrocarbons, and nitrogen oxides in automobile exhaust gases into less harmful substances. Use of catalytic converters has reduced air pollution from automobiles.

Since their invention in the late nineteenth century, automobiles have revolutionized society. However, automobiles represented a significant new source of air pollution, particularly in urban areas. As early as 1943, the effects of automobile emissions on air quality were noted in the Los Angeles area. In 1952 A. J. Haagen-Smit showed that the interaction of nitrogen oxides and hydrocarbons from automobiles with sunlight resulted in the formation of secondary pollutants, such as peroxyacetyl nitrates (PAN) and ozone. This new form of air pollution, termed photochemical or Los Angeles smog, was soon recognized as a major pollution problem in urban atmospheres.

In 1961 the state of California began regulating the release of pollutants from automobiles. Two years later the first federal emission standards were imposed as part of the 1963 Clean Air Act. As further restrictions on automobile emissions were introduced in the 1960's and early 1970's, new methods for emissions reductions were developed. Of these, the catalytic converter was the most important device.

The first catalytic converters consisted of an inert ceramic support material coated with a thin layer of platinum or palladium metal. Carbon monoxide and hydrocarbons from the exhaust gases attach themselves to the metal surface, where they react with molecular oxygen and are converted into carbon dioxide and water. The use of a metal catalyst lowers the temperature and increases the rate at which reaction occurs, making it possible to remove almost all of these two pollutants from the exhaust gases.

Because lead, phosphorus, and other substances can coat the metal surfaces in catalytic converters and prevent them from functioning, the Environmental Protection Agency (EPA) was granted the authority to regulate gasoline composition and fuel additives as part of the 1970 Clean Air Act amendments. A consequence of the introduction of catalytic converters in new automobiles in the mid-1970's was the gradual switch from leaded to unleaded gasoline. As a result, emission of lead from transportation, the major source of lead as an air pollutant, decreased by 97 percent between 1978 and 1987.

While the first-generation catalytic converters dramatically reduced emission of carbon monoxide and hydrocarbons from automobiles, they had no effect on nitrogen oxide emissions. Dual-bed catalytic converters, developed in the early 1980's, use a two-step process to reduce emission of nitrogen oxides, carbon monoxide, and hydrocarbons. Dual-bed catalytic converters were followed by three-way converters, which use platinum and rhodium as the metal catalyst. A feedback system with an oxygen sensor is used to adjust the mixture of air and gasoline sent to the automobile engine to ensure maximum removal of pollutants.

The introduction of catalytic converters significantly reduced the release of pollutants by automobiles. In the United States, emission of carbon monoxide and hydrocarbons from transportation sources decreased by more than 50 percent between 1970 and 1995, even though the number of cars and trucks more than doubled. Similar reductions in automobile pollution have occurred in other countries where catalytic converters are used.

Jeffrey A. Joens

See also: Air pollution; Air pollution policy; Automobile emissions; Clean Air Act and amendments; Smog.

Chalk River nuclear reactor explosion

Date: December 12, 1952
Category: Nuclear power and radiation

The first serious nuclear reactor accident in the world occurred in December, 1952, when an experimental nuclear reactor at Chalk River, Ontario, overheated. The resulting hydrogen-oxygen explosion and release of more than one million gallons of radioactive water forced the evacuation of the reactor building.

The experimental NRX reactor, run by the National Research Council of Canada, began operation at Chalk River, Ontario, in 1947. It was operating at low power on December 12, 1952, when a technician mistakenly opened four air valves in the system used to insert the control rods that slow the rate of reaction. This action caused the control rods to move out of the reactor core, increasing the reaction rate and the amount of heat produced in the core. In the effort to return these control rods to their proper position, miscommunication resulted in the removal of additional control rods, causing the power generated in the reactor core to double every two seconds.

When the operator realized that the power was rapidly increasing, he "scrammed" the reactor, a process that forces the control rods into place, thus halting the nuclear reaction. However, the earlier error of opening the air valves kept some of the control rods from being pushed into place, and the temperature in the reactor continued to increase. To stop the reaction, water rich in deuterium (heavy water) was dumped from the reactor core. Without this water, which slows the neutrons to a point where they can induce fission in the uranium core, the reaction ceased and the core began to cool. However, more than one million gallons of highly radioactive water flooded the basement of the reactor building. The reactor dome, a 4-ton lid on the reactor vessel, rose upward to release pressure from a hydrogen-oxygen explosion, and more radioactive water escaped, flooding the main floor of the building. The reactor operators were forced to evacuate the building, and eventually the entire reactor site, as radioactivity rose above safe levels.

The small size of the experimental NRX reactor minimized the radiation release. Since the Chalk River site was remote, the exposure of the general population to radioactivity was minimal. Nonetheless, a pipeline had to be constructed to divert the radioactive water and avoid contamination of the Ottawa River. The high levels of radioactivity made decontamination and cleanup of the NRX reactor difficult. In some parts of the reactor, workers participating in the cleanup effort could spend only minutes at work before accumulating the maximum permissible annual radiation dose for reactor workers. About three hundred military personnel from the United States and Canada, including future U.S. president Jimmy Carter, volunteered to participate in the cleanup of the NRX reactor. A 1982 study of the health of more than seven hundred workers who participated in the cleanup showed no increase in the death rate from cancer when compared to the general population.

George J. Flynn

SEE ALSO: Antinuclear movement; Chernobyl nuclear accident; Nuclear power; Three Mile Island nuclear accident; Windscale radiation release.

Chelyabinsk nuclear waste explosion

DATE: September 29, 1957
CATEGORY: Nuclear power and radiation

In 1957 a nuclear explosion at Mayak, a weapons production facility in the Chelyabinsk province of the Soviet Union, exposed 270,000 people to high levels of radiation.

The Mayak industrial complex began producing weapons-grade plutonium in 1948. Workers dumped radioactive waste into the nearby Techa River. A waste storage facility was constructed in 1953 after people living near the Techa suffered from radiation poisoning. On September 29, 1957, one waste tank exploded. Although the exact cause of the explosion remains unknown, a cooling system failure contributed to the disaster. A radioactive cloud consisting of between 70 and 90 tons of waste released an estimated 20 million curies of radiation into the environment. Of the waste material that was released, 90 percent fell back on the blast site, while 10 percent drifted through the atmosphere, contaminating 2,000 square kilometers (772 square miles) of territory and exposing 270,000 people to radiation. Eyewitnesses recalled seeing red dust settle everywhere, and the waters of the Techa River turned black for two weeks. Soon thereafter, plants died, and leaves fell off the trees. In fewer than two years, all the pine trees in a 28-square-kilometer (11-square-mile) area around the Mayak complex were dead.

The Soviet government closed all the stores in the area and shipped in food. Some ten thousand people were evacuated from the area, while the government burned houses and demolished entire towns to ensure that the residents could not return. However, many smaller communities continued to use local water sources, and later anecdotal accounts indicated that not all the contaminated crops in the region were destroyed. A dairy farm near the Techa River was allowed to operate until 1959. News of Chelyabinsk was kept secret, and for almost twenty years few people outside the region knew the extent of the disaster. The U.S. Central Intelligence Agency (CIA) learned of the incident but did not make the information public. Zhores Medvedev, a Soviet émigré, published the first account of the accident in 1976.

Evaluating the impact of the explosion proved difficult for two reasons: The Soviet government consistently denied the magnitude of the event, and the region was heavily polluted by other sources, especially the dumping of waste into the Techa River. In 1989 a U.S. official who visited the Mayak complex declared it to be the "most polluted spot on earth." By 1992 nearly one thousand area residents had been diagnosed with chronic radiation sickness. Rates for cancer were higher near Mayak than anywhere else in the Soviet Union, and the general health of the population, especially children, was poor by any standard.

Cleanup efforts were hampered by the secrecy that surrounded the event for nearly three decades, limited funds, and the high levels of contamination. Lake Karachay, located near Mayak, was so radioactive that standing on its shore for more than one hour resulted in a lethal dose of radioactivity. At the close of the twentieth century, Chelyabinsk remained a pressing environmental problem and a legacy of the damage that the Cold War had wrought upon the environment.

Thomas Clarkin

SEE ALSO: Chernobyl nuclear accident; Nuclear accidents; Radioactive pollution and fallout.

Chernobyl nuclear accident

DATE: April 26, 1986
CATEGORY: Nuclear power and radiation

The accident at the Chernobyl nuclear power plant in the Soviet Union drastically changed the lives of thousands of residents in the northern part of Ukraine and the southern portion of Belarus. Questions rose about the future of the plant itself; the ecological, human health, economic, and political repercussions of the incident; and the future of nuclear power programs throughout the world.

On April 26, 1986, nuclear power reactor number 4 exploded at the nuclear plant located about 15 kilometers (9 miles) from the small town of Chernobyl in the republic of Ukraine. As the core of the reactor began to melt, an explosion occurred that blew the top off the reactor and sent a wide trail of radioactive material across large parts of the former Soviet Union and much of Eastern and Western Europe. More than 116,000 people were evacuated within a 30-kilometer (19-mile) radius

In addition to the political, social, and economic aftermath of the explosion, consequences of human inability to prevent widespread damage was evident in the environment within five years of the Chernobyl disaster. The major release of radioactive materials into the atmosphere occurred during the first ten days following the explosion. Radioactive plumes reached many European countries within a few days, increasing radiation levels to between ten and one hundred times normal levels. Over time the contaminated lithosphere created a new biogeographical province characterized by irregular and complicated patterns of radioactivity in Belarus, Ukraine, and Russia, the countries most affected. In spite of some success in efforts to slow the flow of certain soluble radionuclides into the Black and Baltic Seas, as well as into the Pripet, Dnieper, and Sozh Rivers (all of which contribute water to the Kiev water reservoir), much contamination occurred that only time can resolve.

The effects of radioactive contamination on

vegetation has varied depending on the species. Damage to coniferous forests was more than ten times greater than to oak forests and grass communities, and more than one hundred times greater than to lichen communities. Ten years after the explosion, pine forests still had high levels of radionuclides in the uppermost layer of the forest floor, while birch forests had considerably lower levels. Large numbers of highly contaminated trees were felled, and restrictions were placed on cutting and using wood in industry and for fuel. Likewise, the degree of radiation damage and the dose absorbed varied among plant communities. In addition to killing or damaging plant life, radiation has disturbed the function of plant reproductive systems, resulting in sterility or decreases in both seed production and fertility. Other effects involve changes in plant chlorophyll, necessary for removing carbon dioxide from the air.

Among animals, lower life-forms have been found to have higher radiosensitivity levels; that is, mammals are less sensitive to radioactivity than birds, reptiles, and insects. The severity of damage also depends on whether it is external or internal. The impact of environmental contamination may be less on animals than plants because of the ability of animals to move from place to place and make selective contact with the environment. Irradiation can have a wide range of effects among animal species, including death, reproductive disturbances, a decrease in the viability of progeny, and abnormalities in development and morphology.

Contamination of aquatic ecosystems by Chernobyl radiation has been considerable. The contamination of freshwater ecosystems has fallen with time. However, continuing contamination has been determined by factors such as a transfer of radionuclides from bottom sediments and erosion of contaminated soil into water sources. Some restrictions have been placed on fishing in contaminated aquatic ecosystems.

The consequences of Chernobyl on agricul-

Damage caused by an explosion and fire in reactor four of the Chernobyl nuclear power plant on April 26, 1986. (AP/Wide World Photos)

ture have been severe. Drastic changes in land use and farming practices as a result of contamination have been necessary. As with other lifeforms affected by radiation, the degree of contamination varies; in the case of agriculturally related contamination, damage depends on such things as the type of soil and the biological peculiarities of different plant species. Since Ukraine produces 20 percent of the grain, 60 percent of industrial sugar beets, 45 percent of the sunflower seeds, 25 percent of the potatoes, and about 33 percent of all the vegetables used in the former Soviet Union, the problem of soil contamination is serious. Furthermore, the problem extends beyond the borders of Ukraine: The effect on the reindeer herds of Scandinavia and on sheep breeding in mountainous regions of the British Isles has been serious as well.

The scale of the contamination of the environment by the Chernobyl accident was so enormous that the task of protecting the population has not been entirely successful. The long-term health effects of environmental contamination are caused only in part by external radiation. Scientists studying the problem believe that nearly 60 to 70 percent of future health problems will be caused by the consumption of contaminated agricultural products.

If international standards were applied for the use of agricultural land in the affected areas, nearly 1 million hectares would be declared lost for one century, and about 2 million hectares would be lost for ten to twenty years. In terms of the economy, continuing to use heavily contaminated land for food production at the expense of human health could not be justified rationally because whatever was salvaged in the agricultural economy would be lost in future health costs.

Victoria Price

SUGGESTED READINGS: A foundation work that outlines the various aspects of ecological catastrophe and puts the Chernobyl nuclear accident into context is V. K. Savchenko's *The Ecology of the Chernobyl Catastrophe* (1995). *The Social Impact of the Chernobyl Disaster* (1988), by David R. Marples, discusses the cause of the Chernobyl accident, identifies the populations affected, and comments on the environmental, economic, and political repercussions of the event. A somewhat more concise treatment of the Chernobyl nuclear accident is C. C. Bailey's *The Aftermath of Chernobyl* (1993). Alla Yaroshinskaya's *Chernobyl: The Forbidden Truth* (1995) discusses details of the Chernobyl incident that were not immediately announced following the accident. The aftermath of Chernobyl in terms of its impact on agriculture, health, global ecology, and future nuclear power use is discussed in Zhores A. Medvedev's *The Legacy of Chernobyl* (1990).

SEE ALSO: Nuclear accidents; Nuclear regulatory policy; Three Mile Island nuclear accident.

Chipko Andolan movement

DATE: begun April, 1973
CATEGORY: Forests and plants

The Chipko Andolan movement was begun by villagers in northern India who sought to stop lumber companies from clear-cutting mountain slopes and contributing to the rapid destruction of local forests.

The forests of India are a critical resource for the subsistence of rural people throughout the country, especially in the hill and mountain areas. Mountain villagers depend on the forests for firewood, for fodder for their cattle, for wood for their houses and farm tools, and as a means to stabilize their water and soil resources. During the 1960's and 1970's the Indian government restricted villagers from huge areas of forestland, then auctioned off the trees to lumber companies and industries located in the plains. Large lots of trees were sold to the highest bidder by the Forest Department, with the purchase price going to the Indian government.

Because of government restrictions and an ever-growing population, women who lived in mountain villages were forced to walk for hours each day just to gather firewood and fodder. In addition, when mountain slopes were cleared of trees, rains washed away the topsoil, leaving the soil and rocks underneath to crumble and fall in landslides. Much of the soil from the mountain slopes was deposited in the rivers below, raising water levels. At the same time, the bare slopes allowed much more rain to run off directly into the rivers, which resulted in flooding.

As trees were being felled for commerce and industry at increasing rates during the early 1970's, Indian villagers finally sought to protect their lands and livelihoods through a method of nonviolent resistance inspired by Indian leader Mohandas Gandhi. In 1973 this resistance spread and became organized into the Chipko Andolan movement, commonly referred to as the Chipko movement. Chipko comes from a word meaning "to embrace," while Andolan refers to a protest against harmful practices. Together the words literally mean "movement to hug trees." The movement originated in April, 1973, as a spontaneous protest by mountain villagers against logging abuses in Uttar Pradesh, an Indian province in the Himalayas. When contractors sent their workers in to fell the trees, the villagers embraced the trees, saving them by interposing their bodies between the trees and the workers' axes. The movement was largely organized and orchestrated by village women, who became leaders and activists in order to save their means of subsistence and their communities.

After many Chipko protests in Uttar Pradesh, victory was finally achieved in 1980 when the Indian government placed a fifteen-year ban on felling live trees in the Himalayan forests. The Chipko Movement soon spread to other parts of India, and clear-cutting was stopped in the Western Ghats and the Vindyas Mountains. The Chipko protesters staged a socioeconomic revolution in India by gaining control of forest resources from the hands of a distant government bureaucracy that was only concerned with selling the forest in order to make urban-oriented products. The movement generated pressure for the Indian government to develop a natural resources policy that was more sensitive to the environment and the needs of all people.

Alvin K. Benson

SEE ALSO: Deforestation; Logging and clear-cutting; Sustainable forestry.

Chloracne

CATEGORY: Human health and the environment

Chloracne is a skin disease resembling acne that results from exposure to chlorine and chlorine-containing compounds.

Chloracne involves an increase of dry skin and a reduction in the ability to produce sebum, which moistens and protects skin. Chloracne has been linked to exposure to polychlorinated biphenyls (PCBs) and polychlorinated dioxins (PCDDs). PCBs, which are used in industry as lubricants, vacuum pump fluids, and electrical insulating fluids, harm the environment because they evaporate slowly and do not mix well with water. Therefore, they are widely dispersed in water-courses and in the atmosphere. Because they strongly resist degradation, PCBs remain and accumulate in the environment.

Dioxins are a group of chlorinated aromatic hydrocarbons that are formed in trace amounts during production of the many chlorinated compounds used throughout industry. They are highly persistent in the environment. One dioxin commonly linked to chloracne is 2,3,7,8-tetrachlorodibenzodioxin (TCDD), which is a known human carcinogen. The primary source of dioxin in the environment is burning of chlorine-containing compounds. Since dioxins are also formed in the production of chlorinated compounds used as herbicides, chlorine-compound exposure occurs in the workplace and in the environment.

Environmental concerns over the health risks posed to humans and wildlife by dioxin and related chlorine compounds was brought to public attention with the publication of Rachel Carson's *Silent Spring* (1962), which condemned the widespread use of chlorinated pesticides. Through direct skin contact, inhalation, or ingestion of such compounds, chloracne produces skin lesions on the face, with more severe cases involving lesions on the shoulders, chest, and back. Symptoms may include small, colored, blisterlike pimples (pustules) and straw-colored cysts. Chloracne can develop from three to four weeks after exposure and can last up to fifteen years. Side effects from the disease include nausea, bronchitis, and liver disease. It can also have a poisoning effect on the nervous system, which results in extended symptoms such as headache, fatigue, sweaty palms, and numbness in the legs. Chloracne does not respond well to antibiotics. Derivatives of vitamin A (such as gels) and retinoic acid creams may help, and Accutane (isotretinoin) may also be an effective treatment.

In October, 1993, the American Health Association approved a resolution for the gradual phaseout of chlorine. However, the issue is not a simple one because chlorine is used to benefit humans in disinfecting drinking water and making many medicines. Regardless, environmentalists are worried that chlorinated compounds are slowly poisoning the earth by working their way through the air, groundwater, and the food chain, as well as destroying wildlife and causing many diseases such as chloracne in humans. To help fight chloracne and other diseases caused by dioxin and chlorine compounds, a Government Action Plan proposed the phasing out of the remaining PCBs in use by 1999.

Beth Ann Parker and Massimo D. Bezoari

SEE ALSO: Agent Orange; Dioxin; Environmental health; Environmental illnesses; Pesticides and herbicides; Polychlorinated biphenyls.

Chloramphenicol

CATEGORY: Pollutants and toxins

Chloramphenicol is a broad-spectrum antibiotic that is used to combat harmful bacteria and organisms in humans and animals. However, use of the antibiotic is often accompanied by side effects that may lead to severe health problems.

Chloramphenicol (CHPC) is known by many other names, including chloromycetin. CHPC is an effective antibiotic, particularly in those cases that require penetration through purulent material (which either contains or discharges pus) to reach the infecting bacteria, as occurs in infections of pneumonia. It is also a good choice for infections involving the eye, the nervous system, and the prostate gland. The more efficient penetration of CHPC into cells compared to other antibiotics increases its advantage in killing intracellular parasites, such as Chlamydia, mycoplasma, and rickettsia.

The high acidity of CHPC is believed to contribute to its ability to penetrate necrotic material and cellular membranes. It kills bacteria by interfering with the protein manufacturing system occurring in the ribosomes of organisms. The mode of attack is unique in that reptilian, mammalian, and avian ribosomes are unaffected. In addition, the antibiotic does not destroy all the bacteria in infected tissue. Only highly susceptible bacteria are destroyed, with the remainder merely being inhibited from reproducing. This is an advantage because the inactive residual bacteria allow B cells in the immune system to develop an immune response to the bacteria. The result is akin to the use of inactive bacterial vaccinations in fighting disease.

Discovered in 1947, CHPC was released for public use in 1949. In 1959 researchers reported the cardiovascular collapse and death of three infants after they were treated with CHPC. All exhibited a gray pallor (called "gray syndrome") prior to death. A more complete study found that of thirty-one premature infants treated with the drug, 45 percent died exhibiting gray syndrome, while only 2.5 percent of a group of untreated infants died. The common feature among the fatal cases was the high level of the drug used in their treatment: 230 to 280 milligrams of CHPC per kilogram of body weight per day, five to ten times what later became the recommended dose.

CHPC was subsequently found to interfere with the energy-producing capacity of mitochondria in those organs with a high rate of oxygen consumption; the heart, liver, and kidneys are at particular risk. People, particularly infants, who are unable to detoxify the drug efficiently are particularly susceptible to the accumulation of toxic levels in the blood.

CHPC is known to have other severe side effects. Common symptoms include nausea, diarrhea, and loss of appetite. In more severe cases,

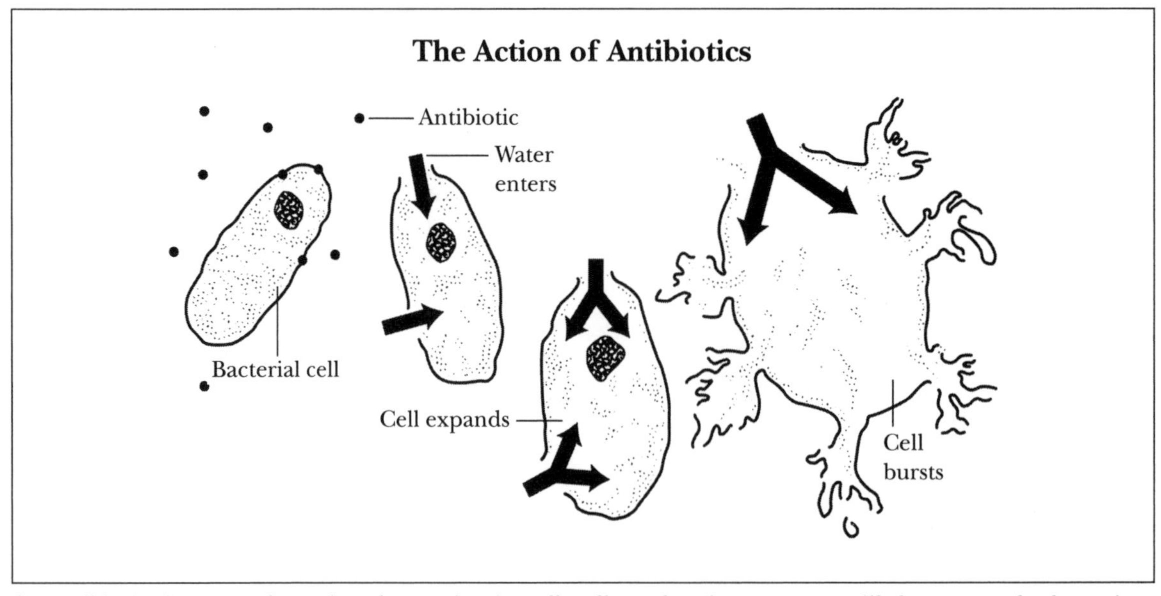

An antibiotic destroys a bacterium by causing its cell walls to deteriorate; water will then enter the bacterium unchecked until it bursts.

blood dyscrasias may result. This is a condition in which abnormal blood cells are produced while normal blood cell production is blocked because of interaction of CHPC on the patient's or animal's bone marrow. If CHPC is taken orally, humans may develop fatal anemia. Patients who recover from such blood abnormalities resulting from the antibiotic have a high incidence of leukemia. Unfavorable interactions with other drugs—including phenytoin (used to treat heart disease), primidone (for seizures), and cyclophosphamide (in chemotherapy)—have also been noted. The antibiotic can also interfere with vaccinations and is a potential carcinogen.

These side effects occur not only in humans, but also in other animals. CHPC may build up to toxic levels in young animals because their liver detoxification mechanisms are not as well developed or efficient as in adults. Therefore, the antibiotic must not be administered to food animals, which can be ingested by the young. Ingestion of CHPC by pregnant or lactating creatures must be avoided for similar reasons. Anorexia in dolphins, whales, and California sea lions has also been a noted side effect.

Although these serious side effects are rare, a study by the California Medical Association in 1967 found the occurrence of cases in humans to be as high as one per twenty-four thousand applications, depending on dose. The accumulation of such data was sufficient for the use of CHPC to be restricted, in the 1960's, to cases for which there is little alternative treatment.

Jacqueline J. Robinson and Massimo D. Bezoari

SEE ALSO: Bacterial resistance and super bacteria.

Chlorination

CATEGORY: Water and water pollution

The practice of disinfecting water by the addition of chlorine is known as chlorination. Although the chlorination of public water in the United States has helped reduce outbreaks of waterborne disease, it has raised concerns about the possible formation of chloro-organic compounds in treated water.

Drinking water, waste waters, and water in swimming pools are the most common water sources where chlorination is used to kill bacteria and prevent the spread of diseases. Viruses are generally more resistant to chlorination, but they can be eliminated by increasing the chlorine levels needed to kill bacteria. Common chlorinating agents include elemental chlorine gas and sodium or calcium hypochlorite. In water these substances generate hypochlorous acid, which is the chemical agent responsible for killing microorganisms by inactivating bacteria proteins or viral nucleoproteins.

Public drinking water was chlorinated in most large U.S. cities by 1914. The effectiveness of chlorination in reducing outbreaks of waterborne diseases in the early twentieth century was clearly illustrated by the drop in typhoid deaths: 36 per 100,000 in 1920 to 5 per 100,000 by 1928. Chlorination has remained the most economical method to purify public water, although it is not without potential risks. Chlorination has also been widely used to prevent the spread of bacteria in the food industry.

In its elemental form, high concentrations of chlorine are very toxic, and solutions containing more than 1,000 milligrams per liter (mg/l) are lethal to humans. Chlorine has a characteristic odor that is detectable at levels of 2-3 mg/l of water. Most public water supplies contain chlorine levels of 1-2 mg/l, although the actual concentrations of water reaching consumer faucets fluctuates and is usually around 0.5 mg/l. Consumption of water containing 50 mg/l has produced no immediate adverse effects.

The greatest environmental concern from chlorination is not usually from the chlorine itself, but from the potential toxic compounds that may form when chlorine reacts with organic compounds present in the water. Chlorine, which is an extremely reactive element, reacts with organic material associated with decaying vegetation (humic acids), forming chloro-organic compounds. Trihalomethanes (THMs) are one of the most common chloro-organic compounds. At least a dozen THMs have been identified in drinking water since the 1970's, when health authorities came under pressure to issue standards for the identification and reduction of THM levels in drinking water.

Major concern has focused on levels of chloroform because of its known carcinogenic properties in animal studies. Once used in cough syrups, mouthwashes, and toothpastes, chloroform in consumer products is now severely restricted. A 1975 study of chloroform concentrations in drinking water found levels of more than 300 micrograms per liter (μg/l) in some water, with 10 percent of the water systems surveyed having levels of more than 105 μg/l. In 1984 the World Health Organization used a guideline value of 30 μg/l for chloroform in

Despite the fact that chlorination has reduced the incidence of waterborne diseases in drinking water supplies, many people remain concerned about the formation of toxic chloro-organic compounds in treated water. (Jim West)

drinking water. Although there are risks associated with drinking chlorinated water, it has been estimated that the risk of death from cigarette smoking is two thousand times greater than that of drinking chloroform-contaminated water from most public sources. However, as water sources become more polluted and require higher levels of chlorination to maintain purity, continual monitoring of chloro-organic compounds will be needed.

Nicholas C. Thomas

SEE ALSO: Drinking water; Water pollution; Water quality; Water treatment.

Chlorofluorocarbons

CATEGORY: Atmosphere and air pollution

Chlorofluorocarbons (CFCs) are a family of chemical compounds used in air conditioners,

Size and Depth of the Antarctic Ozone Hole, 1979-1997

YEAR	MINIMUM VALUE (Dobson Units)	OZONE HOLE SIZE (millions of square kilometers)
1979	209	2.23
1980	205	1.88
1981	205	1.70
1982	189	3.77
1983	169	6.24
1984	154	8.66
1985	146	12.57
1986	159	9.58
1987	120	18.18
1988	173	8.75
1989	124	17.75
1990	128	17.86
1991	117	18.13
1992	124	21.28
1993	94	22.81
1994	88	22.82
1995	No satellite in place	
1996	111	21.59
1997	104	20.89

Source: NASA Goddard Space Flight Center.

Note: Listed are the minimum concentration of ozone (in Dobson units) in the Antarctic ozone hole and the maximum size of the ozone hole for each year (the ozone hole is defined as regions with ozone concentrations below 220 Dobson units) as recorded by NASA satellites from 1979 to 1997.

refrigerators, and aerosol spray cans. Concern over destruction of stratospheric ozone by chlorofluorocarbons led to a worldwide ban on their manufacture and use.

CFCs are organic molecules containing carbon, fluorine, and chlorine atoms. The first CFCs were discovered by Thomas Midgley in 1928. Because these molecules are chemically inert and easily liquefied, CFCs soon became the standard coolant in refrigerators and air conditioners. They also became widely used as propellants in aerosol spray cans. By 1968 2.3 billion aerosol cans containing CFCs had been sold in the United States.

In 1970 the British scientist James Lovelock determined that most CFCs entering the atmosphere remained there without significant decomposition. Three years later Frank Sherwood Rowland and Mario Molina, working at the University of California at Irvine, suggested that CFCs would eventually migrate into the stratosphere. Once there, absorption of ultraviolet light would cause CFCs to release chlorine atoms, which would then react catalytically to remove ozone. Since ozone in the stratosphere prevents high-energy ultraviolet light from reaching the surface of the earth, any decrease in ozone would lead to increased exposure to ultraviolet light, causing higher levels of skin cancer in humans and damage to plants and animals.

Although evidence from laboratory studies suggested that CFCs in the atmosphere would cause depletion of stratospheric ozone, uncertainty remained as to the degree of ozone destruction that would occur. Nevertheless, in 1975 Oregon became the first state to ban CFCs in aerosol spray cans. Several states took similar actions. In 1977 the Food and Drug Administration (FDA) implemented a ban on the use of CFCs as aerosol propellants to be phased in over a two-year period. Continued uncertainties in predictions of ozone loss and the lack of direct evidence for ozone depletion kept most other countries from restricting the use of CFCs. While a total ban on CFCs was discussed by the EPA, no action was taken, in part because of the difficulty in finding adequate substitutes for CFCs.

In 1985 a team of British scientists led by Joseph Farman announced the discovery of significant loss of ozone over the Antarctic. Beginning in the early 1970's, springtime levels of ozone had slowly decreased. By 1985 as much as 40 percent of the ozone usually present in the Antarctic stratosphere during the spring had disappeared. In addition, both the duration and geographic extent of this ozone hole was increasing. Evidence linking ozone hole formation to CFCs in the atmosphere was quickly found.

The discovery of the ozone hole led to further restrictions on CFCs. In 1987 an international agreement called the Montreal Protocol was reached to ban the manufacture and use of CFCs by the year 2010. In the United States, passage of the 1990 Clean Air Act amendments resulted in an accelerated timetable for restrictions on CFCs and related compounds. By the mid-1990's levels of CFCs in the atmosphere had stabilized, and CFCs were expected to gradually disappear from the atmosphere over the next century.

Jeffrey A. Joens

See also: Aerosols; Freon; Molina, Mario; Ozone layer and ozone depletion; Rowland, Frank Sherwood.

Citizen's Clearinghouse for Hazardous Waste

Date: established 1981
Category: Waste and waste management

Founded in 1981, the Citizen's Clearinghouse for Hazardous Waste (CCHW) is a national nonprofit organization that provides information and assistance to the public on hazardous waste issues. The CCHW has worked with community groups to address a broad range of environmental issues, including toxic waste, solid waste, air pollution, incinerators, medical waste, radioactive waste, pesticides, sewage, and industrial pollution.

Headquartered in Falls Church, Virginia, the CCHW was founded by Lois Gibbs, leader of the campaign to provide federal relocation of residents victimized by toxic waste from the Love Canal dump site in Niagara Falls, New York. In the early years, the CCHW focused on helping community groups suffering from the effects of toxic dumps similar to Love Canal. Since the early 1990's, it has expanded its programs to match the expanding concerns of numerous grassroots environmental organizations, working with more than eight thousand community-based groups nationwide. Funding for the organization is provided exclusively by environmental foundations and members of the CCHW.

One of the primary goals of the CCHW is to translate scientific issues into plain language. Consequently, environmental activists can understand the technical scientific language so that they can articulate their concerns and have a voice in the issues and decisions that directly affect their lives. The CCHW reviews and comments on technical reports, cleanup plans, health studies, and alternative technologies. This information helps communities understand why their families may be suffering from environmentally related sickness or what type of cleanup is needed to protect residents' health from further harm.

The CCHW organizes meetings and seminars to bring together grassroots environmental activists, scientists, and representatives of labor unions and national environmental organizations to discuss and plan how to protect public health and the environment. These meetings have led to the implementation of such strategies as the Stop Dioxin Exposure Campaign and the McToxics Campaign, which led McDonalds to stop using styrofoam in packaging its products. The Information Services branch of the CCHW has established an extensive library containing government documents and databases on how to win environmental justice. It has also published more than one hundred self-help guides and fact packs and publishes two periodicals dealing with environmental justice and current research on health and chemical exposures.

The CCHW provides Superfund technical assistance to help communities evaluate cleanup options and site-specific environmental and health studies. In addition, the efforts of the CCHW virtually stopped the development of new hazardous waste landfills, promoted the regulation of solid waste incinerator ash, and persuaded many municipal and state governments to institute environmentally sound and less expensive methods of managing waste by mandating that a percentage of waste be recycled. The CCHW has also been instrumental in assisting groups get laws passed that protect health and the environment. Among these are laws that allow people to access information on chemicals being stored, disposed, and released in their communities; prohibit corporations with felony records from conducting business in a state; and ban the dumping of wastes transported from other states.

Alvin K. Benson

See also: Dioxin; Hazardous waste; McToxics Campaign; Recycling; Solid waste management; Waste treatment.

Clean Air Act and amendments

Dates: passed 1963; amended 1970, 1977, and 1990

Category: Atmosphere and air pollution

The Clean Air Act of 1963, as amended in 1970, 1977, and 1990, federalizes the regulation of air pollution to a large degree. The act provides guidelines for minimum standards of air quality as well as maximum levels for the emissions of pollutants. It has served as a model for other federal environmental legislation.

Since the 1880's state and local governments have put limits on smoke emissions and other forms of air pollution. Federal regulation of the problem, however, did not really begin until 1955, when the Air Pollution Control Act authorized the federal government to conduct research and provide assistance to state and local governments. This act included no national standards, and it ceded responsibility for controlling air quality to the states.

Congress increased the federal role somewhat in the Clean Air Act of 1963. The secretary of the Department of Health, Education, and Welfare (HEW) was authorized to call abatement conferences when air pollution from one state put citizens of another state in danger, but the Clean Air Act failed to include any sanctions for the enforcement of national standards. Meanwhile, evidence was accumulating that air pollution posed a serious threat to public health throughout the country, and President Lyndon Johnson's Great Society looked to federal regulation as the only effective way to deal with such matters.

The 1967 Air Quality Act authorized HEW to consult with the states to determine air-quality standards in regions of particular concern, and the states were then given a year to formulate a plan to implement the guidelines. Environmentalists were disappointed that Congress still had not provided minimum standards of air quality or effective means for forcing the states to achieve their goals. The most significant aspect of the act was the authorization of some federal enforcement of vehicular emissions standards, with criminal fines of up to $1000 for each violation of the standards. Relatively weak requirements based on grams of pollutants emitted per mile took effect for new automobiles in 1968.

1970 Amendments

Widespread support of the environmental movement was demonstrated by the enthusiastic response to Earth Day in 1970, and that same year the first report of the Council on Environmental Quality called on Congress to enact new laws to deal with several problems, including air pollution. Senator Edmund Muskie, a presidential hopeful, was the acknowledged congressional leader in the campaign for tough environmental reform, and he was the chief author of the 1970 clean air bill. President Richard Nixon, also supported an aggressive bill. With this bipartisan support, Congress enacted the far-reaching Clean Air Act amendments of 1970, which initiated the federal government's regulation of air pollution.

Addressing perceived weaknesses in the existing law, the landmark 1970 amendments authorized the newly created Environmental Protection Agency (EPA) to establish standards that were binding on states. Applying a command-and-control approach to regulation, the centerpiece of the legislation was a program for the EPA to determine national ambient air-quality standards (NAAQS) that defined specific levels of air pollution considered harmful to public health. The EPA was also authorized to set emission limits on hazardous pollutants at levels allowing a sufficient margin of safety. Although states might exercise discretion in choosing how to attain federal standards, they were required to develop state implementation plans (SIPs), which utilized appropriate measures to reach those standards. States could maintain an air-quality control program for existing plants, while new plants were required to meet stricter standards based on the best available technology that was economically feasible.

The 1970 legislation stunned automobile manufacturers by requiring them to curtail emissions of the "big three" pollutants—hydrocarbons, carbon oxides, and nitrogen oxides—by 90 percent within six years. The technology did not exist to meet the new standards, although it was hoped that new technology could be developed within the specified time. Most members of Congress understood that it might be necessary to extend the deadline because few were willing to see the collapse of the automobile industry. In fact, the deadlines for meeting the vehicular emission standards turned out to be excessively ambitious, and waivers for the standards were granted in 1971, 1973, 1974, and 1976.

1977 Amendments

With the enthusiastic support of President Jimmy Carter, Congress passed major revisions of the Clean Air Act on August 4, 1977. In addition to making NAAQS more stringent, the amendments required each state to designate "nonattainment" regions based on the NAAQS. The state was then given the choice of either accepting statutory sanctions or revising its SIPs in order to meet the standards in a timely way. The act focused especially on coal-burning power plants, the major source of the sulfur dioxide that contributed to acid rain. Existing stationary sources of pollution were required to provide for "reasonably available control technology," while

The Clean Air Act includes legislation aimed at reducing emissions of pollutants from industrial sources. (Jim West)

new or modified stationary sources were required to utilize technology meeting the "lowest achievable emission rate," which usually meant the use of expensive scrubbers.

In the case of clean-air regions already in attainment, the amendments instituted a Prevention of Significant Deterioration (PSD) program designed to prevent the EPA from allowing deterioration of air quality up to the national standards. Representatives from rural districts had unsuccessfully argued that such a program would unfairly restrict industrial growth in areas where air pollution was not a problem.

The 1977 amendments further extended the required dates for automobiles to achieve emissions control standards. The stricter controls were scheduled for the 1980 model year. American carmakers had insisted that they could not meet the requirements in the existing law, and they had threatened to shut down production lines. This marked the fifth relaxation of the vehicular deadlines

1990 AMENDMENTS

Between 1977 and 1990, there were numerous efforts either to strengthen or to weaken the Clean Air Act, but opposing interest groups prevented major changes in either direction. While there was widespread agreement that the Clean Air Act had been somewhat successful in improving air quality, the administration of President Ronald Reagan strongly opposed any expansion of environmental regulations. During the election of 1988, candidate George Bush pledged to be the "environmental president." When Bush entered the White House, the deadlines for compliance with most air-quality standards had passed, putting noncompliance regions in danger of losing many industrial jobs. In July, 1989, the Bush administration made a sweeping proposal that most Democrats and environmentalists could support, while conservative Republicans were divided on the issue. The resulting amendments were signed into law on November 15, 1990.

The 1990 amendments were extremely complex, requiring more than seven hundred pages. The major regulatory change, modeled on the Clean Water Act, was a requirement that all major sources of air pollution obtain a state operating permit, with the EPA given the authority to veto such permits. The amendments provided additional regulations of emissions that were responsible for acid rain and established an allowance system based on a nationwide limit of 8.9 million tons of sulfur dioxide per year. Other provisions included a phase-out program for chlorofluorocarbons (CFCs) and other ozone-depleting substances, as well as a requirement that industrial plants cut emissions of 189 toxic substances to the level of the cleanest plants within that particular industry.

One of the major goals of the statute was to decrease urban smog, with strict controls on automobile emissions and mandates for cleaner-burning fuels. Beginning with 1994 automobiles, tailpipe exhausts were required to contain 60 percent less nitrogen oxide, and the emission-control equipment was required to last ten years. The EPA was authorized to conduct a study to determine if stricter standards were needed. A pilot program in California required an increasing number of cars and light trucks to run on batteries or nongasoline fuels. Beginning in 1995, oil companies were required to sell only cleaner-burning reformulated gasoline in the smoggiest metropolitan regions, and gasoline stations were mandated to install devices to capture fumes during refueling.

The 1990 amendments considerably strengthened the enforcement provisions under the Clean Air Act. The EPA acquired new powers to issue administrative penalties of up to $25,000 per day, and individuals were empowered to take civil action against polluters. The EPA and the Department of Justice were given new authority for the prosecution of misdemeanors and felonies. The amendments increased maximum sentences for most violations from six months to two years and increased maximum fines from $25,000 to $500,000. An individual who released hazardous air pollutants into the air could henceforth be sentenced to fifteen years in prison and fined up to $250,000, and corporations could be fined up to $1 million.

At an estimated cost of $25 billion per year, the 1990 act is considered the most expensive piece of environmental legislation ever passed. It was expected that the costs would mostly be passed on to consumers in higher prices for cars, gasoline, electricity, and products containing chemicals.

Enforcement

Based on a command-and-control model, the Clean Air Act and its amendments provide a variety of strong mechanisms for enforcing their statutory and regulatory requirements. The EPA has primary responsibility for enforcement at the federal level, and the states share responsibility for regulating SIPs. Citizens are also given broad opportunities to participate in the enforcement process.

When the EPA finds evidence that a violation has occurred, a regional office of the agency issues a notice of violation to both the source and the state. Based on its investigations, the EPA has the discretion to determine whether further action is necessary. The agency may issue an administrative order requiring a person or institution to comply with the applicable statute or regulation. If the recipient of the order fails to comply, the EPA may enforce the order through a civil action. If there is probable cause that a crime has occurred, the EPA will initiate a criminal prosecution, but the Department of Justice usually takes charge of the legal actions.

In formulating its SIPs, each state is required to include a program of legal enforcement. The states are usually given the opportunity to lead in initiating enforcement action if they wish to do so. If the state does not do so, the EPA has the authority to proceed on its own. Any person, moreover, may bring a civil action against an individual or entity alleged to be in violation of the Clean Air Act. If a violation is proven, any monetary awards must be either turned over to the EPA's "penalty fund" or used for "beneficial mitigation projects."

When the landmark amendments of 1970 were passed, proponents of the act tended to be extremely optimistic about the prospects for achieving national air-quality standards without any serious economic costs. The act envisioned full attainment of the standards by 1975, but this expectation turned out to be unrealistic, especially in regard to ozone. In the 1977 amendments, Congress responded to the problem by explicitly recognizing noncompliance regions, which were henceforth required to improve incrementally. It was even more difficult to formulate vehicular emissions standards that were both meaningful and attainable, in part because no one could be certain about the prospects of technological improvement. When automotive technology improved, moreover, no one could be certain about whether Clean Air Act standards were a primary cause.

By the late 1990's, few people denied that the Clean Air Act had helped decrease air pollution and improve the public health. By its nature, however, such legislation does not completely satisfy everyone. Environmental organizations commonly argue that the EPA has not been aggressive enough in its enforcement efforts, while probusiness groups tend to blame the Clean Air Act for forcing plants to close their doors and move to poor countries with a greater toleration for dirty air.

Thomas T. Lewis

SUGGESTED READINGS: An excellent summary of the Clean Air Act is found in the *Clean Air Handbook* (1993), edited by F. William Brownell. An interesting and readable account of the legislative history of the Clean Air Act is provided by Gary Bryner's *Blue Skies, Green Politics: The Clean Air Act of 1990* (1993). Richard Cohen provides a behind-the-scenes perspective of the legislation in *Washington at Work: Back Rooms and Clean Air* (1992). For the important question of vehicular emissions, see Sudhir Rajan's *The Enigma of Automobility: Democratic Politics and Pollution Control* (1996). The *Environmental Law Reporter's Clean Air Deskbook* (1996) is an excellent source for students who want detailed legal information. A short and useful summary of the Clean Air Act is found in Timothy Vanderver, Jr., *Environmental Law Handbook* (1994). For a broad history of environmental policy, see Jacqueline Switzer's *Environmental Politics: Domestic and Global Dimensions* (1994).

SEE ALSO: Acid deposition and acid rain; Air pollution policy; Automobile emissions; Particulate matter; Smog.

Clean Water Act and amendments

DATES: 1965; amended 1972, 1977, 1987, and 1990
CATEGORY: Water and water pollution

The legislation now called the Clean Water Act was largely shaped by the 1972 amendments to the Federal Water Quality Act of 1965. The complex legislation was further strengthened by later amendments. The Environmental Protection Agency, in cooperation with other federal, state, and local agencies, administers the numerous programs established by the legislation.

Before the mid-1960's, the regulation of water pollution was mostly left up to the states. The earliest federal law was the Rivers and Harbors Act of 1899, which prohibited the dumping of debris into navigable waters. Although the purpose of the law was to protect interstate navigation, it became an instrument for regulating water quality sixty years after its passage. The Oil Pollution Act of 1924 prohibited the discharge of oil into interstate waterways, with criminal sanctions for violations. The first Federal Water Pollution Control Act (FWPCA), passed in 1948, authorized the preparation of federal pollution abatement plans, which the states could either accept or reject, and provided some financial assistance for state projects. Although the FWPCA was amended in 1956 and 1961, it still contained no effective mechanisms for the federal enforcement of standards.

By this period, however, many Americans were recognizing water pollution as a national problem that required a national solution. The Federal Water Quality Act of 1965 introduced the policy of minimum water quality standards that could be enforced in federal courts. The standards applied regardless of whether discharges could be proven to harm human health. The act also significantly increased federal funds for sewage plant construction. A 1966 amendment required the reporting of discharges into waterways, with civil penalties for failure to comply. The Water Quality Improvement Act of 1970 established federal licensing for the discharge of pollutants into navigable rivers and provided plans and funding for the detection and removal of oil spills.

Congress and President Richard Nixon agreed that existing programs were ineffective in controlling water pollution, and the resulting 1972 amendments to the Federal Water Quality Act established the basic framework for the Clean Water Act. The centerpiece of the landmark amendments was the national pollutant discharge elimination system (NPDES), which utilized the command-and-control methods earlier enacted in the Clean Air Act. The premise of the legislation was that polluting surface water is an unlawful activity, except for those exemptions specifically allowed in the act. The announced goal was to eliminate all pollutants discharged into U.S. surface waters by 1985.

In addition to standards of quality for ambient water, the amendments also included technology-based standards. Industrial dischargers were given until 1977 to make use of the "best practicable technology" in that industry, and the standard was be increased to the "best technology available" by 1982. The 1972 act also included stringent limitations on the release of toxic chemicals judged harmful to human health. For members of Congress, the most popular part of the act was the grant program for the construction of publicly owned treatment works (POTWs).

The Environmental Protection Agency (EPA), created just two years earlier, was assigned the primary responsibility of regulating and enforcing the legislation. The agency could issue five-year permits for the discharge of pollutants, and any discharge without a license or contrary to the terms of a license was punishable by either civil or criminal sanctions. When faced with a discharge of oil or other hazardous substances, the EPA could go to court and seek a penalty of up to $50,000 per violation and up to $250,000 in the case of willful misconduct. In addition, a discharger might be assessed the costs of removal, up to $50 million. Because of the technical complexity of the law, the EPA for many years relied more on civil penalties than criminal prosecutions.

The 1972 amendments prohibit the discharge of dredged or fill materials into navigable waters unless authorized by a permit issued by the Army Corps of Engineers. Based on the literal wording of the statute, the corps at first regulated only actually, potentially, and historically navigable waters. In 1975, however, the corps revised its regulations to include jurisdiction over all coastal and freshwater wetlands, provided they were inundated often enough to support vegetation adapted for saturated soils. The Supreme Court endorsed the corp's broad construction of the law.

The Clean Water Act amendments of 1977, giving the legislation its present name, focused on a large variety of technical issues. They required industry to use the best available technology to remove toxic pollutants within six years. For conventional pollutants, businesses could seek a waiver from the technology requirements if the removal of the pollutant was not worth the cost. The act further required an environmental impact statement for any federal project involving wetlands, and it extended liability for oil-spill cleanups from 12 miles to 200 miles offshore.

The amendments of 1987, entitled the Water Quality Act, were passed over President Ronald Reagan's veto. In addition to increasing the powers of the EPA, the act significantly raised the criminal penalties for acts of pollution. Individuals who knowingly discharge certain dangerous pollutants can receive a fine of up to $250,000 and imprisonment for up to fifteen years. The maximum prison term for making false statements or tampering with monitoring equipment was increased from six months to two years. The most controversial part of the act was authorization of $18 billion for the construction of wastewater treatment plants. In 1990 Congress further amended the Clean Water Act with the Oil Pollution Act, which strengthened cleanup requirements and penalties for discharges.

In addition to the Clean Water Act, several closely related federal laws also deal with water pollution. In 1972 Congress passed the Marine Protection Research and Sanctuaries Act to regulate ocean dumping, and an amendment of 1988 prohibited ocean dumping of all wastes other than dredge spoil. The Safe Drinking Water Act (1974) authorized the EPA to regulate contaminants in tap water as well as injections into underground sources of drinking water. Amendments of 1986 required the EPA to regulate eighty-three contaminants within three years and authorized the EPA to issue new administrative orders and take enforcement action.

Some of the worst instances of water pollution have been curtailed in the years since the Clean Water Act was overhauled in 1972, even though the act has manifestly failed to achieve its stated goals. It is probably inevitable that economic prosperity and population growth will mean that water in the United States will never be completely free of pollutants. Since 1972, nevertheless, the American public has become increasingly intolerant of dirty and unhealthful water, and Congress, reflecting public sentiment, has continued to strengthen the Clean Water Act.

Thomas T. Lewis

SUGGESTED READINGS: Robert Adler and Jessica Landman provide a useful summary of the

Clean Water Act and its impact in *The Clean Water Act Twenty Years Later* (1993). For a comprehensive yet readable analysis of the Clean Water Act and related laws, see the *Environmental Law Reporter's Clean Water Deskbook* (1993), edited by Kenneth Gray et al. Also useful is the less-detailed *Clean Water Handbook* (1990), edited by J. Gasdon Arckle and Russell Randle. A good summary of the law can be found in Timothy Vanderver, Jr., *Environmental Law Handbook* (1994).

SEE ALSO: Drinking water; Runoff: agricultural; Water pollution policy; Water quality; Water treatment.

Climate change and global warming

CATEGORY: Weather and climate

Climate change refers to a change in the climate of an area over many seasons, while global warming is a gradual increase in surface temperature over the entire world over a period of many years. Many scientists believe that human behavior has greatly increased the rate of these processes.

The earth's global climate is controlled by the amount of solar radiation received at the top of the atmosphere, the amount reflected back into space by clouds and particles, and the loss of infrared radiation to space from the earth's surface and atmosphere. The atmosphere traps a considerable amount of infrared radiation, preventing it from being lost to space. Assuming that the amount of solar radiation received annually at the top of the atmosphere worldwide remains constant over a number of years, slight decreases in the amount of solar radiation reflected back to space by clouds or in the amount of infrared radiation lost to space result in global warming. Clouds and infrared-absorbing gases (sometimes called "greenhouse gases") are the major controls governing the loss of infrared radiation to space.

EVIDENCE FOR CLIMATE CHANGE

Geological records show that large portions of the continents have been covered by ice sheets at various times. Some intervals between glacial epochs called interglacials had climates as warm as or warmer than the current climate, with sea levels about 20 meters (66 feet) higher than they are today. Some interglacials may have been warmer than the maximum warming projected by the year 2050 as a result of the buildup of greenhouse gases in the atmosphere.

The West Antarctic Ice Shelf is currently the only ice sheet having its margins below sea level; evidence recovered from a benthic core taken 700 kilometers (435 miles) inland from the Ross Sea suggests that during a lengthy interglacial about 400,000 years ago, this ice sheet collapsed. Study of a 3,348-meter (10,984-foot) ice core recovered at Vostok Station, Antarctica, indicate that there have been four glacial episodes, each lasting about 90,000 years, in the last 400,000 years. Each glacial epoch is separated by an interglacial period lasting about 10,000 years, with the most recent glaciation reaching a maximum about 18,000 years ago and coming to an end about 10,000 years ago. At the current time, the earth is experiencing an interglacial period, with only Greenland and Antarctica covered by ice sheets. Many high mountain ranges worldwide have peaks that have permanent glaciers.

Historical records indicate that following the end of the Medieval Warm Period (1000 C.E.-1200 C.E..), a rapid decline in agricultural productivity occurred. The rise and decline of the Hanseatic League occurred in the interval between the Medieval Warm Period and the Little Ice Age (1550-1860), during which many European glaciers expanded, sometimes engulfing forests, agricultural land, and buildings. Why the Little Ice Age began and why it ended are not known. However, it roughly corresponded to a period called the Maunder Minimum, during which few sunspots were observed. A small decrease in the amount of solar radiation reaching earth presumably could explain why there was a global cooling and an expansion of glaciers.

Since 1860 glacial retreat has been documented in Switzerland and other mountainous locales. Swiss hikers approaching mountain glaciers such as Vadret da Morteratsch can see the succession of vegetation in the wake of glacial retreat; where the glacier tongue stood one hundred years ago, larches now stand more than 10 meters (33 feet) in height. Scientific evidence indicates that the climate of the Northern Hemisphere warmed by 0.5 degrees Celsius between the 1880's and 1940's. A cooling trend then set in, which lasted until 1980, when global atmospheric temperatures again began to rise. Global sea surface temperatures show this same pattern of warming, cooling, and recent warming.

HUMAN ACTIVITIES AND GREENHOUSE GASES

Fossil fuel combustion produces many unwanted waste products, including oxides of nitrogen and sulfur, and carbon dioxide (CO_2). Before industrialization occurred, atmospheric carbon dioxide levels are estimated to have been around 280 parts per million by volume in air. In 1957 Charles Keeling measured a carbon dioxide concentration of 318 parts per million in air over Mauna Loa volcano in Hawaii. Since that time, carbon dioxide concentrations have shown a continued upward trend, with a mean global carbon dioxide growth rate from 1981 (global average concentration of about 340 parts per million by volume) to 1996 (global average concentration of about 363 parts per million by volume) estimated to be around 1.4 parts per million by volume. Because carbon dioxide is a greenhouse gas, it is widely believed by most scientists and many nonscientists that any prolonged, substantial increase in the atmospheric concentration of carbon dioxide will contribute to global warming. Some estimates suggest

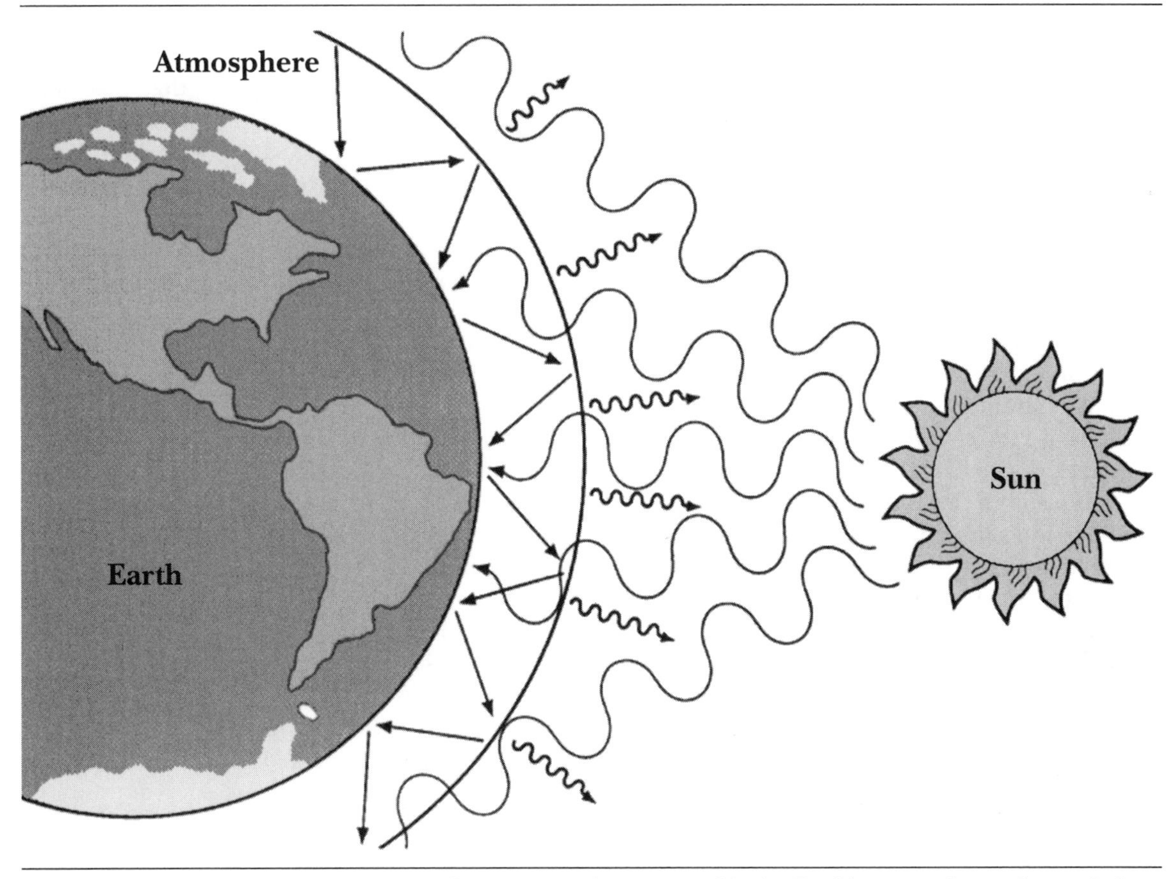

With the increase in "greenhouse gases" such as carbon monoxide in Earth's atmosphere, the sun's heat becomes trapped (straight arrows), leaving less to escape back into space (wavy arrows); as a result, the overall temperature of the planet rises.

that carbon dioxide emissions from the mid-nineteenth century through the end of the twentieth century have already increased the global tropospheric heating rate.

Methane, also a greenhouse gas, has increased from 1.61 parts per million by volume in 1983 to 1.73 parts per million by volume in 1996. Humans have increased the acreage of rice paddies and the size of cattle herds, both major sources of atmospheric methane. Leakage and deliberate venting of methane during mining operations contribute significant amounts of methane to the atmosphere. In addition, methane is a product of incomplete combustion. Nitrous oxide, another greenhouse gas, is also produced as a by-product of combustion. Chlorofluorocarbons such as freon and halon, which are completely human-made contaminants with relatively long atmospheric half-lives, are much more potent greenhouse gases than carbon dioxide and methane.

Although carbon dioxide, methane, nitrogen oxides, freon, and halon all trap infrared radiation that would otherwise escape into space, the greatest effects on climate change would result from changes in the amount of water vapor in the atmosphere and the amount and type of clouds worldwide. Water vapor is itself a greenhouse gas; many climate models factor in feedbacks from a temperature rise caused by other greenhouse gases favoring higher global concentrations of atmospheric water vapor, which adds to global warming. However, most experts predict that worldwide increases in water vapor concentrations in the atmosphere would be coupled with increased cloudiness, which would probably result in greater reflection of solar radiation back to space. This would have a cooling effect.

When coal containing sulfur is burned, various sulfur oxides are formed, which, once released into the atmosphere, are converted to sulfuric acid (SO_4) aerosol. Because of their optical properties, sulfate aerosols reflect solar radiation back into space. Some scientists have postulated that sulfate aerosols produced by burning of fossil fuels have caused global cooling at a sufficient rate to cancel the global warming that would be expected to have occurred with the release of carbon dioxide from that same burning.

Carbon Sequestration

In addition to a steady increase in the amount of fossil fuels burned yearly, which increases the amount of carbon dioxide in the atmosphere, people are destroying vast areas that actively sequester carbon within the biosphere. These carbon sinks include tropical rain forests and old-growth forests throughout the world, which remove carbon dioxide from the atmosphere. Although some prime timber is used to make wood products, tropical rain forest trees are generally burned during land clearance, releasing stored carbon into the atmosphere as carbon dioxide. In developed areas such as the United States, the lumber industry practices controlled reforestation; this is a slow process, and it takes many years to reestablish the carbon sequestration levels present before clear-cutting took place. Between 1860 and 1990 an estimated 120 gigatons of carbon were released to the atmosphere as a result of deforestation.

Atmospheric carbon dioxide that has been incorporated into the terrestrial biosphere is also sequestered as peat, the first step in a slow metamorphosis in which decaying organic material is turned into coal. Peat bogs have been exploited as a traditional fuel source for millennia, and some peat-burning electric power plants have been built.

The oceans are a sink for atmospheric carbon dioxide. The colder the ocean water, the more atmospheric carbon dioxide that dissolves in it. This carbon dioxide is a nutrient for algae, the beginning of the food chain in the oceans. Many organisms living in the oceans incorporate carbon into calcium carbonate ($CaCO_3$), which is then deposited as ocean sediment and stored for lengthy geologic periods. Vast deposits of limestone, dolomite, and marble are found worldwide, representing many gigatons of carbon. Some scientists believe in the Gaia hypothesis, which proposes that the atmosphere, biosphere, and geosphere are linked and self-correcting. They point to the vast terrestrial deposits of peat, lignite, and anthracite, and the vast sedimentary deposits of carbonates as evidence that excess atmospheric carbon dioxide will increase biological productivity on the land and in the oceans. Their critics point out that humankind has perturbed the natural feedback mechanisms of the atmosphere-biosphere-geosphere to such a great extent that natural feedback mechanisms may fail.

Extreme Weather Events

No one doubts that extreme weather events occur and that recent extreme events have been very costly. Sometimes these extreme events have been called "one-hundred-year" or even "five-hundred-year" occurrences. Many environmentalists point to a growing number of extreme weather events worldwide as evidence for global warming. While this argument may have some merit, other explanations must also be considered. A specific example is the regional flooding of the upper Mississippi River during July, 1993. This was an extreme event for farmers in the Mississippi floodplain and for cities from Des Moines to St. Louis. Reasons for this terrible flood included frequent intense precipitation events, the increasing channelization of the river—which decreases the ability of the watershed to retain moisture—and the urban development of marginal floodplains.

Another extreme event occurred in late December, 1996, and early January, 1997, when Northern California, western Nevada, the Pacific Northwest, and British Columbia experienced devastating floods as a result of heavy rain and snow. Similar events have occurred a number of times in recorded history, including major flooding in 1986. However, these previous episodes had been referred to as one-hundred-year events, and development throughout the region had proceeded with little regard for future flood risks. Urbanization increases runoff; heavy precipitation events that may not have been extreme prior to urbanization may suddenly overwhelm recent developments, causing costly floods.

There are many weather systems that recur perhaps once or twice per decade, El Niño being one of the best known. These weather systems and their resultant destruction do not constitute truly extreme events. Similarly, Atlantic hurricanes do not constitute extreme events, although many years may pass before any specific community is struck by a major category 5 storm. The number of Atlantic hurricanes is quite variable from year to year, and it is possible that global warming might lead to an increased number of Atlantic hurricanes occurring over a long period of years; this would constitute a long-term climatic change.

Many meteorologists have suggested that global warming would lead to increased convective activity in temperate climates as well as an increase in tropical storms. Some evidence comparing the number of storm events in the late twentieth century to earlier periods suggests that there has been an increase in convective activity in the Great Lakes and Mediterranean regions.

Consequences of Global Warming

Reliable temperature and precipitation records exist for many sites worldwide since about 1860. During this relatively brief period in the earth's climatic history, there have been some years and even decades in which the seasonal weather for specific regions was better than average, and some periods in which the seasonal weather was worse than average. Some scientists have made the claim that the decade of the 1990's was the warmest of the twentieth century, averaged over the globe. Extrapolations of the consequences of global warming rely heavily on the experiences of specific regions in the 1990's.

What constitutes better and worse climate depends on the region. For farmers in Alaska and Western Canada, the warm summer of 1997 was outstanding; many were able to double crops for the first time in living memory. If a global warming trend established itself, this region of hitherto marginal agriculture would benefit greatly. Agricultural land, selling for around $375 per acre in the late 1990's, would increase substantially in value. Russian agriculture would also benefit from any modest increase in the length of the growing season.

However, melting of permafrost in Alaska, Northern Canada, and Siberia would have adverse effects on infrastructure, especially by undermining roads. Other areas around the world might find global warming and its consequences extremely undesirable. Some regional projections suggest that farming in the U.S. Midwest would be adversely affected. Agriculture is dependent on receiving adequate precipitation throughout the growing season. If global warming caused longer, warmer, and drier summers in the Midwest, crop yields might be limited if the same crops that are currently grown would continue to be cultivated. Cost of production might increase because of increased need for crop irrigation. However, it is possible that more drought-tolerant crops might be introduced, and yield and production costs might remain constant (after adjustment for inflation).

Global agriculture would be expected to benefit from increased concentrations of carbon dioxide. It has been estimated that over the last two centuries, when atmospheric carbon dioxide levels rose from 280 parts per million to 360 parts per million, global agricultural yield increased by about 7.5 percent. Of course, during this time, improved crops, modern farming methods, and human-made fertilizers helped increase productivity. Pessimists point out that productivity might not increase as much in the future because soil nutrients are limited. Great use of irrigation might also contaminate soils with salts, limiting productivity. Models indicate that agriculture in Southeast Asia and Central America would be most adversely affected by global warming. In temperate regions, warmer winters could allow migratory pests to enlarge their ranges northward.

SEA LEVEL INCREASE

Because of the steady increase in greenhouse gases in earth's atmosphere as a result of humankind's activities, the Intergovernmental Panel on Climate Change (IPCC)—sponsored by the United Nations Environment Programme (UNEP)—and the World Meteorological Organization (WMO) have predicted that the globally averaged terrestrial surface temperatures will increase by 1 to 3.5 degrees Celsius by the year 2100. One of the most inescapable of the projected consequences of global warming is an increase in sea level. Estimates of sea-level increase are dependent on partial melting of ice sheets in Greenland and Antarctica. Small increases in the temperatures of the polar oceans could cause increased melting of polar ice. Between 1992 and 1996, the Pine Island Glacier, a major ice stream in West Antarctica, thinned by approximately 3.5 meters (11.5 feet) per year. This thinning has been attributed to basal melting of the floating tongue of the glacier caused by a 0.1-degree-Celsius warming of the seawater of the southeast Pacific.

As a result of the anticipated global warming, the IPCC projects a rise in sea level of between 15 centimeters (6 inches) and 95 centimeters (37 inches) by 2100. This is relatively minor compared with the 5-6 meter (16-20 foot) rise in sea level that would follow the collapse of the West Antarctic Ice Sheet. Any sizable sea level increase would have negative impacts on coastal regions worldwide. Because many people live in coastal areas, any rise in sea level would prove costly. About 40 percent of Bangladesh is at an elevation of 1 meter (3.3 feet) or less above sea level. This country, one of the poorest and most densely populated in the world, is ill-equipped to deal with a modest rise in sea level. Some relatively wealthy cities such as New Orleans, with an elevation about 3 meters (10 feet) below sea level, might need sizable investments in infrastructure to counteract sea level increases.

Anita Baker-Blocker

SUGGESTED READINGS: An excellent, short discussion about global warming and climate change can by found in *Common Questions about Climate Change*, a booklet published by the World Meteorological Organization. *Is the Temperature Rising? The Uncertain Science of Global Warming* (1998), by S. George Philander, is a comprehensive treatment of global change issues. *Biogeochemical Cycles: A Computer-Interactive Study on Earth System Science and Global Change* (1997), by W. L. Chameides and E. M. Perdue, discusses the many feedbacks between the geosphere, biosphere, oceans, and atmosphere and the various elements, including carbon and sulfur, that are important in climate modeling. For a discussion on how climate models are used to predict global change, read *Assessing Climate Change: Results from the Model Evaluation Consortium for Climate Assessment* (1997), edited by Wendy Howe and Ann Henderson-Sellers. The worldwide effects of global warming on agriculture are discussed in *Climate Change and the Global Harvest:*

Potential Impacts of the Greenhouse Effect on Agriculture (1998), by Cynthia Rosenzweig and Daniel Hillel. For a discussion of the possible effects of global warming on water quality and water resources, read *Freshwater Ecosystems and Climate Change in North America: A Regional Assessment* (1997), edited by Colbert E. Cushing, and *Global Warming, River Flows and Water Resources* (1996), by Nigel Arnell. The stability of the West Antarctic Ice Shelf is discussed by E. J. Rignot in "Fast Recession of a West Antarctic Glacier," *Science* 281 (July 24, 1998).

SEE ALSO: Aerosols; Chlorofluorocarbons; Deforestation; Fossil fuels; Freon; Greenhouse effect; Rain forests and rain forest destruction.

Cloning

CATEGORY: Biotechnology and genetic engineering

Cloning is the production of a population of genetically identical cells or organisms derived from a single ancestor, or the production and amplification of identical deoxyribonucleic acid molecules.

The molecular cloning and engineering of deoxyribonucleic acid (DNA) molecules were first made possible with the discoveries of DNA ligase (enzymes that join DNA molecules) in 1967 and restriction endonucleases (enzymes that cut DNA molecules at specific nucleotide sequences) in 1970 by Hamilton Smith and Daniel Nathans. These enzymes allow scientists to cut and join DNA molecules from different species to produce recombinant DNA. For example, the DNA encoding human insulin can be combined with a plasmid, a small piece of DNA often found in bacteria such as *Escherichia coli* (*E. coli*). After the recombinant human-insulin-DNA/plasmid-DNA molecule is constructed, it can be inserted into a host cell such as *E. coli*; the recombinant DNA molecule will then replicate one or more times each time the *E. coli* DNA replicates. Thus, a clone of identical recombinant human-insulin-DNA/plasmid-DNA molecules will result. If the recombinant molecule has been engineered with the requisite signals, the *E. coli* will produce copious amounts of human insulin. In 1972 the first recombinant DNA molecules were made at Stanford University, and in 1973 such molecules were inserted via plasmids into *E. coli.* The first successful synthesis of a human protein by *E. coli* was somatostatin, reported in 1977 by Keiichi Itakura and coworkers. In 1984 insulin was the first human protein made by *E. coli* to become commercially available.

The production of cloned recombinant DNA has been an indispensable tool in biological research and has increased the knowledge and understanding of the structure and function of DNA and the control of gene activity. Cloned DNA has led to the production of important products of medical interest, the production of DNA for gene transfer and genetic engineering experiments, and the identification of mutations and genetic disease. Several different vectors have been developed for the delivery of DNA to a variety of plant, animal, and protistan cells, resulting in the creation of many transgenic species. Nearly one hundred types of transgenic plants that are resistant to certain herbicides, insects, and viruses have already been produced. Larger varieties of many animals have been engineered with the human growth factor gene. Transgenic plants, animals, and protistans now produce a variety of human proteins, including insulin, antithrombin, growth hormone, clotting factor, vaccines, and many other pharmaceuticals and therapeutic agents, as well as molecular probes for the diagnosis of human disease. As a result of this technology, the costs of treating many diseases have declined.

The first human gene transfer experiment using cloned DNA was performed in 1990 on a four-year-old girl with severe combined immune deficiency (SCID). SCID is caused by a mutation in the adenine deaminase (ADA) gene, resulting in white blood cells deficient in their immune response. These cells were removed from the girl, the normal gene was inserted, and the genetically altered white cells were returned to her body, where they repopulated and expressed normal defense mechanisms.

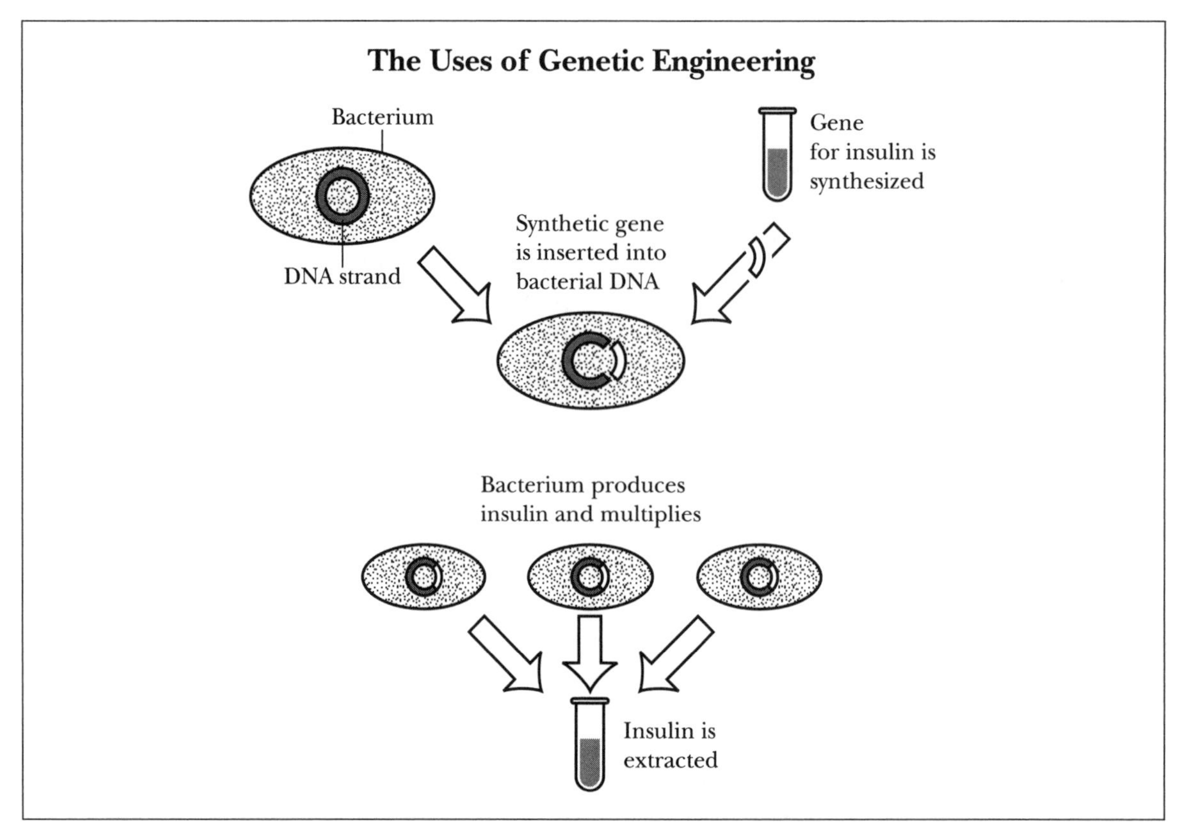

Genetic engineering, the manipulation of genetic material, can be used to synthesize large quantities of drugs or hormones, such as insulin.

Environmental Impact of DNA Cloning

The construction of recombinant DNA molecules and their subsequent cloning have not been without controversy. In 1971 researcher Paul Berg planned an experiment to combine DNA from simian virus 40 (SV40)—a virus that causes tumors in monkeys and transforms human cells in culture—with bacteriophage λ and to incorporate the recombinant molecule into *E. coli*. However, several scientists warned of a potential biohazard. Since *E. coli* is a natural inhabitant of the human digestive system, it was feared that the engineered *E. coli* could escape from the laboratory, enter the environment, become ingested by humans, and cause cancer as a result of its newly acquired DNA. The scientific community imposed a moratorium on recombinant DNA work in 1974 until the National Institutes of Health (NIH) could study the safety of recombinant DNA research and develop guidelines under which such work could proceed. The guidelines, originally published in 1976, were eventually relaxed after it was clearly demonstrated that the work was not nearly as dangerous as initially feared.

In 1983 the NIH granted permission to the University of California at Berkeley to release bacteria that had been engineered to protect plants from frost damage. This was the first experiment intentionally designed to introduce genetically engineered organisms into the environment. Various environmental and consumer protection groups were successful in persuading federal judge John J. Sirica to order the suspension of the planned trial. Environmentalists feared that the bacteria could spread to other plants, prolong the growing season, cause irreparable harm to the environment, or enter the atmosphere and decrease cloud formation or alter the climate. The NIH and the university eventually won approval for their experiment.

Although most scientists believe that the introduction of engineered species into the environment will be harmless and that early concerns were exaggerated, many environmentalists still have reservations since the long-term environmental effects of genetically altered organisms are unknown. It is feared that the introduction of some organisms will negatively impact the environment and have irreversible global effects. Many fear that a genetically engineered organism such as a bacterium or virus could spread throughout the environment, causing human disease or ecological destruction.

The use of genetically engineered organisms could have many adverse effects. The introduction of new genes into an organism could extend the range of that species, causing it to infringe on the natural habitats of closely related or more distant species and thus disrupt the balance of nature. There are already many recorded examples of ecological disruption occurring as a result of plants and animals being introduced into areas with no natural predators. Some environmentalists believe that the use of herbicide-resistant crops will actually prolong, extend, and even increase the use of toxic herbicides, since the crop plants are immune to the effects of the toxins. Others fear that herbicide-resistant genes could be transferred to related plants, producing a population of herbicide-resistant weeds.

The introduction of organisms engineered through the cloning of recombinant DNA into the environment could have a significant positive environmental impact and increase world food production. Genetically altered plants could reduce the input of toxic chemicals into the environment. Researchers could create crops with larger yields; resistance to pests, pesticides, drought, and disease; higher tolerance to cold, heat, and drought; greater shelf life; and greater photosynthetic and nitrogen fixing activity. Such crops may be more cost effective to produce. Plants genetically engineered to produce a natural insecticide would actually protect the environment because the insecticide is confined to the plants themselves, eliminating the need for contaminating the entire area where they are grown. Genetically altered animals are healthier, more productive, and less costly to maintain. Bioremediation of toxic waste dumps, chemical spills, and oil spills could be enhanced by genetically engineered microorganisms.

Animal and Plant Cloning

The cloning of plants from cuttings has been successfully practiced for thousands of years and is commonly used for many important food crops. Successful animal cloning was first reported in 1892 by Hans Adolph Eduard Dreisch. Dreisch separated the first two and four embryonic cells (blastomeres) of the sea urchin and allowed them to develop into complete, genetically identical embryos.

The first report of successful animal cloning by nuclear transplantation was published in 1952 by Robert Briggs and Thomas J. King, who removed nuclei from embryonic frog cells and transplanted them into eggs from which their nuclei had been removed. Pricking the egg with a glass needle induced it to divide and often develop into a complete tadpole. The first reliable reports of successful animal cloning by nuclear transplantation in mammals came in 1986 from Steen Willadsen in Cambridge, England. Willadsen cloned sheep from the nucleus of an early blastula cell. In 1987 Randall Prather and Willard Eyestone cloned cows while working in Neal First's laboratory at the University of Wisconsin.

In 1997 Ian Wilmut and Keith Campbell announced that they had successfully cloned a sheep named Dolly in Edinburgh, Scotland. Dolly was a milestone in cloning research because she was the first mammal cloned from an adult cell. The use of adult cells in cloning is advantageous, since the phenotype of the adult donor animal is known before the animal is cloned. Animal cloning makes genetic engineering more efficient, since an animal would have to be engineered only once and then could be used to donate nuclei for cloning.

Although inbred stocks have been used for centuries in agriculture, it is feared that the extensive use of animal and plant clones could severely reduce the genetic variability of various important crop, forestry, and livestock species, making them more susceptible to disease and extreme environmental factors. On the other hand, the development of successful cloning techniques could rescue endangered species from the brink of extinction or allow for the creation of a population of genetically identical animals valuable in medical research. Moreover, cloning could be used to create an entire population of animals from one individual that has been genetically altered to produce a valuable pharmaceutical.

Charles L. Vigue

SUGGESTED READINGS: *The DNA Story: A Documentary History of Gene Cloning* (1981), by James D. Watson and John Tooze, is an excellent early history of the development of gene cloning techniques. *Clone: The Road to Dolly and the Path Ahead* (1998), by Gina Bari Kolata, chronicles the development of animal cloning and discusses various ethical issues. Marc Lappe's *Broken Code: The Exploitation of DNA* (1984) is a well-written book that discusses many aspects of genetic engineering. *The Genetic Revolution: Scientific Prospects and*

Milestones in Cloning

YEAR	EVENT
1892	Hans Adolph Eduard Dreisch clones sea urchins by separating the first two and four blastomeres.
1902	Hans Spemann successfully repeats Dreisch's experiment using salamanders.
1952	Robert Briggs and Thomas King successfully clone frogs by nuclear transplantation of embryonic nuclei to enucleated eggs.
1967	Deoxyribonucleic acid (DNA) ligase, the enzyme that joins DNA molecules, is discovered.
1969-1970	Danial Nathans, Hamilton Smith, and others discover restriction endonucleases.
1971	Paul Berg plans to combine DNA with a bacteriophage and insert the recombinant DNA molecule into *Escherichia coli* (*E. coli*).
1972	The first recombinant DNA molecules are constructed at Stanford University.
1973	Recombinant DNA is first inserted into *E. coli.*
1974	Scientists call for a moratorium on recombinant DNA research until the National Institutes of Health (NIH) can study the safety of recombinant DNA research and develop guidelines.
1976	The NIH issues its guidelines for recombinant DNA research.
1983	The NIH grants permission to the University of California at Berkeley to release genetically engineered bacteria designed to retard frost formation on plants.
1986	Steen Willadsen clones sheep using early embryonic nuclei.
1987	Randall Prather and Willard Eyestone clone cows using early embryonic nuclei.
1990	The first human gene transfer experiment is performed on a patient with severe combined immune deficiency (SCID).
1997	Ian Wilmut and Keith Campbell announce the successful cloning of Dolly the sheep, the first mammal cloned from an adult cell.

Public Perceptions (1991), edited by Bernard D. Davis, is a compendium of articles covering legal, ecological, environmental, agricultural, and medical issues related to molecular cloning. *Recombinant DNA: Science Ethics and Politics* (1978), edited by John Richards, discusses the history and dangers of recombinant DNA research. *The Recombinant DNA Debate* (1979), edited by David A. Jackson and Stephen P. Stich, presents the scientific background of gene cloning. *The Cloning of Frogs, Mice, and Other Animals* (1985), by Robert Gilmore McKinnell, contains a history of animal and plant cloning with a well-illustrated description of the nuclear transfer technique. *Genetic Engineering* (1988), by J. G. Williams and R. K. Patient, is a short, concise, and excellent description of the principles and techniques of genetic engineering and molecular cloning.

SEE ALSO: Dolly the sheep; Genetically engineered organisms; Wilmut, Ian.

Mike Mathis, a member of the Oklahoma Water Resources Board, discusses the possibility of using cloud seeding to end a drought in Oklahoma in 1996. Tests designed to determine whether cloud seeding techniques actually work have proven inconclusive. (AP/Wide World Photos)

Cloud seeding

CATEGORY: Weather and climate

Cloud seeding is the practice of introducing agents into clouds for the purpose of intentionally modifying the weather.

Cloud seeding is a relatively inexpensive and easily used method that has been the main technique of weather modification. The seeding of clouds is conducted for many purposes, including increasing rainfall for agriculture, increasing snowfall for winter recreational areas, fog and cloud dispersal, and hail suppression.

Most cloud seeding is directed at clouds with temperatures below freezing. These cold clouds consist of ice crystals and water droplets with subfreezing temperatures called supercooled droplets. Initially, supercooled droplets far outnumber ice crystals; however, the ice crystals quickly grow larger at the expense of the droplets and, after reaching sufficient size, fall as precipitation. The objective of seeding cold clouds is to stimulate this process in clouds that are deficient in ice crystals. The seeding agent is either silver iodide (AgI) or dry ice—solid carbon dioxide (CO_2) with a temperature of –80 degrees Celsius (–112 degrees Fahrenheit). Silver iodide pellets act as nuclei on which water droplets freeze to form precipitation, while dry ice pellets are so cold that they cause the surrounding supercooled droplets to freeze and grow into snowflakes. Warm clouds with temperatures above freezing can also be seeded by injecting them with salt crystals, which triggers the development of large liquid cloud drops that fall as precipitation.

Clouds must already be present for such weather modification techniques to work, since clouds cannot be generated by seeding. Most cloud seeding is done from aircraft, although silver iodide crystals can be injected into clouds from ground-based generators. One environmental concern is that seeding may only redistribute the supply of precipitation, so that an increase in precipitation in one area might mean a compensating reduction in another.

Since World War II, considerable research has gone into cloud seeding. However, scientists still cannot conclusively answer the question of whether cloud seeding actually works. One of the first large-scale experiments, conducted over south-central Missouri for five years during the 1950's, actually decreased the rainfall. Apparently, the clouds were overseeded, resulting in too many ice crystals competing for too few water droplets to form precipitation. In the 1970's and early 1980's the National Oceanic and Atmospheric Administration (NOAA) conducted a major experiment over southern Florida to test the effectiveness of seeding cumulus clouds. The initial results showed that under some conditions seeding increases rainfall, but a second, more statistically rigorous set of experiments failed to confirm the earlier results. Of the many cloud-seeding experiments, only one, conducted in Israel in the 1960's and 1970's, has yielded statistically convincing confirmation of an increase in precipitation.

The conflicting results of the Florida and Israel studies point out the uncertainties of cloud seeding and the need for more basic knowledge of cloud and atmospheric processes. A major result of the many decades of investigating cloud seeding is the realization that weather events are quite complex and not yet fully understood. Without further basic understanding of atmospheric processes, large-scale weather modification through cloud seeding cannot be carried out with scientifically predictable results.

Craig S. Gilman

SEE ALSO: Weather modification.

Club of Rome

DATE: established April, 1968
CATEGORY: Ecology and ecosystems

The Club of Rome's influence in solving environmental problems surpasses its low profile. From sponsoring the best-selling Limits to Growth study in 1972 to promoting the benefits of energy efficiency, the informal organization has steadfastly worked for the betterment of humankind.

In April, 1968, thirty-six prominent European scientists, businessmen, and statesmen gathered at the Accademia dei Lincei in Rome and formed the Club of Rome. Since its founding, the club has focused on the *world problematique*, which is characterized by the complex interrelationships of global problems: environmental degradation, poverty, overpopulation, militarism, ineffective governmental institutions, and a global loss of human values. The club commissions studies on important aspects of the *world problematique*. The resulting reports to the Club of Rome become a springboard for behind-the-scenes meetings with decision makers and for initiating projects recommended in the reports. Funding for club reports and activities has come through arrangements with government agencies, academic research centers, and foundations.

By all accounts, Aurelio Peccei has been the most influential Club of Rome member. From the club's founding in 1968 until his death in 1984, Peccei was instrumental in bringing people together, developing ideas into projects, finding funding, and providing logistic support for projects. Despite Peccei's personal influence, the success of the club can be measured by the contributions of its one hundred members from more than fifty countries. Many of the reports to the Club of Rome were made possible or were directly authored by its members.

The first report to the Club of Rome is also its best known. Club members obtained funding from the Volkswagen Foundation to commission a computer modeling team to forecast the future global system. The researchers' results were published in *The Limits to Growth* (1972), by Donella H. Meadows, Dennis L. Meadows, Jørgen Randers, and William W. Behrens. Their alarming results predicted that the human race would collapse by the year 2100 if current trends continued. The book became an international sensation, in large part through the publicity efforts of Peccei. Despite widespread criticism about the team's methodology and results, *The Limits to Growth* succeeded in sparking debate about the fate of humankind. It also generated greater interest in global modeling, including several follow-up studies commissioned by the club.

More than twenty other reports to the Club of Rome have been published, with topics ranging from education to microelectronics. Reports with specific environmental themes include *Factor Four: Doubling Wealth—Halving Resource Use* (1997), by Ernst von Weizsäcker, Amory B. Lovins, and Hunter L. Lovins; *The Future of the Oceans* (1986), by Elizabeth Mann-Borgese; *Energy: The Countdown* (1978), by Thierry de Montbrial; and *Beyond the Age of Waste* (1978), by Dennis Gabor and Umberto Colombo. Although these reports have not achieved the visibility of *The Limits to Growth*, the club has used them as platforms for their quiet environmental protection campaigns.

Andrew P. Duncan

See also: *Limits to Growth, The*; Sustainable development.

Coalition for Environmentally Responsible Economies

Date: established 1989
Category: Philosophy and ethics

The Coalition for Environmentally Responsible Economies (CERES) is an organization made up of investors, pension fund managers, and environmentalists dedicated to injecting environmental considerations into investment decisions and corporate activity.

The 1989 *Exxon Valdez* oil spill in Prince William Sound was the catalyst for the formation of CERES. Meeting in Chapel Hill, North Carolina, fifteen major environmental groups joined with investors and public pension fund managers to encourage greater corporate consideration of the environmental consequences of their actions. With participation by the New York and California state pension funds, the Social Investment Forum, and a coalition of more than two hundred Protestant and Catholic groups, the initial CERES members represented more than $150 billion in invested capital.

CERES is founded on the notion that government regulation alone is not sufficient to materially improve the environment. Progress will require fundamental changes in corporate behavior and a more responsible attitude toward the environment. According to CERES, investors can influence this shift in beneficial ways if the investment decision includes consideration of the corporation's environmental track record. CERES's first project was the development of a ten-point corporate code of environmental ethics. Originally called the Valdez Principles as a reminder of the oil spill, the code is now known as the CERES Principles. CERES uses shareholder resolutions to initiate discussions of environmental responsibility at the highest corporate levels. In some corporations, such resolutions eventually led to formal endorsement of the ten CERES Principles. By endorsing the principles, companies acknowledge their environmental responsibility, actively commit to an ongoing process of continuous improvement, and agree to initiate a comprehensive public reporting of environmental issues.

CERES is also actively involved in the Global Reporting Initiative (GRI), which is the attempt to standardize corporate environmental reporting to generate the equivalent of a financial report. Just as the official corporate financial report includes required information on expenses, revenue, and profitability, each company's GRI would include required information on sustainability, environmental impacts, technological innovation, and other pertinent material. Once developed, the GRIs would improve corporate accountability by ensuring that all stakeholders—investors, fund managers, community groups, environmentalists, and labor organizations—would have access to standardized and consistent information. Armed with this data, environmentally conscious investors could measure corporate compliance with the CERES Principles, thus using capital markets to effectively promote sustainable business practices.

Initially, the CERES Principles were adopted by companies that already had strong "green" reputations. In 1993, following lengthy negotiations, Sun Oil became the first *Fortune* 500 company to endorse the principles. Several other large companies, including BankBoston, Bethlehem Steel, Coca Cola, General Motors, and Polaroid, soon followed Sun Oil's example. By

1998 forty-eight companies and organizations had endorsed the CERES Principles.

Allan Jenkins

SEE ALSO: Environmental economics; Environmental ethics; *Exxon Valdez* oil spill; Green marketing; Valdez Principles.

Commercial fishing

CATEGORY: Animals and endangered species

Commercial fish catches provide consumers with vast amounts of fish and shellfish, the source of 20 percent of the animal protein ingested by humans worldwide. In order to meet growing demand, larger, more efficient fishing fleets began exhausting coastal fishing areas and moving farther from port. By the late twentieth century, such fleets had reached nearly every area of the world's oceans, causing a series of environmental disasters and political crises.

Hooks, nets, and traps had little effect on populations of marine life until Europeans began operating fishing fleets during the late Middle Ages. Mechanization started in the nineteenth century, but world fishing production was still only around 3 million tons at the beginning of the twentieth century and 20 million tons during the 1950's. The subsequent development of better transportation allowed rapid shipment of premium catches, and the fishing industry reacted to the growing market by investing in larger boats and new technologies and techniques, including sonar, aerial spotting, and nylon nets. More important was the introduction of factory ships. Such ships process the catch at sea, so a fishing fleet can work far from its home port. In one hour, a factory ship can harvest as many fish as a sixteenth century fishing boat

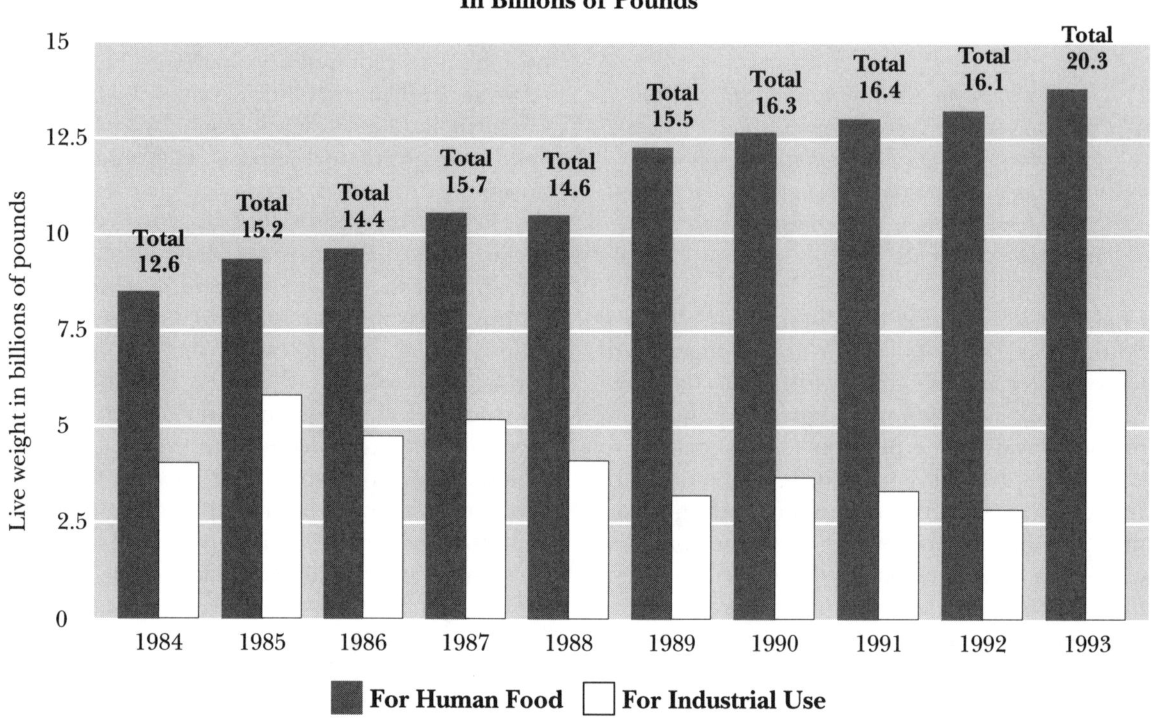

Source: U.S. Department of Commerce, *Statistical Abstract of the United States, 1996*, 1996.

Commercial Fisheries Catch of Major Producers, 1992

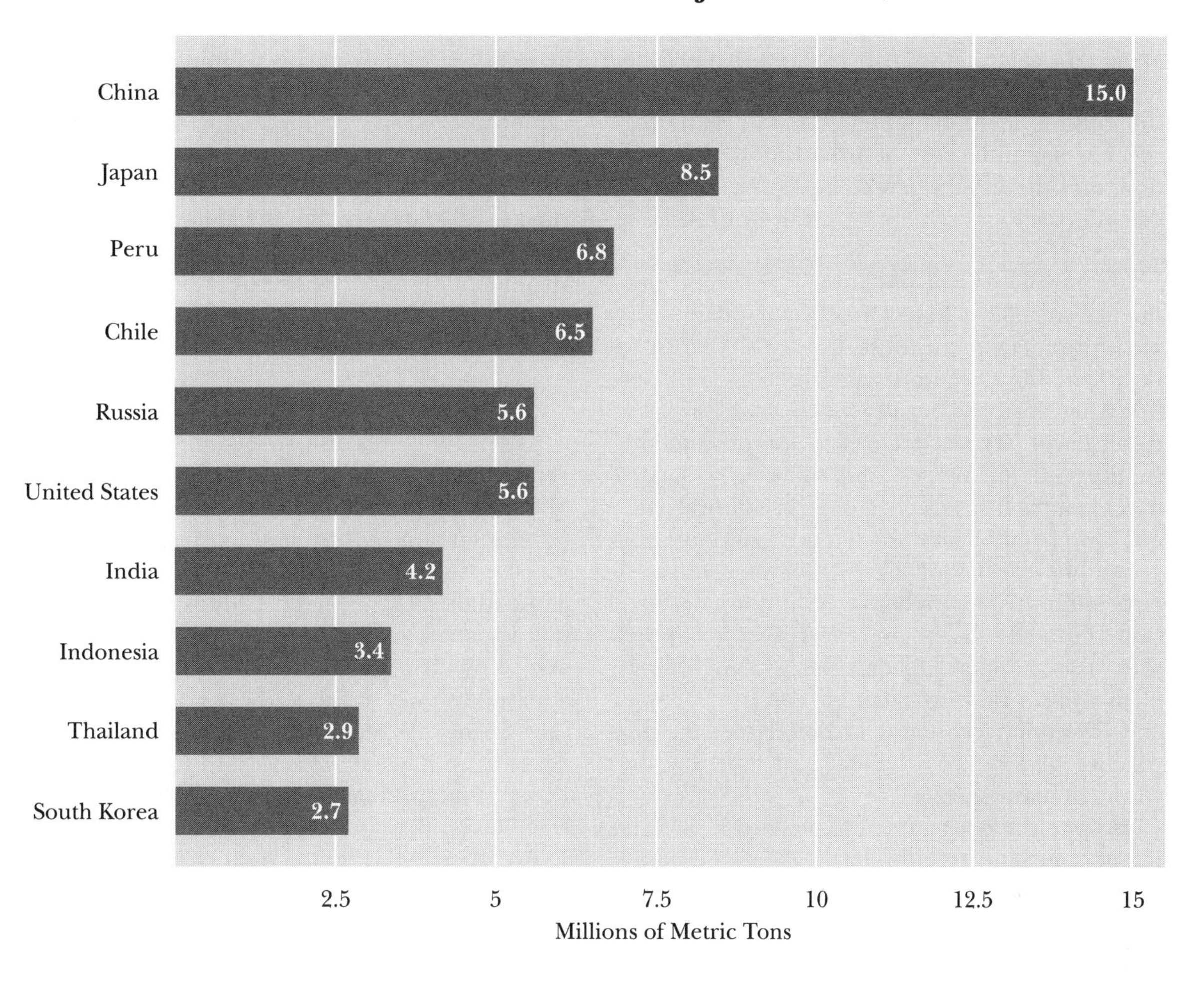

Source: U.S. Department of Commerce, *Statistical Abstract of the United States, 1996*, 1996.
Note: Weights are live weight in thousands of metric tons. Figures include fish, crustaceans, and mollusks.

harvested in one season. Large, high-technology operations and smaller, highly capitalized operations employ about one million people worldwide and take about two-thirds of the catch. These large operations often specialize in a single species and throw back nontarget fish. However, the nontarget by-catch fish that are thrown back are usually dead from being netted, dropped in a hold, and sorted.

Although fishing became industrialized, the old ideas of a limitless supply of fish and an uncontrolled sea remained. The stage was therefore set for a series of environmental disasters and political crises. One problem with fishing is that it is a hunter-gatherer operation rather than an agricultural one. Fishers do not nurture and protect schools of fish as farmers protect herds of cattle. This alone limits productivity. For instance, there is little investment in habitat for fish, such as wetlands or rivers. In addition, fish are considered a common property for which each hunter competes against the others. Any fishers who hold back in catching fish to save breeding stock for the future lose their catch to other boats. American ecologist Garrett Hardin referred to this as "the tragedy of the commons," a concept he originally applied to publicly owned grazing lands. However, cattle on land can be counted, but fish populations are often gauged by catch. Thus, fishers working feverishly to catch the dwindling numbers of fish create the illusion of a stable population. An entire fish species may ultimately be fished to near extinction, and the fishery may collapse, but the worst offenders will gain the most profit until disaster occurs.

Meanwhile, ocean pollution is reducing productivity. Shallow waters close to land are the most productive, but these areas are the first to be affected by pollution. Areas such as the Black Sea, the North Sea, and Chesapeake Bay have experienced serious declines in fish production. In the Gulf of Mexico and the Sea of Japan, excess nutrients, largely from agricultural runoff, have been blamed for red tides (blooms of a poisonous species of algae). Another form of environmental degradation is bottom destruction. Powerful boats use trawl nets equipped with "rock-hoppers" to drag across rough areas of the ocean floor to catch bottom fish, such as cod, flounder, eels, and turbot. However, this practice also destroys habitat and food for the young of many species.

Despite the environmental problems, governments continue to subsidize bigger and more sophisticated boats that travel farther and deeper to catch dwindling fish stocks. The world's fishing fleets get yearly returns of roughly $70 billion but at a cost of roughly $120 billion, and those losses make governments reluctant to enforce limits.

In the 1950's Icelandic gunboats and British warships threatened each other during the so-called Cod War to control fishing access in the North Atlantic. Peru had a similar contest of wills with the United States to control foreign production of anchovies. However, the Peruvians did not limit their own production, and, as anchovy stocks dwindled, the fishery collapsed in 1972. Similarly, Canada and the United States misjudged the carrying capacity of the Grand Banks shoals southeast of Newfoundland, and by 1992 cod fishing in the area was halted.

Total fish production would probably have declined during the 1990's had fish farming not rapidly increased during the same period. Fisheries can produce indefinitely if they limit catches to practical levels. Even devastated fisheries can eventually recover. Doing the same for the open ocean, however, would require major diplomatic efforts. Voluntary bans on whaling and drift nets have been routinely ignored by some countries. Beyond management of wild stocks, fish farming has the potential for great production increases. However, it raises new environmental concerns of pollution and genetic weakening of wild stocks if large numbers of domestic stocks escape.

Roger V. Carlson

Suggested readings: *Scientific American* 9 (Fall, 1998) includes Carl Safina's overview, "The World's Imperiled Fish," as well as related articles on aquaculture and ocean law. In "Fishing the Commons," *Natural History* 85 (August-September, 1976), Garrett Hardin describes economic dynamics of a resource owned by no one and protected by no one. Other articles that address declines in fish stocks worldwide include Carl Safina, "Where Have All the Fishes Gone?" *Issues in Science and Technology* 10 (Spring, 1994); Peter Weber, "Abandoned Seas: Reversing the Decline of the Oceans," *Worldwatch Paper* 116 (November, 1993); Peter Weber, "Net Loss: Fish, Jobs, and the Marine Environment," *Worldwatch Paper* 120 (July, 1994); and Robert Kunzig, "Twilight of the Cod: Atlantic Cod in Danger of Extinction," *Discover* 16 (April 25, 1997). Raymond A. Rogers, *The Oceans Are Emptying* (1995), provides a detailed overview of the subject.

See also: Drift nets and gill nets; Food chains; Law of the Sea Treaty.

Commoner, Barry

Born: May 28, 1917; Brooklyn, New York
Category: Nuclear power and radiation

Barry Commoner's interest in becoming an environmental activist was sparked in the 1950's by the testing of nuclear weapons. Since then he has spoken out on a number of important environ-

mental issues, including the science and information movement, energy resources, organic farming and pesticides, waste management, and toxic chemicals.

Commoner graduated from Columbia University in 1937 and received an M.A. (1938) and a Ph.D. (1941) from Harvard University. He has been awarded a number of honorary degrees from prominent institutions, and his biography has appeared in *Who's Who in Science and Engineering*. In 1965 Commoner was president of the St. Louis Commission on Nuclear Information, and in 1966 he founded the Center for the Biology of Natural Systems (CBNS) at Washington University in St. Louis, Missouri. In 1981 the CBNS moved to Queens College, in the City University of New York (CUNY) system. The CBNS program has continually informed the public and government on environmental issues, including waste recycling, energy resources, and the extensive analysis of the generation and fate of dioxins.

Barry Commoner's environmental activism began during the 1950's when he became aware of the destructive potential of nuclear weapons testing. (Jim West)

Commoner is the author of ten books on the environment, including *Science and Survival* (1963), *The Closing Circle* (1971), *Ecology and Social Action* (1973), *The Politics of Energy* (1979), and *Making Peace with the Planet* (1990). He publicizes environmental problems and relates them to modern technology. Commoner claims that modern methods of production cause human illness and that modern technologies that have resulted in environmental crises have safer alternatives. He believes that transformations to more benign methods, such as eliminating the use of plastic wrap and recycling instead of incineration, are needed to restore the environment. He has promoted the view that there is economic value and profitability in the replacement technologies. He has drawn attention to the affiliations of environmental scientists and questioned their opinions, stating that environmental hazards are reliably recognized only through studies done by independent scientists. He has encouraged the public to become more informed and achieve a greater understanding of the effects of modern technology on the environment before using that technology.

In recognition of his contributions to public awareness of environmental concerns, Commoner has been hailed by *Time* magazine as the "Paul Revere of Ecology" and by *The Earth Times* as "the dean of the environmental movement, who has influenced two generations." In 1980 Commoner was the presidential candidate of the Citizens' Party, a liberal political party he helped found whose interests included an end to nuclear power, a switch to solar energy, and public control of the energy industry.

Commoner continues to lead the movement to eliminate pollution at its source, stating his view that prevention works while controls do not, and publicly advocates the abandonment of fossil fuels as a primary source of energy. Having served society and the environment since the 1950's, he has succeeded in elevating the importance of considering environmental issues in all aspects of life.

Marcie L. Wingfield and Massimo D. Bezoari

SEE ALSO: Antinuclear movement.

Composting

CATEGORY: Waste and waste management

Composting is the recycling of organic waste, such as food, yard waste, and paper products. As these items rot or decay, bacteria help break the organic matter into mulch, a rich soil conditioner suitable for organic gardening. Composting helps reduce the amount of waste that goes to landfills.

All living things are part of a natural recycling process. For example, a dead tree slowly breaks down into a soil-like substance that nurtures the growth of new forest plants. Composting is the same process in a slightly controlled environment.

To make a backyard or garden compost bin, a box with holes or slats, to let in air, is filled with yard and kitchen waste. A fifty-fifty balance of items containing nitrogen and items containing carbon is needed. Items containing nitrogen include grass clippings, egg shells, coffee grounds, and vegetable and fruits peelings. Items containing carbon include dead leaves, evergreen needles, bark chips, and dryer lint. Items are added in layers and topped with dirt to hold in moisture. As the waste decays, the bacteria present thrive. They feast on the waste, generating energy and more bacteria. The decomposing pile is occasionally turned with a shovel, and more yard or kitchen waste is added. The waste products from the bacteria, together with what is left after they feast on scraps, form a dry mulch that can be used as a natural garden fertilizer.

A backyard compost bin. (Ben Klaffke)

Garbage collected from the public by sanitation workers is sometimes composted in municipal processing plants. Rather than add it to landfills or burn it, decomposable materials are sorted from glass, metal, and other inorganic materials. The organic items are then shredded or broken down into smaller pieces.

Two methods are used for mechanical composting. An open-window facility requires a large area of land. It is similar to backyard compost bins but on a large scale. Mounds of organic waste are piled in long, low rows called windows. They are turned or mixed every few days so the aerobic (oxygen-requiring) bacteria that are digesting the waste have adequate oxygen. The aerobic bacteria may take up to eight weeks to completely digest the organic waste. The activity of the microbes creates temperatures in the piles of up to 65 degrees Celsius (150 degrees Fahrenheit). Any bacteria or organisms carrying disease die at this temperature. The odor produced as the material decomposes makes open-window facilities unpopular.

The other type of mechanical composting facility encloses the waste in tanks. Paddles mix the material, adding air at the same time. Tank composting facilities use 85 percent less space than open-window facilities, and the compost is fully digested in about one week. Digested compost is dried, screened, and made into pellets before it is used as a soil conditioner or fertilizer. Mechanically processed compost has low marketability, however, because of the cost of transport and competition from chemical fertilizers available to commercial farmers.

Lisa A. Wroble

SEE ALSO: Landfills; Organic gardening and farming; Waste treatment.

Conservation

Category: Resources and resource management

Conservation is a management policy in which natural resources are used in a way that will benefit the maximum number of people for as long as possible. As such, it is often in conflict with the ethic of preservation, which promotes the indefinite preservation of natural resources in their natural, undisturbed state.

For centuries few people recognized that nature's resources were finite. Only since the mid-nineteenth century have conservation issues been taken seriously. In North America, the eighteenth and nineteenth centuries were largely devoted to conquering the wilderness. As pioneers cleared forests, little or no thought was given to the idea that they could ever be exhausted. Bison were hunted nearly to extinction, and bounties were placed on animals. In the seas, millions of seals were killed, while whales were hunted relentlessly.

Almost too late, it became apparent to many that soils were being depleted by erosion and native plants and that wildlife was threatened. Even the quality of air and surface water deteriorated. An increasing number of people began calling for an end to destructive practices. An important influence was that of George Perkins Marsh. His travels had made him keenly aware of the results of centuries of land abuse in Europe. His book *Man and Nature; Or, Physical Geography as Modified by Human Action* (1864) defined the link among soil, water, and vegetation. From such influences came the park movement, including the establishment of Yellowstone National Park.

During the early twentieth century, living standards of most Americans generally improved, but at the same time it was recognized that consumption of natural resources must be controlled. Out of this realization emerged a human-centered conservation philosophy called the wise-use philosophy, which dictated that nature should be utilized in such a way that its resources could continue to be used over long periods of time.

Among the conservationists of the era were U.S. president Theodore Roosevelt and forester Gifford Pinchot. Roosevelt increased the extent of national forests and created many wildlife refuges. Pinchot, a friend and associate of Roosevelt, became chief forester of the U.S. Department of Agriculture and greatly influenced the conservation movement. He applied European methods of managing forests that were consistent with the wise-use philosophy. This put him in conflict with naturalist John Muir and others who wished to preserve wilderness areas in their natural state. Two opposing camps developed: On one side were preservationists, who argued that nature deserved to be protected for its own sake; on the other were conservationists, who believed in regulated exploitation.

During the 1920's and 1930's, scientific advances and economic problems influenced conservation views and policies. The ecosystem concept, the principle that nature is composed of local units with interacting living and nonliving components, developed. It was to become the cornerstone of the science of ecology and an important tool of conservation. During this same period a new, scientific approach to wildlife management emerged. Aldo Leopold, regarded as the father of the new science, wrote the influential textbook *Game Management* (1933) but is better known for *A Sand County Almanac* (1949). Paul B. Sears achieved prominence with *Deserts on the March* (1935), which vividly dramatized the problems of the Dust Bowl of the 1930's.

The conservation legacy of President Franklin D. Roosevelt includes his attempts to restore the economy by creating various work programs, including the Civilian Conservation Corps (CCC). Many of these programs were aimed at soil conservation, reforestation, and flood control.

In the two decades following World War II, concern with the Cold War and economic expansion kept conservation far from the forefront of American thought. As the population rapidly expanded, air and water pollution worsened, grasslands were overgrazed, and agricultural chemicals were used in increasing amounts. The complacency of the times was shattered by the publication of Rachel Carson's *Silent Spring* (1962). Her warnings that dichloro-diphenyl-trichloroethane (DDT) and other pesticides threatened the lives of animals and humans caused widespread public concern. Laws were soon passed that outlawed DDT.

Under President John F. Kennedy's administration, attention was given once more to conservation issues. Funds were appropriated to improve air quality, and new land was acquired for parks. Steward L. Udall, secretary of the interior, advocated a new positive attitude toward environmental concerns. The 1960's were a time of great scientific activity and social ferment. Ecologists conducted studies in ecosystem analysis using increasingly sophisticated technologies. Birth-control advocates had increased support for their views. Among the social movements of the decade came a new environmentalism, which culminated in the first Earth Day celebration in 1970.

The 1970's saw the expansion of "environmentalism," a term which by then had become almost synonymous with conservation. The distinction between the latter and preservationism was blurred. Environmental organizations continued to elaborate and increase their visibility. More action-oriented alternative groups used sabotage tactics to stop development in wilderness areas. In contrast, the Nature Conservancy became known for its businesslike policy of acquisition, protection, and management of natural areas. In response to the growing environmental awareness of citizens, a new round of federal legislation was passed. Under President Richard Nixon, the Environmental Protection Agency (EPA) was created in 1970. The revised Clean Air Act of 1970 was followed by the Clean Water Act of 1972. One year later, Congress enacted the Endangered Species Act.

The 1980's were marked by new conservation problems and a general indifference at the federal level. Among global problems identified were stratospheric ozone depletion and global warming. Reversals of environmental protection occurred during the administration of President Ronald Reagan after he appointed James Watt as secretary of the interior. Watt attempted to dismantle the environmental protection programs that had been put in place during the previous decades.

The 1990's had their own environmental challenges and successes. After twelve years of apathy, conservationists were reinvigorated by the election of Bill Clinton as president and high-profile environmentalist Albert Gore, Jr., as vice president. Despite continued optimism and significant gains, a growing antienvironmental movement developed. Many individuals and organizations with a conservative agenda were concerned that environmentalism had gone too far. Business leaders and property owners complained of overregulation. Even some religious authorities felt that nature had been elevated in importance above that of humans. Nevertheless, the majority of citizens continued to feel that nature deserved to be protected. An encouraging development of the late twentieth century was a growing awareness of the importance of biodiversity: All species must be protected against extinction. From this realization emerged the value-laden science of conservation biology.

Thomas E. Hemmerly

SUGGESTED READINGS: *First Along the River: A Brief History of the U.S. Environmental Movement* (1997), by Benjamin Kline, provides a useful overview of the history of environmentalism in the United States. *Conservation Biology: Concepts and Applications* (1997), by George Cox, explores scientific investigation based on the concept of biodiversity. Another excellent book on the subject is *Principles of Conservation Biology* (1997), by Gary K. Meffe and E. Ronald Carroll.

SEE ALSO: Forest management; Pinchot, Gifford; Renewable reources; Roosevelt, Theodore; Wise-use movement.

Conservation policy

CATEGORY: Resources and resource management

Conservation policy involves laws and regulations designed to limit the economic exploitation of natural resources in the public interest. Some conservation policies, such as the establishment of national parks and wilderness areas, prohibit most economic uses of the affected resources, preserving them in an undeveloped state. Other conservation policies, such as the establishment of national forests and laws against polluting air and water, encourage the wise use of resources rather than their preservation in a natural state.

First use of the term "conservation" was claimed by Progressive intellectuals in the early twentieth century. In his autobiography *Breaking New Ground* (1947), Gifford Pinchot, first chief of the U.S. Forest Service and a close friend of President Theodore Roosevelt, recalled realizing that all the natural resource problems were really one problem: the use of the earth for the permanent good of humans. The idea needed a name. Presidential adviser Overton Price suggested "conservation," and the matter was settled.

Origins of Conservation Policy

The antecedents of modern conservation policies go back centuries. Aboriginal cultures around the world developed taboos that governed behavior on the hunt. Venetians established reserves for deer and wild boar in the eighth century. Hunting reserves were common in Europe and Asia, and in colonial America those trees thought best for ships' masts were preserved by decree.

Conservation policy, as the term is now understood, was a product of the nineteenth century, when population growth, urbanization, and industrialization created unprecedented opportunities for people to influence the natural world in which they lived, both for good and for ill. By the end of the nineteenth century, more Americans lived in cities than on farms. The nation was connected from coast to coast by telegraph and rail, and economic growth was rapid.

Most natural resource policies of the nineteenth century were designed to facilitate economic development. Best known among these policies was the Homestead Act of 1862, which gave free land to settlers, but similar policies provided free land to railroads and states. Other laws stimulated growth by providing free use of timber and minerals on public lands.

The economic progress of the nineteenth century came at high cost to the environment, typified by the profligate use of forests and the near-extermination of North American buffalo, and some prominent Americans took notice. In 1832 the artist and journalist George Catlin wrote of the probable extinction of buffalo and American Indians, and he advocated a large national park where both might be preserved. Henry David Thoreau echoed Catlin's concerns in 1858, calling for national preserves. In 1864 George Perkins Marsh published *Man and Nature,* the earliest important text with an ecological perspective.

As economic exploitation diminished the supply of natural resources, public attitudes began to change, and with them governmental policies. Although many laws continued to encourage economic growth, others reflected the growing desire for conservation. In 1864 the U.S. Congress sought to preserve the Yosemite Valley and Mariposa Big Tree Grove by giving them to the state of California for a public park. Eight years later Congress established the world's first national park at Yellowstone. In 1884 additional legislation prohibited all hunting and commercial fishing within Yellowstone National Park. In 1891 Congress established what would eventually become the national forests system when it authorized the president to set aside forest reserves on public lands.

Preservation Versus Wise Use

These early conservation policies stressed preservation. Both parks and forest reserves were simply set aside; neither was effectively managed. The lack of management—especially forest management—displeased the advocates of scientific forestry, who also considered themselves conservationists. So early in its history the American conservation movement was divided, with some conservationists preaching "preservation" and others "wise use."

These contradictory tendencies were epitomized in the conflict between John Muir, founder of the Sierra Club, and Gifford Pinchot, principal architect and first chief of the U.S. Forest Service. Muir was the intellectual heir of Thoreau and Catlin. A perceptive scientist and popular author, he devoted his life to the exploration, enjoyment, and preservation of natural ecosystems worldwide. Pinchot had studied scientific forestry at its source in Germany. A gifted politician, his passion was not for preservation but for wise use. Muir's conservation was aesthetic and spiritual: He believed that people could not improve on nature. Pinchot's conservation was economic and utilitarian: He was committed to maximizing the human benefits from resource use through science. Although friends for a time, Muir and Pinchot eventually parted ways, with Muir becoming an advocate for preservation and national parks and Pinchot an advocate for wise-use and national forests.

In the United States the legacy of Pinchot is alive and well in the multiple-use management principles of the Forest Service and the Bureau of Land Management and in membership groups such as the Society of American Foresters, the International Society of Fish and Wildlife Managers, the National Rifle Association, and the Soil Conservation Society of America. Muir's emphasis on preservation has been institutionalized in the National Park Service, the Fish and Wildlife Service, and in membership organizations such as the National Audubon Society, the Nature Conservancy, the Sierra Club, and the Wilderness Society.

Progressive Era

Many historians emphasize three eras of American conservation policy corresponding roughly to the Progressive Era, the New Deal, and the so-called environmental decade of the 1970's. The Progressive Era, epitomized by the presidency of Theodore Roosevelt, was the first golden age of American conservation policy. During this era Congress passed a number of path-breaking conservation laws. The Lacey Act of 1900 put the power of federal enforcement behind state game laws, criminalizing the interstate transport of wildlife killed or captured in violation of state regulations. Another milestone of wildlife conservation was marked by ratification of a migratory bird treaty with Canada and passage of a law to enforce the treaty. With the Migratory Bird Treaty Act of 1918, the federal government asserted national authority to manage wildlife for conservation purposes, authority that was upheld by the Supreme Court in the case of *Missouri v. Holland* (1920).

Two critically important governmental agencies were created during this era: the Forest Service and the Park Service. In 1905 advocates of wise use and scientific forestry were rewarded with a Forest Service in the Department of Agriculture. The new agency's first director was Gifford Pinchot, the nation's foremost advocate of multiple-use forest management based on scientific principles. Under his leadership the concepts of multiple use and sustained yield were applied in the rapidly growing national forest system. There were 46 million acres of national forest when Roosevelt became president. By the end of his term of office, Roosevelt had increased the total size of the national forest system to 194 million acres.

During the Progressive Era advocates of preservation suffered at the hands of wise-use conservation, but in the end they also had a victory. The most painful loss came in Yosemite National Park, where advocates of wise use joined forces with the city of San Francisco to dam the Hetch Hetchy Valley for a municipal water supply, forever destroying a natural valley some regarded as comparable to Yosemite Valley itself. The public outcry over damming Hetch Hetchy contributed to pressure for better park protection, and in 1916 preservationists achieved a long-sought goal: creation of a National Park Service to manage a growing system of fourteen national parks, including Yellowstone (1872), Yosemite and Sequoia (1890), Mount Rainier (1899), Crater Lake (1902), Wind Cave (1903), Mesa Verde (1906), Glacier (1910), Rocky Mountain (1915), and Hawaii and Lassen Volcanic (1916).

Two new forms of conservation reservation made their debut during this era: national wildlife refuges and national monuments. Roosevelt regarded wildlife sanctuaries as critical to the survival of game species. In 1903 he acted on his belief, creating the nation's first national wildlife refuge on Pelican Island in Florida. He had no specific legal authority to create a national wildlife refuge, but his usurpation was accepted at the time and later approved in principle. In 1910 the Pickett Act authorized the president to set aside land for any public purpose. National monuments began on a firmer foundation, but here too Roosevelt pushed conservation to the limit. The Antiquities Act of 1906 authorized the president to establish national monuments. As the name suggests, the law anticipated relatively small reservations to protect archaeological sites, but Roosevelt's monuments included 85,000 acres at Petrified Forest, 298,000 acres at Mount Olympus, and 806,000 acres at the Grand Canyon. Each is now a national park.

New Deal Era

The New Deal was, for the most part, a response to disaster. The primary disaster was the Great Depression, but the 1930's was also the decade of the Dust Bowl, a minor climatic change that produced disastrous results on the Great Plains. New Deal conservation policies were responsive to the economic and ecological crises of the era, and they stressed wise use through scientific management rather than preservation. The Tennessee Valley Authority was created in 1933 to stimulate employment and economic growth in Appalachia through scientific management of the area's natural resources. The Taylor Grazing Act of 1934 was designed to end overgrazing of western public lands by imposing a system of permits based on principles of scientific management. The Civilian Conservation Corps and the Soil Conservation Service were both established during this era, and each contributed to repairing environmental damage. Greater concern for the management rather than the disposal of western public lands was also reflected in the creation in 1946 of the Bureau of Land Management to replace the General Land Office.

Environmental Decade

The so-called environmental decade lasted almost twenty years. It began with the inauguration of President John F. Kennedy, persisted through the presidential administrations of Lyndon Johnson, Richard Nixon, Gerald Ford, and Jimmy Carter, and ended with the inauguration President Ronald Reagan. The conservation policies of this era were responsive to postwar economic growth that seemed to assure economic prosperity while threatening quality of life. Interior Secretary Stewart Udall warned of a *Quiet Crisis* (1963), Rachel Carson of a *Silent Spring* (1962), Barry Commoner of *The Closing Circle* (1971), and Paul Ehrlich of *The Population Bomb* (1968). Conservation policy matured into environmental policy during this era. Conservation was still about husbanding natural resources, but to the historic concerns of conservation—such as forests, wilderness, and wildlife—were added clean water, clean air, energy supply, and hazardous and toxic waste. During this era Congress passed most of the major laws that continued to shape conservation policy at the end of the twentieth century.

Preservation policy was strengthened. In 1964 Congress passed the Wilderness Act and the Land and Water Conservation Fund Act. The former established a National Wilderness Preservation System, which has since grown to more than 100 million acres. The latter facilitated acquisition of land for parks and open space. Four years later Congress established a national system of wild and scenic rivers that were protected from certain kinds of development and a national system of trails. Both the Forest Service and the Bureau of Land Management were given new statutory direction emphasizing planning and preservation, sometimes at the expense of economic development. At the end of the era Congress passed the Alaska National Interest Lands Conservation Act (1980), making conservation withdrawals of more than 100 million acres of public lands and doubling the size of the national park and national wildlife refuge systems nationwide.

Wise-use conservation was also well served as Congress radically increased federal regulation of resource use. President Johnson established a presidential commission on natural beauty and addressed world population and resource scarcity in his 1965 state of the union speech. Environmental management was nationalized through a series of far-reaching statutes addressing air pollution, water pollution, marine resources, noise pollution, biological diversity, toxic chemicals, and hazardous waste. New burdens were placed on government and private citizens. The National Environmental Policy Act of 1970 required all governmental agencies to study the probable environmental effects of

Milestones in Conservation Policy

Year	Event
1864	Yosemite Valley is ceded to California to create a park.
1872	The Yellowstone National Park Act establishes the world's first national park.
1891	The Forest Reserve Act authorizes the U.S. president to establish national forests.
1894	The Yellowstone Game Protection Act closes parks to hunting and commercial fishing.
1897	The Forest Management Act mandates that national forests be managed to perpetuate water supplies and wood products.
1900	The Lacey Act prohibits interstate shipment of wildlife that has been killed illegally.
1902	The Newlands Act establishes a national reclamation policy.
1903	The first National Wildlife Refuge is created at Pelican Island, Florida.
1905	The U.S. Forest Service is created within the Department of Agriculture to manage national forests.
1906	The Antiquities Act authorizes the creation of national monuments by presidential proclamation.
1910	The Pickett Act authorizes presidential land withdrawals for any public purpose.
1911	The Weeks Act provides for governmental purchase of national forestlands.
1913	The Hetch Hetchy Dam is authorized in Yosemite National Park.
1916	National Park Service is created in the Interior Department to manage national parks.
1918	The Migratory Bird Treaty Act restricts the hunting of migratory birds.
1933	The Tennessee Valley Authority is created.
1934	The Taylor Grazing Act regulates grazing on public lands.
1937	Civilian Conservation Corps Act is passed.
1937	The Federal Aid in Wildlife Restoration (Pittman- Robinson) Act provides federal aid to states for wildlife management.
1946	The U.S. Bureau of Land Management is created.
1950	The Federal Aid in Fish Restoration (Dingell-Johnson) Act provides federal aid to states for sport fish management.
1956	The Fish and Wildlife Act creates the U.S. Fish and Wildlife Service in the Interior Department.
1956	Echo Park Dam proposal in Dinosaur National Monument is defeated.
1960	The Multiple-Use and Sustained Yield Act clarifies the purposes of national forests.

Year	Event
1964	The Wilderness Act establishes the National Wilderness Preservation System.
1964	The Land and Water Conservation Fund Act provides a trust fund for parkland acquisition.
1968	The National Wild and Scenic Rivers Act establishes a national river conservation system.
1968	The National Trails System Act establishes a national system of recreational trails.
1970	The National Environmental Policy Act requires environmental impact statements for federal activities that affect the environment.
1970	The Environmental Protection Agency (EPA) is created.
1970	Clean Air Act amendments establish stricter air-quality standards.
1971	The United Nations Educational, Scientific, and Cultural Organization (UNESCO) Biosphere Reserve Program recognizes areas of global environmental significance.
1972	The Clean Water Act establishes stricter water-quality standards.
1972	The United Nations Environmental Conference in Stockholm, Sweden, is attended by 113 nations.
1972	Federal Water Pollution Control Act amendments provide protection for wetlands.
1972	The Federal Environmental Pesticides Control Act requires pesticide registration.
1972	The Marine Mammal Protection Act imposes a moratorium on hunting or harassing of marine mammals.
1973	The Convention on International Trade in Endangered Species of Wild Fauna and Flora (CITES) prohibits international trade in endangered species.
1973	The Endangered Species Act commits the United States to the preservation of biological diversity.
1974	The Safe Drinking Water Act sets federal standards for public water supplies.
1976	Toxic Substances Control Act authorizes the EPA to ban substances that threaten human health or the environment.
1976	The Federal Land Policy and Management Act directs the Bureau of Land Management to retain public lands and manage them for multiple uses.
1976	The Resource Conservation and Recovery Act directs the EPA to regulate waste production, storage, and transportation.
1976	The National Forest Management Act gives statutory protection to national forests and sets standards for management.
1977	The Surface Mining Control and Reclamation Act establishes environmental standards for strip mining.

(continued)

YEAR	EVENT
1977	Clean Air Act amendments set high standards for air quality in large national parks and wilderness areas.
1980	The Fish and Wildlife Conservation Act provides federal aid for the protection of nongame wildlife.
1980	The Alaska National Interest Lands Conservation Act establishes more than 100 million acres of national parks and wildlife refuges in Alaska.
1980	Comprehensive Environmental Response, Compensation, and Liability Act (CERCLA) establishes the Superfund hazardous waste cleanup program.
1982	The Nuclear Waste Policy Act establishes a process for siting a permanent nuclear waste repository.
1987	The Montreal Protocol limits the production and consumption of chlorofluorocarbons (CFCs).
1988	The Ocean Dumping Act prohibits the dumping of sewage sludge and industrial waste.
1990	Clean Air Act amendments strengthen the Clean Air Act.
1992	The Earth Summit in Rio de Janeiro, Brazil, is attended by 179 nations.
1997	The Kyoto Protocol on Climate Change encourages global reduction in greenhouse gas emissions.

their actions before moving forward. A large number of environmental enforcement programs were reorganized in 1970 into the newly created Environmental Protection Agency (EPA). The EPA is an independent agency, but presidents have routinely regarded its director as having cabinet status.

CONSERVATION POLICY FOR A NEW MILLENNIUM

At the policy level, the era following the environmental decade was one of consolidation rather than new initiatives. The Reagan administration was hostile to environmental policy and attempted to tilt public policy toward less environmental regulation. President Reagan was able to prevent the adoption of major new conservation policies, but his administrative efforts—led by Interior Secretary James Watt—to roll back environmental laws were successfully resisted by Congress. A major new air pollution statute was passed during the presidential administration of George Bush, but this was exceptional for the era. The Clinton administration gave greater attention to conservation policy, but in the years following the 1994 elections there was little cooperation between the president and Congress on environmental issues.

Americans have come to expect government to practice conservation and protect environmental quality. Doing so is increasingly difficult. Beyond the policy gridlock of the 1980's and 1990's, the issues themselves have become more difficult. The most pressing issues—such as biological diversity, acid precipitation, stratospheric ozone depletion, and climate change—are beset by scientific uncertainty. They are also global in scope and thus beyond the ability of any nation to address independently. The future of conservation policy appears to be in the international arena, where extant institutions lack the authority to govern. Treaties on chlorinated fluorocarbons, biological diversity, and greenhouse gas emissions demonstrate that nations are giving increasing attention to conservation issues, but international achievements remain modest.

Craig W. Allin

SUGGESTED READINGS: Modern conservation policy is effectively summarized in David Howard Davis's *American Environmental Politics* (1998) and Walter A. Rosenbaum's *Environmental Politics and Policy* (1998). A leading textbook in conservation is *Natural Resource Conservation: Management for a Sustainable Future* (1997), by Oliver S. Owen, Daniel D. Chiras, and John P. Reganold. Craig W. Allin's *Politics of Wilderness Preservation* (1982) deals with public land preservation through 1980. *The Ends of the Earth* (1989), edited by Donald Worster, explores the relationship between natural resources and people on a global scale. Samuel P. Hayes's *Conservation and the Gospel of Efficiency* (1959) and Roderick Nash's *Wilderness and the American Mind* (1982) are conservation classics.

SEE ALSO: Conservation; Forest Service, U.S.; National forests; Pinchot, Gifford; Roosevelt, Theodore; Wise-use movement.

Convention on International Trade in Endangered Species

DATE: 1975

CATEGORY: Animals and endangered species

The Convention on International Trade in Endangered Species resulted from an international conference on endangered species held in Washington, D.C., in 1973. The 144 signatories made legal commitments to conserve endangered animal and plant species.

Until the 1970's the international agreements that dealt with preservation of species did not include a binding legal commitment on the part of the countries signing them. They were ineffectual in protecting the species that they were written to protect. In 1969, however, the United States passed the Endangered Species Conservation Act (ESA), which contained a provision that gave the secretaries of interior and commerce until June 30, 1971, to call for an international conference on endangered species.

Although it went beyond the ESA's time limit, the international conference was held in Washington, D.C., in March, 1973, resulting in the Convention on International Trade in Endangered Species of Wild Fauna and Flora (CITES). The United States was the first country to ratify the convention, which entered into effect on July 1, 1975; 143 other countries have also ratified. CITES is intended to conserve species and does this by managing international trade in those species. It was the first international convention on the conservation of wildlife that constituted a legal commitment by the parties to the convention and also included a means of enforcing its provisions. This enforcement includes a system of trade sanctions and an international reporting network to stop trade in endangered species.

However, the system established by CITES does contain loopholes through which states with a special interest in a particular species can opt out of the global control for that species. The major aspect of CITES is its creation of three levels of vulnerability of species. Appendix I includes all species that are threatened with extinction and whose status may be affected by international trade. Appendix II includes species that are not yet threatened but might become endangered if trade in them is not regulated. It also includes other species that, if traded, might affect the vulnerability of the first group. Appendix III lists species that a signatory party identifies as subject to regulation in order to restrict exploitation of that species. The parties to the treaty agree not to allow any trade in the species on the three lists unless an exception is allowed in CITES.

The species listed in the appendices may be moved from one list to another as their vulnerability increases or decreases. According to the convention, states may implement stricter measures of conservation than those specified in the convention or may ban trade in species not included in the appendices. CITES also established a series of import and export trade permits within each of the categories. Each nation designates a management authority and a scientific authority to implement CITES. Exceptions to the ban on trade are made for scientific and museum specimens, exhibitions, and movement of a species under permit by a national management authority.

The parties to CITES maintain records of trade in specimens of species that are listed in the appendices and prepare periodic reports on their compliance with the convention. These reports are sent to the CITES secretariat in Switzerland, administered by the United Nations Environment Programme (UNEP), which issues notifications to all parties of state actions and bans. The secretariat's functions are established by the convention and include interpreting the provisions of CITES and advising countries on implementing those provisions by providing assistance in writing their national legislation and organizing training seminars. The secretariat also studies the status of species being traded in order to assure that the exploitation of such species is within sustainable limits.

The CITES Conference of Parties meets every two or three years in order to review implementation of the convention. The meetings are also attended by nonparty states, intergovernmental agencies of the United Nations, and nongovernmental organizations considered "technically qualified in protection, conservation or management of wild fauna and flora." The meetings are held in different signatories' countries: The first took place in Berne, Switzerland, on November 2-6, 1976. At the conference, the parties may adopt amendments to the convention and make recommendations to improve the effectiveness of CITES.

CITES has been incorporated into Caring for the Earth: A Strategy for Sustainable Living. The strategy was launched in 1991 by UNEP, the International Union for the Conservation of Nature (IUCN), and the World Wildlife Fund (WWF). Other nongovernmental groups working to support CITES are Fauna and Flora International (FFI), Trade Records Analysis of Flora and Fauna in Commerce (TRAFFIC International), and the World Conservation Monitoring Centre (WCMC).

Some of the species protected by CITES have received additional protection under later agreements. In certain cases, however, states have allowed trade in listed species to continue for economic purposes or have refused to sign CITES because of the extent to which they trade in a species or species part, such as ivory. Others have signed because they needed help in stop-

Monetary Value of Endangered Species and Animal Parts, 1990

International Species	Price (U.S. Dollars)	North American Species	Price (U.S. Dollars)
Olive python	1,500	Bald eagle	2,500
Rhinoceros horn	12,500/lb.	Golden eagle	200
Siberian tiger skin	3,500	Gila monster	200
Tiger meat	130/lb.	Peregrine falcon	10,000
Leopard	8,500	Grizzly bear	5,000
Snow leopard	14,000	Grizzly claw necklace	2,500
Elephant tusk	250/lb.	Polar Bear	6,000
Mountain Gorilla	150,000	Black Bear paw pad	150
Giant panda	3,700	Mountain Lion	500
Ocelot coat	40,000	Mountain Goat	3,500
Amazon Macaw	30,000	Saguaro Cactus	15,000

Source: U.S. Fish and Wildlife Service.

ping illegal trade and poaching of species within their borders. Whales have proven to be a difficult species to protect. Whales were given protection under CITES according to the status of the specific whale species. The moratorium on commercial whaling by the International Whaling Commission (IWC) was intended to strengthen the CITES protection by species, but the whaling states have disagreed on the numbers of whale populations, and some have withdrawn from the IWC and resumed their whaling activities.

Colleen M. Driscoll

SUGGESTED READINGS: Paul Ehrlich and Anne Ehrlich, *Extinction: The Causes and Consequences of the Disappearance of Species* (1981), provides an understanding of the basic problem and potential effects of the disappearance of species. For an overview of the problem CITES seeks to correct, see *Conserving the World's Biological Diversity* (1990), by the International Union for Conservation of Nature.

SEE ALSO: Endangered species; Endangered Species Act; Pets and the pet trade.

Convention on Long-Range Transboundary Air Pollution

DATE: November 13, 1979
CATEGORY: Atmosphere and air pollution

In 1979 thirty-two nations signed an agreement to limit air pollution, including pollution created in one country that affected the environment in another. Although the treaty had little direct effect on air quality, it was the first agreement among nations of Eastern Europe, Western Europe, and North America regarding the environment.

Until the 1970's most local, regional, and national regulations regarding industrial air pollution were concerned only with pollution generated in the immediate area. For example, regulations in a particular community might call for taller industrial smokestacks to carry pollution farther away, but there was little official concern about where that pollution might eventually return to earth. Similarly, local assessments and treatments of pollution tended not to consider pollution that might come to an area from distant generators. The only exceptions were a small number of treaties between two countries, such as the United States and Canada, or Germany and France.

In 1972 the United Nations Environmental Conference in Stockholm, Sweden, drew attention to the harmful effects of acid rain, including damage to forests, crops, surface water, and building and monuments, especially in Europe. Data revealed that while all European nations produced alarming levels of air pollution, several nations were receiving more pollution from beyond their borders than they were generating on their own. It became clear that pollution is both imported and exported, that sulfur and nitrogen compounds can travel through the air for thousands of miles, and that any serious attempt to deal with air pollution must reach beyond political boundaries. Two major studies of the long-range transport of air pollutants (LRTAP) were conducted under United Nations (U.N.) sponsorship in 1972 and 1977, conclusively proving that air pollution was an international—even a global—problem primarily caused by fossil fuel combustion. However, the pollution caused harm to both industrial and nonindustrial nations around the world.

In 1979 the United Nations Environment Programme organized a convention in Geneva, Switzerland, for the thirty-four member countries of the United Nations Economic Commission for Europe (ECE), a group that includes all European nations, the United States, and Canada. Significantly, the gathering had the participation of Eastern European nations under the Soviet Union, marking the first time that these nations had collaborated with Western Europe to solve an international environmental problem. The Convention on Long-Range Transboundary Air Pollution was signed by thirty-two nations on November 13, 1979, and went into effect on March 16, 1983. It called upon signatory nations to limit and eventually reduce air pollution, in particular sulfur emissions, using

The Convention on Long-Range Transboundary Air Pollution was designed to seek international solutions to problems caused by industrial air pollution. Tall smoke stacks carry pollution away from the immediate vicinity and deposit it elsewhere, often across international boundaries. (Jim West)

the best and most economically feasible technology; share scientific and technical information regarding air pollution and its reduction; permit transboundary monitoring; and collaborate in developing new antipollution policies. Under the terms of the convention, an international panel would undertake a comprehensive review every four years to determine whether goals were being met, and an executive body would meet each year.

The convention did not include any specific plan for the reduction of air pollution; there was no language calling for particular amounts by which emissions would be reduced, nor was there a schedule by which the reductions would occur. Scandinavian nations, which were among the countries most affected by acid rain, urged the other participants to adopt these kinds of policies, but other countries, led by the United States, the United Kingdom, and West Germany, defeated the proposal.

In the years following the convention, however, several nations did make commitments to reduce emissions by specific amounts, including West Germany, which changed its position as further information was revealed about deforestation caused by acid rain. At the 1983 executive body meeting, eight nations, including Canada, West Germany, and the Scandinavian countries, made a formal commitment to reduce their emissions by 30 percent by 1993, using 1980 levels as the baseline. Over the next two years thirteen more nations announced similar goals, and in 1985 the commitment to a 30 percent reduction was formally adopted as an amendment to the convention that was signed by nineteen nations.

Neither the United States nor the United Kingdom agreed to the 30 percent reductions, and neither country signed the 1985 protocol. The United Kingdom informally agreed to attempt to reduce emissions by 30 percent but was unwilling to commit the financial resources to guarantee it, especially since the benefits were uncertain. In fact, many scientists felt that 30 percent reductions would not be enough to yield significant improvement. The United States argued that it had already taken major steps to reduce its emissions prior to 1980, so using 1980 data as a baseline would subject the United States to unrealistic and unfair demands for further reduction. This refusal to ratify the protocol caused tension between the United States and Canada, because much of the air pollution that affects Eastern Canada comes from the Great Lakes industrial belt in the United States.

Cynthia A. Bily

suggested readings: Much of the writing about transboundary air pollution is technical and bureaucratic, written for scientists and policymakers. A notable exception is Derek Elsom's *Atmospheric Pollution: Causes, Effects, and Control Policies* (1987), an overview of air pollution issues that is thorough and accessible to nonspecialist readers. Jutta Brunnee's *Acid Rain and Ozone Layer Depletion: International Law and Regulation* (1988) is excellent but slightly more difficult. *Global Alert: The Ozone Pollution Crisis* (1990), by Jack Fishman, is readable and reliable but borders on the sensational. Levels of international cooperation in Europe are explored in Peter H. Sand, "Air Pollution in Europe: International Policy Responses," in the journal *Environment* (1987). G. S. Wetstone and A. Rosencranz focus on the limits to cooperation in *Acid Rain in Europe and North America: National Responses to an International Problem* (1983). The complete treaty, as well later protocols, is available as a small book issued by the United States Government Printing Office under the title *Long-Range Transboundary Air Pollution: Protocols Between the United States of America and Other Governments* (1996).

see also: Acid deposition and acid rain; Air pollution; Air pollution policy; United Nations Environment Programme; United Nations Environmental Conference.

Convention on the Conservation of Migratory Species of Wild Animals

Date: June 23, 1979
Category: Animals and endangered species

The Convention on the Conservation of Migratory Species of Wild Animals is an international treaty designed to protect vulnerable species of wild animals that migrate across national boundaries.

On June 23, 1979, several sovereign nations and regional economic organizations signed the Convention on the Conservation of Migratory Species of Wild Animals in Bonn, Germany. After receiving the necessary fifteen ratifications, the treaty went into effect on November 1, 1983. Within ten years it had been signed and ratified by more than sixty countries and qualified organizations. The signatories of the treaty agreed to protect any endangered species whose entire population (or a significant portion of the population) "cyclically and predictably cross one or more national jurisdictional boundaries." The term "endangered" was defined as meaning that "the migratory species is in danger of extinction throughout all or a significant portion of its range." In addition, the treaty obligated the signatories to take appropriate action to prevent unfavorable-status species from becoming endangered.

The treaty established a Conference of the Parties, which makes decisions concerning the obligations of the signatories. The conference determines its budget and formulates a scale for assessing the contribution of each party to the conference. The treaty also provided for a Secretariat, which is appointed by the executive director of the United Nations Environment Programme (UNEP). The Secretariat has several executive functions, which include listing endangered species, arranging meetings of the conference, promoting liaisons between the parties, and performing duties entrusted to it by the conference. Finally, the treaty established a Scientific Council for the purpose of providing scientific recommendations to the conference.

Appendix I of the treaty lists those migratory species that are considered endangered. The list is regularly updated; in 1992 it included more than fifty species, including four species of whales, twenty-four bird species, six species of marine turtles, and one species of gorilla. In regard to such species, the parties to the treaty agreed to conserve and, where feasible, restore the habitats that were important "in removing the species from danger of extinction." The parties agreed to prohibit the taking of members of species on the list, with exceptions for scientific purposes and a few extraordinary circumstances.

Appendix II lists those migratory species that have an "unfavorable conservation status," as well as additional "conservation status" species that would benefit from an international agreement. In regard to the species of Appendix II, the parties agreed to endeavor to conclude agreements that would promote survival of the species, especially those in an unfavorable conservation status. The Secretariat is provided with a copy of each such agreement.

The convention of 1979 was primarily a declaration of principles, and it delegated almost no enforcement powers to the Conference of the Parties or the Secretariat. Although sometimes referred to as an example of "soft law," the convention's importance is enhanced by other international agreements that deal with related problems, and it has the potential of making a significant impact on public opinion.

Thomas T. Lewis

SEE ALSO: Convention on International Trade in Endangered Species; Endangered species; International Whaling Ban.

Convention on Wetlands of International Importance

DATE: 1971
CATEGORY: Preservation and wilderness issues

The Convention on Wetlands of International Importance is an intergovernmental treaty that provides the framework for the conservation and wise use of wetlands. The treaty, signed in Ramsar, Iran, in 1971, is commonly referred to as the Ramsar Convention.

During the 1960's there was growing concern about the decline in populations of waterfowl in many parts of the world. One of the major factors causing this trend was the decline in the number and size of wetlands, which are habitats that are heavily used by waterfowl. In order to address these interrelated environmental issues,

The Convention on Wetlands of International Importance protects wetlands that provide essential stopovers for migrating waterfowl. (Ben Klaffke)

the Ramsar Convention was developed. It broke new ground in global environmental efforts, serving as the first international treaty on the conservation and wise use of a natural resource.

The importance of wetlands to the global environment and the needs of humans, as well as the need for an international approach to deal with wetlands, gradually became apparent. Wetlands are important because of their high biodiversity and their role in water purification, water storage, flood abatement, and groundwater recharge. Many wetlands extend across national boundaries, which often results in disagreements on wetland conservation and use. For example, fish may hatch in the wetlands of one country but be caught as adults in those of another country. Also, many birds migrate hundreds or thousands of kilometers twice each year and need wetlands of many countries to rest, feed, and breed.

The Ramsar Convention has now been ratified by more than 110 countries with the designation of almost one thousand wetland areas as "wetlands of international importance." Under the treaty, each country is obligated to implement the convention in four basic ways: designate at least one wetland for inclusion in the list of wetlands of international importance, include wetland conservation and wise use as a major focus within its national land-use planning, promote wetland conservation by establishing nature reserves on wetlands and promoting wetland education, and consult with other countries concerning the implementation of the convention.

The policy-making body of the convention is called the Conference of Parties. Each member nation sends representatives to a conference every three years for the purpose of receiving and reviewing reports on the work of the convention and for approving the work and budget of the convention for the next three years. The Ramsar Convention is administered by the Ramsar Bureau, located in Gland, Switzerland. The bureau is advised on a regular basis by a Standing Committee and a Scientific and Technical Review Panel, each composed of representatives from various member nations. The work of the convention is funded by contributions from the member nations.

Key issues of the convention include urging member nations to develop management plans for wetlands that include the involvement of local communities and indigenous people, wetland education, and establishing monitoring programs that have the ability to detect changes in the ecological character of wetlands.

Roy Darville

SEE ALSO: Biodiversity; Wetlands; Wildlife refuges.

Convention Relative to the Preservation of Fauna and Flora in Their Natural State

DATE: 1933
CATEGORY: Preservation and wilderness issues

An international treaty signed in London, England, by nine nations in 1933 established preservation policies for European colonies in Africa. The treaty recognized that most of the danger to Africa's animals was posed by Westerners, in particular by European and American trophy hunters, and sought to protect animals from extinction by regulating such hunting.

In 1900 the European nations that had recently divided sub-Saharan Africa among themselves and established colonial governments signed the first international conservation treaty, the Convention for the Preservation of Animals, Birds, and Fish in Africa. The men who drafted this document did not recognize any inherent value in living creatures—they were not protecting animals because they felt animals had a right to live. The intention of the treaty was to preserve animals that were popular trophies for hunters, such as elephants and giraffes, and encourage the eradication of animals harmful to agriculture, including lions, leopards, and wild dogs.

In 1930 a surveying expedition sponsored by the British Society for the Protection of the Fauna of the Empire made it clear that the 1900 treaty was ineffective from a conservation standpoint. Elephants and other animals were still being overhunted. Several animal and plant species were drawing closer to extinction. In order to protect species without substantially limiting human activity, an expanded system of national parks was proposed for East and Central Africa. National parks would be under the control of the colonial government. The public would be encouraged to visit the national parks to observe the plants and animals, but no "hunting, killing, or capturing" would be permitted within park boundaries.

Several nations that held substantial amounts of land in Africa met in London in 1933 to discuss the issues. The resulting Convention Relative to the Preservation of Fauna and Flora in Their Natural State was signed by South Africa, Belgium, the United Kingdom, Egypt, Spain, France, Italy, Portugal, and the Sudan. It established national parks for public enjoyment and "strict natural reserves" for the exclusive use of scientists. One plant species and twenty animals—including gorillas, white rhinoceroses, and shoebill storks—were fully protected by the treaty. New rules for hunters outside the parks forbade the use of cars and aircraft to chase or herd animals and also prohibited poison and traps.

However, neither the treaty nor the discussions leading up to it considered the role black Africans might play in preserving or endangering the fauna and flora. The treaty was made by Europeans to ensure that white people would have enough animals to hunt. Much of the land newly dedicated to national parks had been home to Africans who were now forbidden to hunt, farm, or live on that land. Animals were protected or not protected according to their usefulness for or danger to white hunters and settlers, without consideration of which animals provided food or presented a danger to native villagers.

Cynthia A. Bily

SEE ALSO: Convention on International Trade in Endangered Species; Endangered species and animal protection policy; National parks.

Corporate average fuel economy standards

DATE: enacted 1975
CATEGORY: Energy

The corporate average fuel economy standards consist of U.S. government regulations concerned with automobile fuel efficiency.

In response to the energy crisis of the 1970's, the U.S. Congress enacted legislation intended to reduce American dependence on oil imports. The 1975 corporate average fuel economy (CAFE) standards mandated fuel efficiency levels that automobile manufacturers were required to meet. Each manufacturer's annual automobile output had to meet the assigned average for that year. If a fleet exceeded the CAFE standards, the manufacturer faced a substantial fine. To ensure that manufacturers did not import fuel-efficient foreign cars to offset low averages in their domestic output, import and domestic fleets were evaluated separately. The 1978 standard for passenger cars was 18 miles per gallon (mpg). Averages gradually increased over the years, and by the 1990's the passenger car standard was 27.5 mpg, while the standard for light trucks (which included vans and sports utility vehicles) was 20.7 mpg.

During the late 1990's many environmentalists pressed for more stringent standards. They argued that improved fuel economy reduced the introduction of greenhouse gases and other harmful automobile by-products into the atmosphere. It also afforded protection to the Arctic National Wildlife Refuge and other wilderness areas in which the threat of oil drilling remained a possibility. In 1995 one group advocated an immediate increase to 45 mpg for passenger cars and 35 mpg for light trucks. These proposals met with little support from politicians or the general public. Opponents argued that CAFE standards had no impact on foreign oil imports, which had continued to rise after the regulations were enacted. After a period of concern during the Persian Gulf War in 1991, most consumers regarded the continuing decline in gasoline prices that marked the mid-1990's as evidence that the issue was not critical.

Critics also contended that the CAFE standards hurt the economy and endangered vehicle passengers. They claimed that the costs of creating vehicles with increased fuel efficiency were passed on to consumers in the form of higher vehicle prices, which was unfair to small businesses. More important, critics asserted that manufacturers achieved better performance by building smaller cars from lighter materials. Such vehicles provided less protection to passengers in the event of an accident. This argument gained ground in 1991 when the U.S. Department of Transportation released a study indicating that higher CAFE standards were directly related to increases in traffic injuries and fatalities.

Environmentalists rejected these claims, noting that higher vehicle costs were offset by savings in fuel expenses. As for the increase in traffic deaths, they maintained that automobile exhaust caused environmental damage that also resulted in deaths. They pointed to global warming and its consequences as proof that CAFE standards were necessary. However, these arguments made little impression on most politicians. Efforts to raise the CAFE standards during the 1990's consistently failed. As long as energy prices remained low, it appeared that increases in the CAFE standards would not garner significant popular support.

Thomas Clarkin

SEE ALSO: Automobile emissions; Energy conservation; Energy policy.

Jacques-Yves Cousteau wears an Aqua-Lung, which he invented, in a still from Louis Malle's film The Silent World. (Library of Congress)

Cousteau, Jacques-Yves

Born: June 11, 1910; Saint-André-de-Cubzac, France
Died: June 25, 1997; Paris, France
Category: Animals and endangered species

Jacques-Yves Cousteau was one of the twentieth century's best known explorers and conservationists. He also developed, with Emile Gagnan, the Aqua-Lung, the first commercially available self-contained underwater breathing apparatus (scuba).

Jacques-Yves Cousteau was, in his own mind, neither scientist nor adventurer. Rather, he considered himself a filmmaker. In 1971 Cousteau said, "I am not a scientist, I am rather an impresario of scientists." Cousteau, however, probably did recognize the vital role he played in exciting public interest in the ocean and revealing the intricacies of humanity's relationship to it.

Cousteau's long career began in 1933 with a stint in the French navy. In 1950 he bought the retired mine sweeper *Calypso* for use as a research ship. The many films and books Cousteau authored in his lifetime helped finance the *Calypso*'s expeditions, which brought the wonders of the oceans to an international audience. He is perhaps best known from his television series, *The Undersea World of Jacques-Yves Cousteau*, which aired from 1968 to 1976.

By 1956 Cousteau's research activities had become a full-time career, so he resigned from the navy. In 1957 Prince Rainier III named Cousteau director of the Oceanographic Museum of Monaco, a post he held for thirty-one years. During the 1950's and 1960's Cousteau sometimes explored the same sites several times. As a result of these trips, he eventually recognized that human activities were degrading the aquatic environment.

In 1958 Cousteau helped establish the world's first undersea marine reserve off the coast of Monaco. In 1960 he spoke out against France's plans to dispose of nuclear waste in the Mediterranean Sea. One decade later, he summed up his fears about ocean pollution by stating, "The oceans are in danger of dying."

In 1974 Cousteau founded the Cousteau Society, a nonprofit organization "dedicated to the protection and improvement of the quality of life for present and future generations." Among other achievements, this group has provided logistical support and facilities for hundreds of scientists, helped develop postgraduate environmental science and policy programs at universities worldwide, and worked toward United Nations (U.N.) adoption of a Bill of Rights for Future Generations—all causes to which Cousteau himself was strongly committed. In addition to revealing the ocean's majesty through books, films, and television, Cousteau strove to alert the public to the dilemma he recognized during his explorations:

> The pursuit of technology and progress may today endanger the very survival of . . . practically all life on earth. . . . [However] the technology that we use to abuse the planet is the same technology that can help us to heal it.

During his lifetime, Cousteau received numerous awards, honors, and titles for his environmental efforts, including the United Nations International Environmental Prize (1977), the United Nations Global 500 Roll of Honor (1988), Advisor for Environmental Sustainable Development to the World Bank, and membership on the United Nations High-Level Board on Sustainable Development (1992). In 1992 France's president appointed Cousteau chairperson of the newly created Council on the Rights of Future Generations. Three years later he resigned in protest over resumed nuclear weapons testing in the South Pacific.

Clayton D. Harris

SEE ALSO: Environmental education; Ocean pollution; Sustainable development.

Cross-Florida Barge Canal

CATEGORY: Preservation and wilderness issues

Originally designed as a shortcut for shipping across Florida that would link the Gulf of Mexico with the Atlantic Ocean, the Cross-Florida Barge Canal caused decades of controversy among environmentalists, government officials, and business interests until it was finally decommissioned in 1990 and converted into a state recreation and conservation area.

Florida's unique peninsular shape and its 6,115 kilometers (3,800 miles) of tidal shoreline have long frustrated military, industry, and shipping interests. Since no river cuts across the state, ships and barges have had to travel around Florida's southern tip. In earlier times, this trip was often a perilous undertaking because of the existence of many dangerous reefs, shoals, and turbulent storms.

Even before Florida became a state, various interest groups and individuals began calling for the creation of a transpeninsula waterway. Bowing to pressure, the Florida legislature created the Florida State Canal Commission in 1821 to explore the possibilities of building such a canal. Five years later, Congress adopted the cause and authorized the first of twenty-eight surveys that were carried out to find a convenient and safe route for an inland ship canal across the state.

One proposal after another either proved to be impractical or failed to gain political support. In 1935 President Franklin Roosevelt, intending to ease unemployment problems in Florida, used $5 million in federal relief money as startup funds for the construction of a ship canal. The proposal called for a 9-meter (30-foot) sea-level ship canal that would stretch across the north-central part of the state from the Atlantic Ocean to the Gulf of Mexico. On September 19, 1935, Roosevelt pressed a telegraph key from his office in Washington, D.C., and set off an explosive blast in Florida that officially began the construction.

Roosevelt's action received popular support across north-central Florida, a region where many communities welcomed the new jobs and economic boost the canal was expected to generate. However, there was also intense opposition to the canal. Railroad interests, fearing competition from shipping interests, lobbied in Congress and hotly objected to the canal construction. Many south Floridians also joined the protest, fearing that a ship canal would allow salt-water intrusion to jeopardize the state's downstream supply of underground water. Yielding to political pressure, Roosevelt cut off funding three years later and called on Congress to help. Congress, however, failed to appropriate any money for the project, and construction ground to a halt.

World War II rekindled interest in the canal. German submarine attacks against U.S. ships along the Florida coast prompted Congress to ask the Army Corps of Engineers to reexamine the canal to meet the nation's wartime needs. The corp responded with scaled-back plans for a barge canal, but Congress failed to provide funding, and the project stagnated for years. In 1964 Congress finally authorized $1 million to get construction underway and promised more funds later. The Army Corps of Engineers now had responsibility for the project. Its engineers came up with plans for a five-lock waterway that would stretch 296.8 kilometers (184.4 miles) from Port Inglis on Florida's west coast to the Intracoastal Waterway at the St. Johns River on the east.

Opposition to the project quickly developed. Environmentalists argued that the canal would disrupt the natural flow of rivers and creeks in the region, flood woodland areas, and destroy many endangered and threatened plants and animals. In 1969 the Environmental Defense Fund (EDF), along with the Florida Defenders of the Environment, sued in a U.S. District Court to stop construction. Nearly two years later, on January 15, 1971, the court granted the plaintiffs an injunction. Four days later, President Richard Nixon, citing environmental and economic concerns, issued a presidential order that suspended construction. By now, $74 million had been spent to build less than one-third of the canal. Great stretches of trees had been leveled, rivers and streams altered, two locks built, a dam constructed, and much earth moved.

In March, 1974, the Middle District Court of Florida overruled Nixon's action, but it upheld the injunction. Supporters of the canal received another blow in 1977 when both the Army Corps of Engineers and the Florida cabinet went on record calling for an end to construction. Despite these actions, canal proponents continued to lobby in Congress and succeeded in postponing a complete dismantling of the transwaterway project for years. Finally, in 1990, both the U.S. Congress and the Florida legislature officially and permanently deauthorized the barge canal.

Environmentalists hailed the defeat of the Cross-Florida Barge Canal as a major environmental victory, but they soon faced new problems. Debates arose over questions of what was to be done with completed sections of canal and the adjacent canal lands. In 1993 the Florida legislature resolved the issue when it authorized the conversion of 177 kilometers (110 miles) of the defunct canal zone into a huge nature preserve named the Cross-Florida Greenway State Recreation Area, or the Cross-Florida Greenway. In addition to providing hiking, canoeing, and horseback riding for humans, the 70,000-acre corridor also serves as a permanent wildlife refuge—one of the largest in the southern United States.

One controversial issue remained unresolved. It focused on the fate of the Rodman Dam—a reservoir built in the center of the state along the Ocklawaha River prior to the final decommission of the canal. Environmentalists argued that since the reservoir was no longer needed, it should be demolished so that the Ocklawaha River could be restored to its natural flow pattern. On the other hand, supporters of the dam—including fishers, fish camp owners, and local merchants—wanted to keep the reservoir in place, citing its economic benefits to the local community as a recreational bass fishing area. Even though Florida governor Lawton Chiles, the state cabinet, and the Department of Environmental Protection expressed support for restoration efforts of the Ocklawaha River and elimination of the Rodman Dam, the state legislature withheld the necessary funds.

John M. Dunn

SUGGESTED READINGS: A detailed history of the various attempts to create a ship canal in Florida from the 1820's to the 1970's appears in George E. Buker's *Sun, Sand, and Water: A History of the Jacksonville District U.S. Army Corps of Engineers, 1821-1975* (1980). Charlton W. Tebeau's *A History of Florida* (1981) is a classic rendering of Florida's history and includes a concise explanation of the barge canal controversy. *The Ocala Star Banner*, a Florida newspaper located in the canal vicinity, provides scores of articles covering a host of issues associated with the Cross-Florida Barge Canal. Its special twenty-page 1992 report, "Florida's Greenway: How to Create and Preserve It," provides a comprehensive look at the canal issues and the Florida Greenway. Lucy Tobias's "Cross-state Canal Long a Dream, Nightmare for Floridians" provides a succinct and informative summary of the canal controversy.

SEE ALSO: Dams and reservoirs; Wildlife refuges.

Cultural eutrophication

CATEGORY: Water and water pollution

Cultural eutrophication is an unwanted increase in nutrient concentrations in sensitive waters caused by human activities. Eutrophication causes the degradation of productive aquatic environments, which has prompted state and federal governments to regulate point- and non-point-source pollution in surrounding watersheds.

Eutrophication (from the Greek term meaning "to nourish") is the sudden enrichment of natural waters with excess nutrients, such as nitrogen, phosphorus, and potassium, which can lead to the development of algae blooms and other vegetation. In addition to clouding otherwise clear water, some algae and protozoa (namely *Pfiesteria*) release toxins that harm fish and other aquatic wildlife. When the algae die, their decomposition produces odorous compounds and depletes dissolved oxygen in these waters, which causes fish and other organisms to suffocate.

Eutrophication is a naturally occurring process as environments evolve over time. Cultural eutrophication is distinct because the process is accelerated by human activities such as wastewater treatment disposal, runoff from city streets and lawns, deforestation and development in watersheds, and agricultural activities such as farming and livestock production. These activities contribute excessive amounts of available nutrients to otherwise pristine waters and promote rapid and excessive plant growth.

Eutrophication of the Great Lakes, particularly Lake Erie, was one of the key factors that prompted passage of the Clean Water Act and various amendments during the 1970's. This act specifically addressed the disposal of sewage into public waters, a major contributor to cultural eutrophication. However, it did not specifically address non-point-source pollution, which comes from sources that are not readily identifiable. Agricultural activities such as farming, logging, and concentrated livestock operations all contribute to non-point-source pollution through fertilizer runoff, soil erosion, and poor waste disposal practices that supply readily available nutrients to surrounding watersheds and lead to eutrophication in these environments.

The Chesapeake Bay is an excellent case study in cultural eutrophication. As development surrounding the bay dramatically increased, wetland and riparian buffers that helped reduce some of the impact of additional nutrients were destroyed. Eutrophication in the bay during the 1980's threatened the crabbing and oyster industry. Consequently, in 1983 and 1987, Maryland, Pennsylvania, and Virginia, the three states bordering the Chesapeake Bay, agreed to a 40 percent reduction of nutrients by the year 2000 from point and nonpoint sources in all watersheds contributing to the bay. These reductions included such things as banning phosphate detergents, requiring management plans to control soil erosion, protecting wetlands, and putting controls on production and management of animal wastes. Although these steps reduced phosphorus levels in the Chesapeake Bay and kept nitrogen levels constant, regulators remained unsure how much nutrient reduction must take place for the bay and its surroundings to resemble their original condition.

Mark Coyne

SEE ALSO: Agricultural chemicals; Erosion and erosion control; Lake Erie; Runoff: agricultural; Sewage treatment and disposal; Watersheds and watershed management.

Cuyahoga River fires

DATE: 1959 and June 22, 1969
CATEGORY: Water and water pollution

In 1969 the oil-slicked Cuyahoga River caught fire near Cleveland, Ohio, as a result of waste discharged from waterfront industries. The fire, which demonstrated the poor environmental condition of the Great Lakes, turned into a major media event that served to sway public opinion toward supporting the cleanup of Lake Erie.

The Cuyahoga River divides the city of Cleveland into an east and west side. Originating on the Appalachian Plateau 56 kilometers (35 miles) east of Cleveland, it meanders 166 kilometers (103 miles) to Lake Erie. About 8 kilometers (5 miles) from its mouth, it becomes a sharply twisting but navigable stream that forms part of Cleveland's harbor. Industrial development took place along the river in the early nineteenth century, and by 1860 docks and warehouses lined the ship channel. Industry had claimed virtually all of Cleveland's riverfront by 1881 when, according to Cleveland mayor Rensselaer R. Herrick, the discharge from factories and oil refineries made it an open sewer running through the center of the city.

In 1951 the Ohio Department of Natural Resources reported that the Cuyahoga River was heavily polluted with industrial effluents at its mouth, creating conditions that were unsatisfactory for the existence of aquatic life. In September of that year, thousands of dead fish were

The Anthony J. Celebrezze *fireboat combats flames on the Cuyahoga River in June of 1969. The fire, which severely damaged two railroad bridges over the river, brought attention to the poor environmental condition of nearby Lake Erie.* (State of Ohio Environmental Protection Agency)

washed ashore just west of the river's mouth, and observers noted that the area gave off strong river odors.

An oil slick burned on the Cuyahoga River for eight days in 1959, causing an estimated $1.5 million damage, without attracting national attention. Ten years later, at approximately noon on Sunday, June 22, 1969, the Cuyahoga River again caught fire. The fire was brought under control by approximately 12:20 P.M., but not before it had done some $50,000 worth of damage to two key railroad trestles over the river in the Flats area of Cleveland. An oil slick on the river had caught fire and floated under the wooden bridges, setting fire to both. Witnesses reported that the flames from the bridges reached as high as a five-story building. The fireboat *Anthony J. Celebrezze* rushed upstream and battled the blaze on the water while units from three fire battalions brought the flames on the trestles under control. Responsibility for the oil slick was placed on the waterfront industries, which used the river as a dumping ground for oil wastes instead of reclaiming the waste products.

The railroad trestles that burned were not all that sustained damage in the fire. Cleveland's reputation as "the best location in the nation" was severely damaged by the occurrence. The city became the brunt of numerous jokes from comedians and the media across the country, which characterized it as the only city with a river so choked with pollution that it had burned. No river in the United States had a more notorious national reputation than the Cuyahoga River. Media coverage of the fire helped galvanize nationwide public support for efforts to clean up not only Lake Erie in particular but also the environment in general.

Charles E. Herdendorf

SEE ALSO: Hazardous and toxic substances regulation; Lake Erie; Oil spills; Water pollution.

D

Dams and reservoirs

CATEGORY: Preservation and wilderness issues

Dams are designed for a number of purposes, including conservation, irrigation, flood control, hydroelectric power generation, navigation, and recreation. However, since dams obstruct the natural flow of water, they have the potential to significantly affect stream and river ecosystems.

Dams are structures that obstruct the flow of water in streams or rivers. A reservoir is a body of water created by the impoundment of water behind or upstream of a dam. Not all dams create reservoirs of significant size. Low dams or barrages have been used to divert a portion of stream flow into canals or aqueducts since the first attempts at irrigation thousands of years ago. Canals, aqueducts, and pipelines are used to change the direction of water flow from a stream to agricultural fields or areas with high population concentrations.

Dams and reservoirs provide the chief, and in most cases the sole, means of storing stream flow over time. Small dams and reservoirs are capable of storing water for weeks or months, allowing water use during local dry seasons. Large dams and reservoirs have the capacity to store water for several years. As urban populations in arid regions have grown and irrigation agriculture has dramatically expanded, dams and reservoirs have increased in size in response to demand. They are frequently located hundreds of kilometers from where the water is eventually used. The construction of larger dams and reservoirs has resulted in increasingly complex environmental and social problems that have affected larger numbers of people. This has been particularly true in tropical and developing nations, where most of the large dam construction of the last three decades of the twentieth century was concentrated.

SIZE AND PURPOSE OF DAMS

Early dams and their associated reservoirs were small and remained small, for the most part, until the twentieth century. The first dams were simple barrages constructed across streams

Dams utilize the flow of water to provide electricity to nearby communities, but they also affect the environment in adverse ways. (Ben Klaffke)

to divert water into irrigation canals. Water supply for humans and animals undoubtedly benefited from these diversions, but the storage capacity of most dams was small, reflecting the limited technology of the period. Early dams were constructed five thousand years ago in the Middle East and became common two thousand years ago in the Mediterranean region, China, Central America, and South Asia.

The energy of falling water can be converted by water wheels into mechanical energy to perform a variety of tasks, including the grinding of grain. Dams create a higher "head" or water level, increasing the potential energy, and thus served as the earliest energy source for the beginnings of the Industrial Revolution during the nineteenth century. The most significant contribution of dams to industrialization was the development of hydroelectricity in 1882, which permitted energy to be transferred to wherever electric power lines were built, rather than being confined to river banks.

During the nineteenth century, large-scale settlement of the arid regions of western North America and Asia soon exhausted the meager local supplies of water and prompted demands for both exotic supplies from distant watersheds and storage for dry years. Big dams for storage and big projects for transportation of the water were thought to be the answer. Small dams and projects could be financed locally; grander schemes required the assistance of the federal or national governments. To justify expenditures on larger dams and projects, multiple uses for reservoir water were listed as benefits, offsetting the project's cost. Benefit-cost ratios thus became the tool by which potential projects were judged. In order to raise the ratio of benefits to costs, intangible benefits—those to which it is difficult to assign universally agreeable currency values—became far more important. While dams in arid regions were originally justified chiefly for irrigation, public water supply, and power, dams in wetter areas were usually based upon projected benefits from flood control, navigation, and recreation, in addition to power generation and public water supply.

Complicating the equation is the fact that multiple uses are frequently conflicting uses. While all dams are built to even out the uneven flow of streams over time, flood control requires an empty reservoir to handle the largest floods; conversely, power generation requires a high level of water in the reservoir to provide the highest head. Public water supply and navigation benefit most from supplies that are manipulated in response to variable demand. Recreation, fishing, and the increasingly important factor of environmental concerns focus upon in-stream uses of the water. By the last two decades of the twentieth century, environmental costs and benefits and the issue of American Indian water rights in the American West dominated decisions concerning dam projects in the United States, and few dams were constructed. Most of the best sites for the construction of large dams in the developed nations had been utilized, and the industry turned its attention to the developing nations. Most of the large projects of the last quarter of the twentieth century were constructed or proposed for developing nations and the area of the former Soviet Union.

Human Impact

Small dams have small impacts upon the environment; they affect small watersheds and minor tributaries and usually have only a single purpose. Farm ponds and "tanks," as they are known in many parts of the world, generally cover a fraction of 1 hectare in area and are only a few meters in height. These tiny ponds are designed to store water for livestock and occasionally for human supply. They frequently serve a recreational purpose as well, such as fishing. During dry spells they become stagnant and subject to contamination by algae and other noxious organisms, which can threaten the health of humans and livestock. Otherwise, they have little negative impact upon the environment or nearby people and animals.

Large dams and reservoirs are responsible for environmental and social impacts that often appear to be roughly related to their size: The larger the dam or reservoir, the greater the impact. Geographical location is also important in assessing a project's impact. Scenic areas in particular, or those with endangered species of plants or animals or irreplaceable cultural or

United States Dams Built During the Twentieth Century

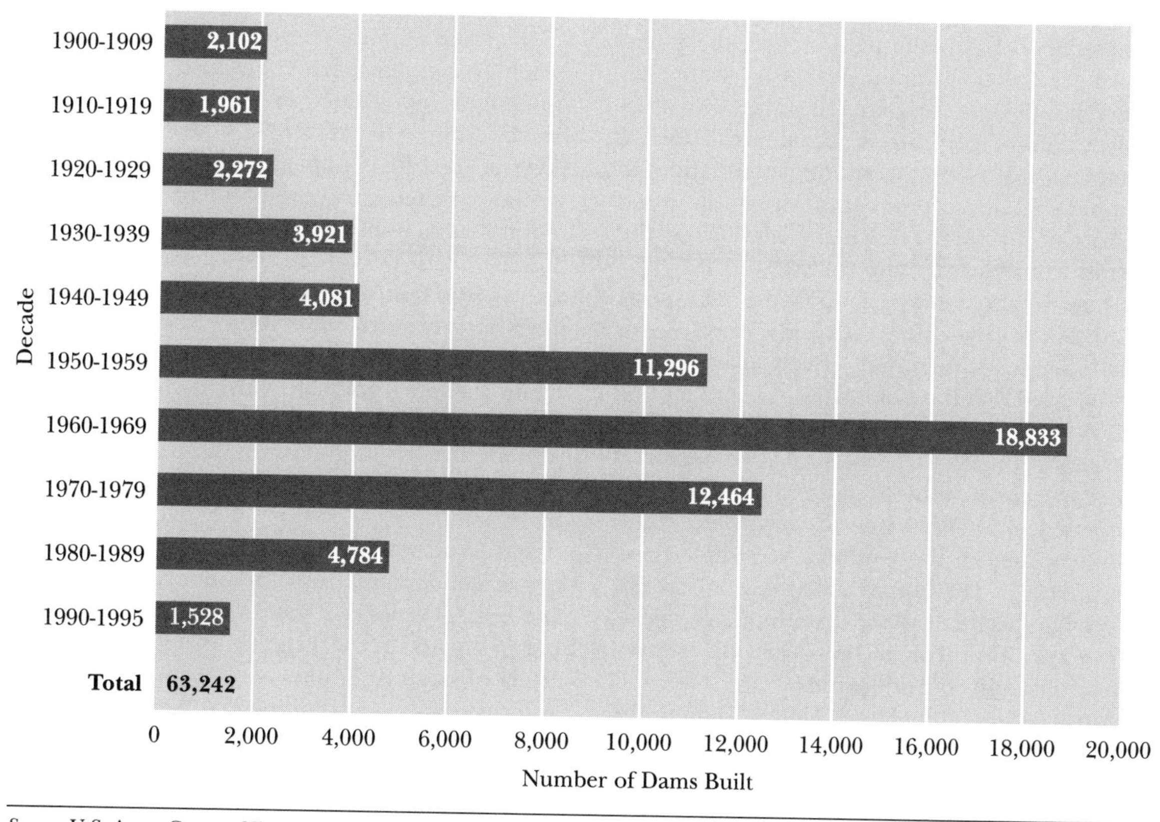

Source: U.S. Army Corps of Engineers National Inventory of Dams, 1996.

archeological features, raise more controversy and litigation if they are chosen as potential sites for dam and reservoir projects.

People in tropical regions suffer proportionately greater health-related impacts from dams than those in corresponding nontropical areas. The large number of workers required for the construction of big dams and associated irrigation projects carries disease into unprotected populations. Stagnant or slow-moving waters in reservoirs and irrigation canals, as well as fast-moving waters downstream of dams, are associated with particularly vicious tropical health risks. Snails in slow-moving water carry schistosomiasis, parasites that infect intestinal and urinary tracts, causing general listlessness and more serious consequences, including failure of internal organs and cancer. Estimates of the number of people infected range into the hundreds of millions. Malaria, lymphatic filariasis (including elephantiasis), and other diseases are carried by mosquitoes that breed in water; the incidence of such insect-borne diseases dramatically expands near irrigation projects and reservoirs. River blindness, which results from the bite of black flies, is associated with fast-flowing water downstream of dams and affects hundreds of thousands of humans.

The flooding of densely populated river valleys by reservoirs displaces greater numbers of people, with attendant health problems and social impacts, than similar projects in sparsely populated areas. Population displacement in developing countries, especially those in the tropics, causes greater health and social problems than in developed nations, where remedial measures and compensation are more likely to assuage the loss of homestead and community.

ENVIRONMENTAL IMPACT

Many of the environmental problems created by large dams are associated with rapid changes in water level below the dam or with the ponding of stream flow in the reservoir, which replaces fast-flowing, oxygenated water with relatively stagnant conditions. Indigenous animal species, as well as some plant life, are adapted to seasonal changes in the natural stream flow and cannot adjust to the postdam regime of the stream. Consequently, the survival of these species may be threatened. In 1973 the Tennessee Valley Authority—the worldwide model for many large, integrated river basin projects—found that the potential demise of a small fish, the snail darter, stood in the way of the completion of the Tellico Dam. After considerable controversy and litigation, the dam was completed in 1979, but few large projects have been proposed in the United States since then, particularly in the humid East. Since the early 1970's, the arguments for abandoning projects have been more likely to be backed up by laws, regulations, and court decisions.

Construction of the Hetch Hetchy Dam in the Sierra Nevada of California in the early twentieth century sparked vigorous dissent, which is said to have led to the growth of the Sierra Club and organized environmental opposition to dam building. This opposition successfully challenged the construction of the Echo Park Dam on the Green River in Colorado in the early 1950's but was unsuccessful in stopping the construction of the Glen Canyon Dam on the Colorado River, which was completed in 1963. Glen Canyon, however, was the last of the big dams constructed in the American West. The preservationists, whose arguments chiefly concerned scenic and wilderness values, with attendant benefits to endangered species, lost the Glen Canyon battle but won the war against big dams. The controversy surrounding the Glen Canyon Dam continued for more than three decades after its completion, pitting wilderness and scenic preservationists against powerboat recreationists, who benefit from the access accorded by its reservoir to the upstream canyonlands.

All reservoirs eventually fill with silt from upstream erosion; deltas form on their upstream ends. Heavier sediments, mainly sands, are trapped behind the dam and cannot progress downstream to the ocean. The Atlantic coastline of the southeastern United States suffers from beach erosion and retreat because the sands are no longer replenished by the natural flow of nearby rivers. The Aswan High Dam on the Nile River in Egypt has impacted the Nile Delta in a similar fashion. Moreover, the natural flow of sediments downstream historically replenishes the fertility of floodplain soils during floods. To the extent that the flood-control function of dams is successful, new fertile sediment never reaches downstream agricultural fields. While irrigation water provided by the dam may permit the expansion of cropland, water in arid regions is often highly charged with salts, which accumulate in the soils and eventually become toxic to plant life.

Reservoir waters release methane from decaying organic matter into the atmosphere. Methane is a greenhouse gas that promotes global warming, and some estimates suggest that the effect of large reservoirs is roughly equal to the greenhouse gas pollution of large thermal-powered electrical generation plants. The weight of the water in large reservoirs has also been implicated in causing earthquakes, which may lead to failure of the dam. Dam failure may also occur because of inadequate knowledge of the geology of the site or poor construction or design. Tens of thousands of lives have been lost as a consequence of such failures.

Neil E. Salisbury

SUGGESTED READINGS: Joseph E. Stevens, *Hoover Dam: An American Adventure* (1988), is an account of the planning and construction of one of the first major dams of the American West. Kenneth D. Frederick's chapter on water resources in *America's Renewable Resources* (1991) places dams in the total picture of water resource development in the United States and provides a succinct account of legislation affecting dam construction, including environmental constraints. A thorough overview of the scientific aspects of the impact of dams and reservoirs upon the environment is provided by Geoffrey E. Petts's *Impounded Rivers: Perspectives for Ecologi-*

cal Management (1984). The relation between social and environmental impacts of dams is documented in *The Social and Environmental Effects of Large Dams* (1984), by Edward Goldsmith and Nicholas Hildyard. This message is updated in Patrick McCully's *Silenced Rivers: The Ecology and Politics of Large Dams* (1996). The conflict between conservationist and preservationist viewpoints with respect to dam building is discussed in *TVA and the Tellico Dam: 1936-1979* (1986), by William Bruce Wheeler and Michael J. McDonald.

See also: Flood control; Hydroelectricity; Irrigation; Tennessee Valley Authority.

Darling, Jay

Born: October 21, 1876; Norwood, Michigan
Died: February 12, 1962; Des Moines, Iowa
Category: Animals and endangered species

Jay Darling was a cartoonist and wildlife conservationist who initiated the Federal Duck Stamp Program.

Jay Darling was an editorial cartoonist by profession. He was known among his peers as Ding, the name he penned to his drawings. His biting cartoons, which won him two Pulitzer Prizes and national recognition, often depicted the destruction of America's waterfowl and their habitat by overhunting and periodic drought—particularly during the Dust Bowl of the 1930's, which dried the wetlands that the birds required.

The passage of the Migratory Bird Conservation Act by the U.S. Congress in 1929 laid the groundwork for Darling's major contribution to waterfowl conservation. That act authorized the U.S. Department of Agriculture to acquire wetlands and preserve them as waterfowl habitat, but it provided no permanent source of funding for the purpose. In 1934 President Franklin D. Roosevelt appointed a committee to look into the need for waterfowl refuges. Among its members were Darling, who had been a fierce critic of Roosevelt's wildlife policies, wildlife conservationist Aldo Leopold, and publisher Thomas Beck. The committee's recommendation that $50 million be spent for new refuges rekindled an idea that had lain dormant for years: to require waterfowl hunters to buy duck stamps. The revenue generated from the stamps would be used to buy new refuge lands authorized by the Migratory Bird Conservation Act.

In March, 1934, Congress passed the Migratory Bird Hunting Stamp Act, requiring waterfowl hunters sixteen years or older to buy an annual duck stamp. That same month, Roosevelt appointed Darling chief of the Department of Agriculture's Bureau of Biological Survey, forerunner of the U.S. Fish and Wildlife Service. While serving as chief, Darling carried out the 1934 act's mandate by initiating the Federal Duck Stamp Program.

Darling designed the first duck stamp. It took some time for duck stamp revenue to start flowing, however, so Darling began raising money from other programs within the Department of Agriculture to purchase refuge land. He is credited with obtaining $20 million for wildlife conservation and setting aside 4.5 million acres as refuge land during his twenty-month tenure at the Bureau of Biological Survey. He resigned in November, 1935, dismayed by conservationists' lack of collective strength in focussing attention on wildlife issues.

Darling then helped to organize the United States' 36,000 wildlife societies into a national body called the General Wildlife Federation at the first government-sponsored North American Wildlife Conference in 1936. Darling was unanimously chosen president of the federation. Two years later, the group's name was changed to National Wildlife Federation, and Darling was reelected president.

In 1942 Darling was awarded the Theodore Roosevelt Medal for distinguished service in wildlife conservation and, in 1960, a National Audubon Society medal for distinguished service in natural resource conservation. A 4,975-acre wildlife refuge established on Sanibel Island, Florida, in 1945 was dedicated to Darling in 1978. Now known as the J. N. "Ding" Darling National Wildlife Refuge, it supports a wide diversity of birds and other animals.

Jane F. Hill

See also: Conservation; Fish and Wildlife

Service, U.S.; Wildlife management; Wildlife refuges.

Darwin, Charles

BORN: February 12, 1809; Shrewsbury, Shropshire, England
DIED: April 19, 1882; Downe, Kent, England
CATEGORY: Ecology and ecosystems

Darwin's theory of evolution through natural selection, the dominant paradigm of the biological sciences, underlies the study of ecosystems.

Charles Robert Darwin was born on February 12, 1809, the fifth of six children. His mother died when he was eight, leaving him in the care of his elder sisters. His father was a country doctor with a wide practice.

In 1825, Darwin was sent to Edinburgh to study medicine. He proved to be a poor student of anatomy, however, and he was sent to Christ's College, Cambridge, to prepare for the ministry. Though not a distinguished student, Darwin took an interest in natural science. He met John Stevens Henslow, a botany professor who encouraged his interest in natural history and helped to secure for him a position as naturalist aboard HMS *Beagle*, soon to depart on a five-year scientific expedition around the world. Darwin's experiences during the voyage from 1831 to 1836 were instrumental in shaping his theory of evolution.

The voyage took Darwin along the coast of South America. Darwin kept detailed journals in which he carefully observed differences among the South American flora and fauna, particularly on the Galapagos Islands. He would later draw on these extensive field observations to formulate his theory of natural selection.

After his return to London, Darwin began a study of coral reefs, and he became secretary to the Geological Society and a member of the Royal Society. He married his first cousin, Emma Wedgwood, in January, 1839.

Darwin worked for the next twenty years on his journals from the *Beagle*'s voyage, gathering information to support his theory of evolution through natural selection. His preliminary work might have continued indefinitely if he had not received on June 18, 1854, an essay from Alfred Russel Wallace, a field naturalist in the Malay archipelago, outlining a theory of evolution and natural selection similar to his own. Darwin immediately wrote to his friends, Sir Charles Lyell and Joseph Hooker, explaining his dilemma and including an abstract of his own theory of evolution. Lyell and Hooker proposed that in order to avoid the question of precedence, the two papers should be presented simultaneously. Both were read before a meeting of the Linnean Society in Dublin on July 1, 1858, and were published together in the society's journal that year.

Darwin then began writing an abstract of his

Charles Darwin introduced his theory of evolution through natural selection to the general public in On the Origin of Species. *The theory became a key element in later ecosystem studies.* (Library of Congress)

theory, which he entitled *On the Origin of Species.* All 1,250 copies sold out on the first day of publication in London on November 24, 1859. Darwin argued that since all species produce more offspring than can possibly survive, and since species populations remain relatively constant, there must be some mechanism working in nature to eliminate the unfit. Variations are randomly introduced in nature, some of which will permit a species to adapt better to its environment. These advantageous adaptations are passed on to the offspring, giving them an advantage for survival. Darwin did not understand the genetic mechanisms by which offspring inherit adaptations. It would take another seventy years before the forgotten work of the Austrian geneticist Gregor Mendel was rediscovered and integrated with Darwin's theories to provide a fuller view of the evolutionary process.

Darwin was surrounded by a storm of controversy after the publication. Objection came both from orthodox clergy and unconvinced scientists. For the rest of his life, Darwin worked at home on successive editions of *On the Origin of Species*, further studies on plants and animals, and his famous *The Descent of Man and Selection in Relation to Sex* (1871). He died on April 19, 1882, and was buried with full honors in the scientists' corner at Westminster Abbey, next to Sir Isaac Newton.

Darwin has had an immeasurable influence on the development of modern biology, ecology, morphology, embryology, and paleontology. His theory of evolution established a natural history of the earth and enabled humans to see themselves for the first time as part of the natural order of life. A lively debate continues among scientists about revisionist theories of evolution, including Stephen Jay Gould's notion of "punctuated equilibria," or sudden and dramatic evolutionary changes followed by long periods of relative stability. While they disagree about details, however, modern biologists agree that neoevolutionary theory remains the only viable scientific explanation for the diversity of life on earth.

Alexander Scott

See also: Biodiversity; Extinctions and species loss.

Debt-for-nature swaps

Category: Preservation and wilderness issues

Debt-for-nature swaps are a strategy for reducing foreign debt in developing nations by trading debt forgiveness or debt reduction for guarantees of environmental activities by debtor nations.

An international finance crisis began during the late 1980's when many developing nations found that they had borrowed more from international lending institutions, mostly private banks, than they could repay. To recover some of the principle on the loans, banks began to sell the loans in financial markets, usually discounted to a fraction of their principle value because of the threat of default. Several options to relieve this debt burden on developing nations were explored. The debt could be refinanced to lower the interest rates and extend the time for repayment. Another strategy was to encourage domestic financial reforms by increasing domestic investment, expanding the domestic economy, raising taxes, reducing nondebt-related expenditures, or inflating the currency in order to eventually pay the debt. Finally, creditor nations and institutions could partially forgive the debt. Debt-for-nature swaps combine all three methods.

In debt-for-nature swaps, environmental organizations buy discounted debt in the financial markets from the banks. Instead of collecting the full amount of interest and principle from the debtor nations, environmental organizations agree to forgive all or a portion of the debt if debtor nations invest an amount up to the principle value of the debt in local preservation efforts, often the purchase of land for national parks or investment in skilled staff and improvements for existing parks. Conservation organizations benefit through an increase in local funding for conservation. Banks benefit because they have a new market for the debt. Debtor nations benefit because they are able to invest their funds in their own nation rather than transferring funds to the lending nations; use inflated local currency rather than high value, scarce, dollar-based foreign exchange; and reduce their outstanding debt and interest payments on that debt, thus allowing them to continue making payments on the remaining debt and maintain their international credit ratings.

Debt-for-nature swaps erase the "debt overhang," that portion of debt that, if forgiven, allows the remaining debt to continue to be assumed by creditor nations or institutions with satisfactory levels of burden on the debtor nations and satisfactory debt payment risks for the creditor nations. Because environmental organizations purchase the debt at a discount, they are able to multiply their impact on the environment. For example, in the first debt-for-nature swap in 1987, Conservation International purchased $650,000 in Bolivian debt for $100,000, then required the Bolivian government to establish a $250,000 endowment fund in local Bolivian currency to pay operating costs for a biosphere reserve in the Bolivian Amazon before erasing the debt.

In debt-for-nature proposals, debtor nations buy back a portion of their outstanding debt with an investment in a portion of their own natural capital. Natural capital includes caches of nonrenewable resources such as mineral or oil reserves, natural resources such as old-growth forests or endangered species habitat, historical artifacts such as prehistoric ruins or fossil deposits, or cultural resources such as the homelands of primitive indigenous peoples or significant architectural sites. The preservation of this natural capital has positive benefits for the ecology, sustainable development, and biodiversity. The swap is also likely to improve the debtor nation's economic ability to repay the remaining debt, since the preserved cultural and natural resources often serve as tourist attractions, cultural centers, and locations for academic and commercial research.

Since some private lending institutions will not voluntarily participate in a debt relief process that reduces the financial institution's capital or potential profit from loans, governments in creditor nations either mandate the financial institution's participation or provide financial incentives, such as tax deductions and tax credits, to encourage participation. Creditor governments justify their action as a component of their foreign economic development programs, as a component of their environmental programs, as philanthropic support for preservation of the earth's cultural and natural heritage, or as economically justifiable domestic self-interest. For example, encouraging nations in tropical zones to protect rain forests ensures nations in the northern temperate zones that existing climate patterns supported by those rain forests will be maintained. Maintaining these climate patterns is necessary to prevent natural and human-made disasters, reduce the demands on industrialized nations to provide humanitarian relief from increased numbers of natural disasters, ensure continued rainfall in agricultural zones in temperate regions, prevent desertification, maintain air quality and reduce the greenhouse effect, ensure global biodiversity, and improve the living conditions for every person on the planet.

Some nations, including Costa Rica, welcome the opportunities presented by the swaps. Other nations, including Brazil, see the swaps as a form of environmental imperialism in which foreign environmental organizations shape domestic government policy. Placing lands into parks and reserves reduces the remaining acres available for economic production and access by poor subsistence farmers.

Gordon Neal Diem

SUGGESTED READINGS: Gerald M. Meier, *The International Environment for Business* (1998), discusses the basic principles of international trade and finance and the interplay between finance and public policy. Jeffrey Sachs, "Making the Brady Plan Work," *Foreign Affairs* (Summer, 1989), argues for increases in debt forgiveness programs. Diana Page, "Debt-for-Nature Swaps: Experience Gained, Lessons Learned," *International Environmental Affairs* (Fall, 1989), reviews the first programs established in Bolivia, Ecuador, the Philippines, and Costa Rica. The article "Swapping Debt for Nature," *The Nature Conservancy* (July, 1993), describes the Nature Conservancy's program.

SEE ALSO: Biosphere reserves; Ecotourism; Environmental economics; National parks; Rain forests and rain forest destruction; World Heritage Convention.

Deep ecology

CATEGORY: Philosophy and ethics

Deep ecology is a school of environmental philosophy based on environmental activism and ecological spirituality. The term was first used by Norwegian philosopher Arne Naess in 1972 to suggest the need to go beyond the anthropocentric view that nature is merely a resource for human use.

The term "deep ecology" has been used in three major ways. First, it refers to a commitment to deep questioning about environmental ethics and the causes of environmental problems. Such questioning leads to critical reflection on the fundamental worldviews that underlie specific environmental ideas and practices. Second, deep ecology refers to a platform of generally agreed upon values that a variety of environmental activists share. These values include an affirmation of the intrinsic value of nature, the recognition of the importance of biodiversity, a call for a reduction of human impact on the natural world, greater concern with quality of life rather than material affluence, and a commitment to changing economic policies and the dominant view of nature. Third, deep ecology refers to particular philosophies of nature that tend to emphasize the value of nature as a whole (ecocentrism), an identification of the self with the natural world, and an intuitive and sensuous communion with the earth.

Because of its emphasis on fundamental worldviews, deep ecology is often associated with non-Western spiritual traditions such as Buddhism and Native American cultures, as well as radical Western philosophers such as Baruch Spinoza and Martin Heidegger. It has also drawn on the nature writing of Henry David Thoreau, John Muir, Robinson Jeffers, and Gary Snyder. Deep ecology's holistic tendencies have led to associations with the Gaia hypothesis, and its emphasis on diversity and intimacy with nature has linked it to bioregionalism. Deep ecological views have also had a strong impact on environmental activism, including the Earth First! movement.

Deep ecologists have sometimes criticized the animal rights perspective for continuing the traditional Western emphasis on individuals while neglecting whole systems, as well as for a revised speciesism that still values certain parts of nature (animals) over others. Some deep ecologists have also been critical of mainstream environmental organizations such as the Sierra Club for not confronting the root causes of environmental degradation.

On the other hand, deep ecology has been criticized by ecofeminists for failing to consider gender differences in the experience of the self and nature, the lack of an analysis of the tie between the oppression of women and nature, and promoting a holism that supposedly disregards the reality and value of individuals and their relationships. Social ecologists have criticized deep ecology for a failure to critique the relationship between environmental destruction on the one hand and social structure and political ideology on the other. In addition, a distrust of human interference with nature has led some thinkers to present the ideal as pristine wilderness with no human presence. In rare and extreme cases, deep ecologists have implied a misanthropic attitude. In some instances, especially early writings by deep ecologists, such criticisms have considerable force. However, these problematic views are not essential to deep ecology, and a number of thinkers have developed a broadened view that overlaps with ecofeminism and social ecology.

David Landis Barnhill

SEE ALSO: Earth First!; Ecofeminism; Environmental ethics; Naess, Arne; Social ecology; Speciesism.

Deforestation

CATEGORY: Forests and plants

Deforestation is the loss of forestlands through encroachment by agriculture, industrial development, or nonsustainable commercial forestry. Concerns about deforestation, particularly in tropical regions, have risen as the role that tropi-

cal forests play in moderating global climate has become better understood.

Radical environmental activists have long decried the apparent accelerating pace of deforestation in the twentieth century because of the potential loss of wildlife and plant habitat and the negative effects on biodiversity. By the 1990's research by mainstream scientists had confirmed that deforestation was indeed occurring on a global scale and that it posed a serious threat to global ecology.

Deforestation as a result of expansion of agricultural lands or nonsustainable timber harvesting has occurred in many regions of the world at different periods in history. The Bible, for example, refers to the cedars of Lebanon. Lebanon, like many of the countries bordering the Mediterranean Sea, was thickly forested several thousand years ago. A growing population, overharvesting, and the introduction of grazing animals such as sheep and goats decimated the forests, which never recovered.

Similarly, the forests of Europe and North America have shifted in total acreage as human populations have changed over the centuries. When the European colonists arrived in the New World, they immediately began clearing the forests. Trees were harvested for building materials and export back to Europe or were simply felled and burnt to clear space for farming. In North America, however, as agriculture became increasingly mechanized and farming shifted to the prairies, abandoned farms reverted to woodland. Environmental historians believe, in fact, that a greater percentage of land area in North America is now forested than was covered with trees prior to the arrival of European colonists. This is becoming true in many northern European countries, too, as their populations become increasingly urbanized.

As the European industrialized nations have gained forestland, however, the less developed countries in Latin America, Asia, and Africa have lost woodlands. While some of this deforestation is caused by a demand for tropical hardwoods for lumber or pulp, the leading cause of deforestation in the twentieth century, as it was several hundred years ago, is the expansion of agriculture. The growing demand by the industrialized world for agricultural products such as beef has led to millions of acres of forestland being bulldozed or burnt to create pastures for cattle. Researchers in Central America have watched with dismay as large beef-raising operations have expanded into fragile ecosystems in countries such as Costa Rica, Guatemala, and Mexico.

A tragic irony in this expansion of agriculture into tropical rain forests is that the soil underlying the trees is often unsuited for pastureland or raising other crops. Exposed to sunlight, the soil is quickly depleted of nutrients and often hardens. The once-verdant land becomes an arid desert prone to erosion that may never return to forest. As the soil becomes less fertile, thorny weeds begin to choke out the desirable forage plants, and the cattle ranchers move on to clear a fresh tract.

Slash-and-Burn Agriculture and Logging

Apologists for the beef industry often argue that their ranching practices are simply a form of slash-and-burn agriculture and do no permanent harm. It is true that many of the indigenous peoples in tropical regions have practiced slash-and-burn agriculture for millennia with only a minimal impact on the environment. These farmers burn the understory, or low-growing shrubs and trees, to clear small plots of land. Any large trees that survive the fire are cut down with axes and then burnt.

Anthropological studies have shown that the small plots these peasant farmers clear can usually be measured in square feet, not hectares like cattle ranches, and are used for five to ten years. As fertility declines, the farmer then clears a small plot next to the depleted one. The farmer's family or village will gradually rotate through the forest, clearing small plots and using them for a few years, and then shifting to new ground, until they eventually come back to where they began one hundred or more years before. As long as the size of the plots cleared by peasant farmers remains small in proportion to the forest overall, slash-and-burn agriculture does not contribute significantly to deforestation. If the population of farmers grows, however, and more land must be cleared with each succeeding generation, as has been happening in many tropical countries, then even traditional slash-and-burn agriculture can be as ecologically devastating as the more mechanized cattle ranching operations.

Although logging is not the leading cause of deforestation, it remains a significant factor. Tropical forests are rarely clear-cut, as they typically contain hundreds of different species of trees, most of which may have no commercial value. Loggers may select only a few trees for harvesting from each stand. Selective harvesting is a standard practice in sustainable forestry. However, just as loggers engaged in the disreputable practice of high-grading across North America in the nineteenth century, so are loggers high-grading in the late twentieth century in Malaysia, Indonesia, and other tropical forests. High-grading is a practice in which loggers cut over a tract to remove the most valuable timber while ignoring the damage being done to the residual stand. The assumption is that, having logged over the tract once, the timber com-

Population Increase, Deforestation, and Results

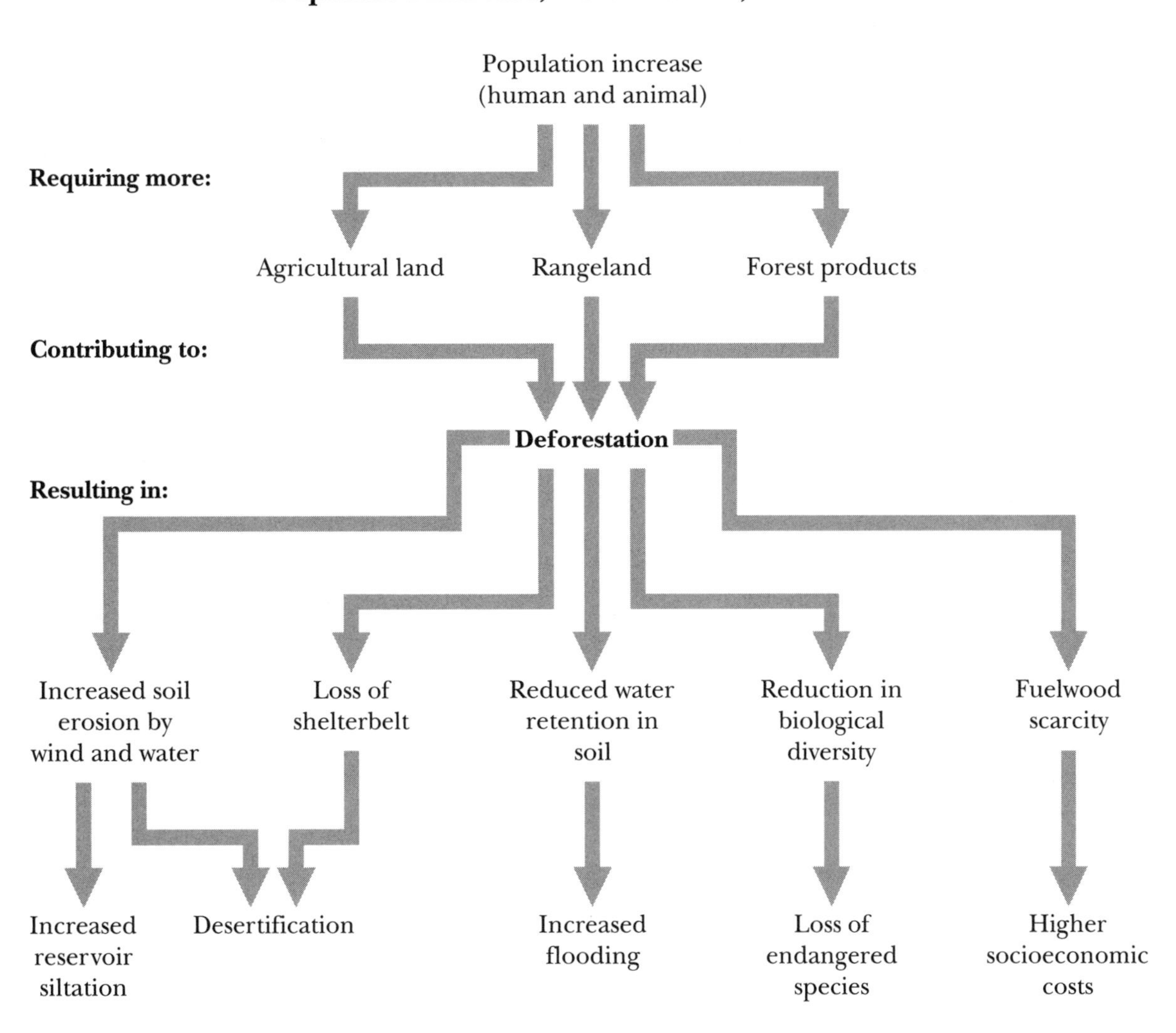

Source: Adapted from A. K. Biswas, "Environmental Concerns in Pakistan, with Special Reference to Water and Forests," in *Environmental Conservation,* 1987.

Loss of Tropical Rain Forest, 1980 and 1990

Region	Tropical Rain Forest Lost (millions of hectares)	
	1980	1990
Africa	569	524
Asia and Pacific	350	310
Latin America	992	918
Total	1,911	1,752

Source: United Nations Food and Agriculture Organization.

pany will not be coming back. This practice stopped in North America not because the timber companies voluntarily recognized the ecological damage they were doing but because they ran out of easily accessible, old-growth timber to cut. Fear of a timber famine caused logging companies to begin forest plantations and practice sustainable forestry. While global satellite photos indicate significant deforestation has occurred in tropical areas, enough easily harvested old-growth forest remains in some areas that there is no economic incentive for timber companies to switch to sustainable forestry.

Logging may also contribute to deforestation by making it easier for agriculture to encroach on forestlands. The logging company builds roads for use while harvesting trees. Those roads are then used by farmers and ranchers to move into the logged tracts, where they clear whatever trees the loggers have left.

Environmental Impacts of Deforestation

Despite clear evidence that deforestation is accelerating, the extent of the problem remains debatable. The United Nations Food and Agriculture Organization (FAO), which monitors deforestation worldwide, bases its statistics on measurements taken from satellite images. These data indicate that in the ten years between 1980 and 1990 at least 159 million hectares of land became deforested. These data also reveal that, in contrast to the intense focus on Latin America by both activists and scientists, the most dramatic loss of forestlands occurred in Asia and the Pacific. The deforestation rate in Latin America was 7.45 percent, while in Asia 11.42 percent of the forests vanished. Environmental activists are particularly concerned about forest losses in Indonesia and Malaysia, two countries where timber companies have been accused of abusing or exploiting native peoples in addition to engaging in environmentally damaging harvesting methods.

Researchers outside the United Nations have challenged the FAO's data, with some scientists claiming the numbers are much too high while others provide convincing evidence that, if anything, the FAO numbers are too low. Few researchers, however, have tried to claim that deforestation on a global scale is not happening. In the 1990's the reforestation of the Northern Hemisphere, while providing an encouraging example that it is possible to reverse deforestation, was not enough to offset the depletion of forestland in tropical areas. The debate among forestry experts centers on whether deforestation has slowed, and, if so, by how much.

Deforestation affects the environment in a multitude of ways. The most obvious is in a loss of biodiversity. When an ecosystem is radically altered through deforestation, the trees are not the only thing to disappear. Wildlife decreases in number and in variety, and other plants also die. As forest habitat shrinks through deforestation, many plants and animals become vulnerable to extinction. Many biologists believe that numerous animals and plants native to tropical forests will become extinct from deforestation before humans have a chance to become aware of their existence.

Other effects of deforestation may be less obvious. Deforestation can lead to increased flooding during the rainy season. Rain water that once would have been slowed or absorbed by trees instead runs off denuded hillsides, pushing rivers over their banks and causing devastating floods downstream. The role of forests in regulating water has long been recognized by both engineers and foresters. Flood control was, in fact, one of the motivations behind the creation of the federal forest reserves in the United States during the nineteenth century. More recently, disastrous floods in Bangladesh have been blamed on logging tropical hardwoods in the mountains of Nepal and India.

Conversely, trees can also help mitigate against drought. Like all plants, trees release water into the atmosphere through the process of transpiration. As the world's forests shrink in total acreage, fewer greenhouse gases such as carbon dioxide will be removed from the atmosphere, less oxygen and water will be released into it, and the world will become a hotter, dryer place. Scientists and policy analysts alike are in agreement that deforestation is a major threat to the environment. The question is whether effective policies can be developed to reverse it, or if short-term economic greed will win out over long-term global survival.

Nancy Farm Männikkö

SUGGESTED READINGS: Leslie E. Sponsel, Robert Converse Bailey, and Thomas N. Headland, editors, *Tropical Deforestation: The Human Dimension* (1996), is an anthology that shows how deforestation affects native populations. John F. Richards and Richard P. Tucker, editors, *World Deforestation in the Twentieth Century* (1990), is one of the few books to discuss the simultaneous reforestation of many industrialized nations and deforestation of developing nations. Thomas K. Rudel and Bruce Horowitz, *Tropical Deforestation: Small Farmers and Land Clearing in the Ecuadorian Amazon* (1994), provides a case study that gives an in-depth analysis of a specific area between 1920 and 1990. Warren Dean's *With Broadax and Firebrand: The Destruction of the Brazilian Atlantic Forest* (1997) is a solidly researched study of the destruction of one of the world's most endangered forests. William W. Bevis's *Borneo Log: The Struggle for Sarawak's Forests* (1995) is both disturbing and enlightening as it describes the exploitation of Third World resources by industrialized nations. Marcus Colchester and Larry Lohmann, editors, *Struggle for Land and the Fate of the Forests* (1993), presents case studies from around the world.

SEE ALSO: Logging and clear-cutting; Rain forests and rain forest destruction; Slash-and-burn agriculture; Sustainable forestry.

Desalination

CATEGORY: Water and water pollution

Since 97 percent of all water on earth is too salty to drink or use for plants, the removal of dissolved salts from seawater to provide potable water would theoretically provide an unlimited supply of fresh water. However, the high energy costs of desalination have restricted its use to arid countries with abundant energy resources.

Attempts to make fresh water from seawater go back several thousand years. For example, the ancient Greeks and Romans may have used simple evaporation and condensation of heated seawater for emergency water supplies. The British navy experimented with desalination on sailing vessels as early as the late seventeenth century. Ships used desalination when steam propulsion became common during the nineteenth century. The twentieth century witnessed the spread of the process to many countries, particularly those in the drier parts of the world.

Desalting treatment is now applied to other forms of water besides seawater, such as fresh water, brackish water, and wastewater. The total number of desalination plants in the world is estimated to be about ten thousand, of which 70 percent are in the Middle East. The total capacity of the plants is estimated to be 18 million cubic meters per day (4,752 million gallons per day), of which 65 percent process seawater.

Desalination methods include thermal (evaporation) and membrane (reverse osmosis, nanofiltration, and electrodialysis). The particular method selected depends on the amount of salts in the water and the source of energy. The salt content of seawater and brackish water is about 3.5 and 0.5 percent, respectively. Distillation and evaporation are better suited for high-salinity water (seawater), whereas membrane methods are generally better for low-salinity water (brackish).

Multistage flash distillation (MSF) and reverse osmosis (RO) account for 60 and 25 percent of the desalination plants, respectively. In the MSF process, saline water is evaporated, and fresh water is obtained by condensation. The RO method uses membranes that are selectively permeable to water rather than salts. By increasing the pressure to as high as 105.5 kilograms per square centimeter (1,500 pounds per square inch), fresh water is pushed through the membrane as salts are left behind.

Although the technology is well developed, high energy costs restrict major desalination efforts to arid countries that have abundant energy resources, such as Saudi Arabia, the United Arab Emirates, and Kuwait. Another factor that makes desalination a less viable option is the high expense of pumping fresh water to inland areas, which may be some distance from the coast and also at a higher elevation. Consequently, desalination is able to furnish only a fraction of the total world need for fresh water. However, there are many locations, including inland ones, where local sources of groundwater are brackish. Thus, widespread application of desalting will probably occur first with brackish water, which has considerably less salt than seawater.

Robert M. Hordon

SEE ALSO: Drinking water; Environmental engineering; Water quality; Water treatment.

Desertification

CATEGORY: Land and land use

Desertification is the degradation of arid, semiarid, and dry, subhumid lands as a result of human activities and climatic variations, such as a prolonged drought. Desertification is recognized by scientists and policymakers as a major economic, social, and environmental problem that afflicts more than one hundred countries and impacts about one billion people throughout the world.

Deserts are climatic regions that receive fewer than 25 centimeters (10 inches) of precipitation per year. They constitute the most widespread of all climates of the world and occupy 25 percent of the world's land area. Most deserts are surrounded by semiarid climates referred to as steppes, which occupy 8 percent of the world's lands. Deserts occur in the interior of continents, on the leeward side of mountains, and along the west side of continents in subtropical regions. All of the world's deserts risk further desertification.

Scientists use various methods to determine the historical climatic conditions of a region. These methods include studies of the historical distribution of trees and shrubs determined by the deposit patterns in lakes and bogs, patterns of ancient sand dunes, changes in lake levels through time, archaeological records, and tree rings (dendrochronology).

The largest deserts occur in North Africa, Asia, Australia, and North America. Four thousand to six thousand years ago, these desert areas were less extensive and were occupied by prairie or savanna grasslands. Rock paintings found in the Sahara show that humans during this era hunted buffalo and raised cattle on grasslands where giraffes browsed. The region near the Tigris and Euphrates Rivers was also fertile. In the desert of northwest India, cattle and goats were grazed, and people lived in cities that have long since been abandoned. The deserts in the southwestern region of North America appear to have been wetter, according to the study of tree rings from this area. Ancient Palestine, which includes the Negev Desert of present-day Israel, was lush and was occupied by three million people.

The United Nations claims that future desertification will claim a combined area the size of the United States, the former Soviet Union, and

Human activity threatens to transform once-fertile areas into sandy expanses of land unsuitable for agriculture or grazing. (McCrea Adams)

Australia. The earth's creeping deserts supported approximately 720 million people, or one-sixth of the world's population, in the late 1970's. It has been stated that "the forests came before civilization, the deserts after." Climate has created the deserts, but humankind has aided their growth across the grasslands of the steppe and savanna climatic boundaries.

According to the United Nations, the world's hyperarid or extreme deserts are the Atacama and Peruvian Deserts (located along the west coast of South America), the Sonoran Desert of North America, the Takla Makan Desert of central Asia, the Arabian Desert of Saudi Arabia, and the Sahara Desert of North Africa, which is the largest desert in the world. The arid zones surround the extreme desert zones, and the semiarid zones surround the arid zones. Areas having a high risk of becoming desert surround the semiarid zones. By the late 1980's the expanding deserts were claiming about 15 million acres of land per year, or an area approximately the size of the state of West Virginia. The total area threatened by desertification equaled about 37.5 million square kilometers (14.5 million square miles).

Causes of Desertification

Desertification results from a two-prong process: climatic variations and human activities. First, the major deserts of the world are located in areas of high atmospheric pressure, which experience subsiding dry air unfavorable to precipitation. Subtropical deserts have been experiencing prolonged periods of drought since the late 1960's, which causes these areas to be dryer than usual.

The problem of desertification came to the attention of the world in the late 1960's and early 1970's as a result of severe drought in the Sahel Desert, which extends in an east-west direction along the southern margin of the Sahara in West Africa. Rainfall has declined an average of 30 percent in the Sahel, and scientific research has been conducted to study the natural mechanisms that are causing the drought. One set of studies is related to changes in global circulation patterns associated with changes in the heat distribution in the ocean. A correlation has been found between sea surface temperatures and the reduction of rainfall in the Sahel. It has been determined that the Atlantic

Ocean's higher surface temperatures south of the equator and lower temperatures north of the equator west of Africa are associated with lower precipitation in northern tropical Africa. However, the cause for the change in sea surface temperatures pattern has not been determined.

Another set of studies is associated with land-cover changes, such as desertification. A lack of rain causes the ground and soils to get extremely dry, which causes the thin soil to blow away. As the water table drops from the lack of the natural recharge of the aquifers and the withdrawal of water by the inhabitants of the desert, inhabitants are forced to migrate to the grasslands and forests at fringes of the desert. Overgrazing, overcultivation, deforestation, and poor irrigation practices (which can cause salinization of soils) eventually lead to a repetition of the process, and the desert begins to encroach. These causes are influenced by such factors as changes in population, climate, and social and economic conditions.

The fundamental cause of desertification, therefore, is human activity. This is especially true when environmental stress occurs because of seasonal dryness, drought, or high winds. Many different forms of social, economic, and political pressure can cause the overutilization of these dry lands. People may be "marginalized" and pushed onto unsuitable agricultural land because of land shortages, poverty, and other uncontrollable forces, while farmers overcultivate the fields in the few remaining fertile land areas.

Consequences of Desertification

A reduction in vegetation cover and soil quality may impact the local climate by causing a rise in temperatures and a reduction in moisture. This can, in turn, impact the area beyond the desert by causing changes in the climate and atmospheric patterns of the region. It is predicted that by the year 2050 substantial vegetation cover changes in humid and subhumid areas will occur and cause substantial regional climatic changes. Desertification is a global problem because it can cause the loss of vegetation and animal diversity, as well as the pollution of rivers, lakes, and oceans. As a result of excessive rainfall and flooding in subhumid areas, fields lacking sufficient vegetation may be eroded by runoff.

Studies have been conducted to determine how rising greenhouse gas levels will impact the rate of desertification. Desertification and even the efforts to combat it may be impacting climatic change because of the emission and absorption of greenhouse gases. The decline in vegetation and soil quality can result in the release of carbon, while revegetation can influence the absorption of carbon from the atmosphere. The use of fertilizer to reclaim dry lands may cause an increase in nitrous oxide emissions. However, scientists involved in these studies have not been able to gather evidence conclusive enough to support such theories.

As a result of the Sahelian drought, which lasted from 1968 to 1973, representatives from various countries met in Nairobi, Kenya, in 1977 for a United Nations conference on desertification. The conference resulted in the Plan of Action to Combat Desertification. The plan listed twenty-eight measures to combat land degradation by national, regional, and international organizations. A lack of adequate funding and commitment by governments caused the plan to fail. When the plan was assessed by the United Nations Environment Programme (UNEP), it found that little had been accomplished and that the desertification problem had worsened.

As a result of the 1977 United Nations conference, several countries developed national plans of action to combat desertification. One example is Kenya, where local organizations have worked with primary schools to plant five thousand to ten thousand seedlings per year. One U.S.-based organization promotes reforestation by providing materials to establish nurseries, training programs, and extension services. Community efforts to combat desertification have been more successful, and UNEP has recognized that such projects have a greater success rate than top-down projects. The Earth Summit, held in Rio de Janeiro, Brazil, in 1992, supported the concept of sustainable development at the community level to combat the problem of desertification.

Roberto Garza

SUGGESTED READINGS: A thorough overview of desertification is provided in *Desertification: Environmental Degradation in and Around Arid Lands* (1977), edited by Michael H. Glantz. A short book on the causes of desertification and how to stop it is Alan Grainger's *Desertification: How People Make Deserts, Why People Can Stop, and Why They Don't* (1982). Reid A. Bryson and Thomas J. Murray's *Climates of Hunger: Mankind and the World's Changing Weather* (1977) provides an overview of climatic change and how it has impacted humankind. "Land's End," *Worldwatch* (May-June, 1989), by Sandra Postel, discusses small-scale efforts to curtail desertification. "Exploring the Links Between Desertification and Climate Change," *Environment* (July-August 1993), by Mike Hulme and Mick Kelly, discusses the link between desertification and global warming. "What's Happening to Our Climate," *National Geographic* (November, 1976), by Samuel W. Matthews, discusses climatic changes for the last 850,000 years. "The Desert: An Age-old Challenge Grows," *National Geographic* (November, 1979), discusses the expanding deserts of the world.

SEE ALSO: Climate change and global warming; Deforestation; Grazing; Soil, salinization of.

Detoxification

CATEGORY: Human health and the environment

Detoxification is the reduction or elimination of the toxic properties of a substance to make it less harmful or more compatible with the environment.

Increasing industrialization during the twentieth century led to the release of large amounts of hazardous waste and by-products into the environment. Another source of toxins was the use of pesticides, which were required to maintain the crop yields necessary to feed the growing population of the world. Some toxins are analogues of harmful substances that occur naturally and may degrade rapidly by natural means. Others are more persistent in the environment and produce unwanted effects. Detoxification is the term used to describe various processes by which toxins are removed from the environment or are rendered less harmful.

A substance is considered hazardous if it poses a threat to human health or the environment when it is spread, treated, disposed of, or transported. Toxic and hazardous substances often occur as a result of manufacturing processes that produce materials designed to protect and improve the quality of life. Sources of hazardous waste include chemicals and allied products, the primary metals industry, petroleum and coal products, and fabricated metals. Environmental releases of toxic chemicals may occur unintentionally through emissions from compressors, pump seals, valves, spills, pipelines, and storage tanks, or intentionally as discharges or discarded solid wastes into air and water. The Environmental Protection Agency (EPA) has reported that Americans generate 1.6 million tons of household hazardous wastes each year.

The disposal of hazardous substances is not a simple matter. Many toxic substances are not suitable for disposal in regular landfills used for trash. A hazardous substance may be water soluble and could leach through the soil into rivers, lakes, and groundwater supplies and pollute sources of potable water. Some wastes have a significant vapor pressure and can be spread over wide areas by wind and air currents. Corrosive wastes require the use of containers that will not decompose.

Public concern regarding toxic substances in the environment has elicited different approaches to solving these problems. The use of natural pesticides and agricultural chemicals has been advocated by environmental activists. The United States Congress has addressed environmental issues with detoxification procedures that regulate wastewater treatment, soil contamination, and landfills. Such regulations include the Federal Water Pollution Control Act (1974), the Safe Drinking Water Act (1974), and the Pesticides Control Act (1972).

Many natural processes cause detoxification of harmful substances in the environment. Gaseous pollutants or toxins that are exposed to sunlight are subject to photochemical decomposition, in which ultraviolet light causes bonds within the compounds to break. The resulting fragments react with oxygen (oxidation) or water (hydration), forming less toxic compounds. These may undergo repeated degradation in the same manner. Microbial degradation, in which organisms metabolize a wide variety of organic compounds to carbon dioxide and water or convert them into less harmful substances, promotes detoxification of many organic toxins. Some newer pesticides, such as organophosphates, are designed to degrade on repeated exposure to water, forming relatively harmless products. Older pesticides, such as polychlorinated biphenyls (PCBs), were found to degrade slowly in the environment. Toxins with slow detoxification pathways bioaccumulate in organisms, causing harmful effects on fish and wildlife. Such effects may be magnified in the food chain.

Efforts to supplement natural detoxification processes include enzymatic (biological) and other chemical methods. Many microorganisms capable of metabolizing toxins have been isolated and cultured in order to treat hazardous wastes. Such treatments are usually carried out at regional waste-treatment centers. One type of chemical treatment involves chelation or precipitation. This is useful for eliminating metals, either in ionic or elemental form, from water and soil. In this method, an organic compound forms an insoluble precipitate with the metal. Filtration removes the precipitate, which can then be subjected to further disposal methods in concentrated form. Composting, or land farming, involves spreading waste materials over a large land area, where they decompose. Pesticides and paper mill wastes have been detoxified this way. Land farming requires monitoring to ensure that waste does not leach into groundwater.

Thermal treatment is considered a safer process. An example of this type of detoxification method is incineration, during which high temperatures oxidize the solid and liquid organic wastes to carbon dioxide and water in the presence of oxygen. However, people living in communities near incinerators fear possible emissions or leaks. One solution to this concern is the incineration of wastes on ships. *Vulcanus*, a Dutch ship, was used to incinerate large quantities of Agent Orange, a hazardous herbicide contaminated with toxic dioxins.

Another method of detoxification is vitrification, in which toxic materials are converted to glass. This method offers a new way to dispose of asbestos, which is considered to be a highly hazardous material. According to Robert E. Prince, president of GTS Duratek of Columbia, Maryland, vitrification can work with almost any kind of wastes, including industrial sludges, soil contaminated by lead, and medical wastes.

Beth Ann Parker and Massimo D. Bezoari

SUGGESTED READINGS: For information on environmental impacts of toxicity and hazardous substances, as well as environmental cleanup, see "The Quest for Answers: Advancing the Health Debate," *Chemical Week* 159 (February 19, 1997). For an in-depth look into detoxifying enzymes, detoxification pathways, and detoxification in different organisms, see Samuel K. Shen and Patrick F. Dowd, "Detoxifying Enzymes and Insect Symbionts," *Journal of Chemical Education* 69 (October, 1992). For more information about new options in hazardous waste disposal and detoxification, see "A Glass Melange: New Options for Hazardous Wastes," *Science News* 147 (January 21, 1995), by Adrienne C. Brooks. For an understanding of detoxification and its environmental effects, see Lester B. Lave and Arthur C. Upton, *Toxic Chemicals, Health, and the Environment* (1987).

SEE ALSO: Environmental health; Hazardous and toxic substance regulation; Waste treatment.

Diamond v. Chakrabarty

DATE: 1980

CATEGORY: Biotechnology and genetic engineering

Diamond v. Chakrabarty was a pivotal U.S. Supreme Court decision in which the Court determined that genetically engineered microorganisms were patentable products of human ingenuity.

In 1972 Ananda Chakrabarty, a microbiologist at the General Electric Research and Development Center in Schenectady, New York, attempted to patent a genetically engineered bacterium that could decompose compounds such as camphor and octane in crude oil. Chakrabarty's patent application was initially rejected because the patent office had a long history of excluding living organisms from patent protection. Chakrabarty, through General Electric, successfully appealed this decision. In 1979 the acting commissioner of Patents and Trademarks appealed the reversal. The case was argued before the U.S. Supreme Court on March 17, 1980.

In a 5-4 decision, the U.S. Supreme Court ruled that living things were patentable if they represented novel, genetically altered variants of naturally occurring organisms. The majority decision held that Chakrabarty's organism was manufactured since he had inserted new genetic information into it and that the organism was new because a similar organism was unlikely to occur in nature without human intervention. Thus, the organism fell within the meaning of the patent statute. It was a product of human ingenuity with a distinctive name, character, and use. The minority opinion held that previous congressional acts that specifically excluded living organisms from patent protection were clearly intended to apply in this case.

This decision let emerging biotechnology companies get patent protection for their living products and potentially capitalize on the revolution in genetic engineering. The Supreme Court realized the ramifications of their decision in terms of its impact on the ethics of patenting living things and the potential to accelerate the release of potentially harmful genetically engineered organisms. However, the basis of their decision was fundamentally narrow: Did Chakrabarty's work constitute patentable material? The further development of biotechnology or its restrictions was held to be a congressional and executive concern, not a judicial one.

Diamond v. Chakrabarty did not greatly influence the extent to which genetically altered organisms were released into the environment. Public opposition to the release of genetically engineered organisms played the dominant role. Instead, the lasting impact of *Diamond v. Chakrabarty* was that it extended the definition of patentable products to compounds or organisms that exist in nature but can be further manipulated by biotechnological means. This had only been true for certain hybrid plants developed by conventional breeding techniques. Furthermore, it became the judicial basis for attempts to patent genetic sequences that may be common to living organisms but require human ingenuity to extract, sequence, replicate, and reinsert into new organisms with their properties intact.

Mark Coyne

See also: Biotechnology and genetic engineering; Genetically altered bacteria; Genetically engineered organisms.

Dichloro-diphenyl-trichloroethane (DDT)

Category: Pollutants and toxins

Dichloro-diphenyl-trichloroethane (DDT) is a synthetic insecticide that has been used extensively in agriculture and for control of insect-borne diseases worldwide. However, DDT's toxicity, persistence in the environment, and ability to accumulate in the food chain has resulted in devastating consequences to wildlife. The harmful effects of DDT became a major focus for the emerging environmental movement during the 1960's.

During the 1930's scientists began searching for nonmetal insecticides. Prior to that time, insecticides were mainly derived from toxic metals, such as arsenic and mercury. In 1939, while experimenting with chlorinated hydrocarbons, Swiss chemist Paul Hermann Müller noted that DDT had insecticidal properties. This led to the development of the first synthetic, organic (carbon-based) insecticide, which was introduced commercially by the Swiss chemical company J. R. Geigy A. G. in 1942. DDT was initially used to provide protection against typhus to civilians and Allied troops during World War II by killing body lice. When the DDT story was re-

Milestones in DDT History

YEAR	EVENT
1874	The first synthesis of DDT is reported.
1939	Paul Müller discovers DDT's insecticidal properties.
1942	The first commercial DDT formulations are introduced by the Swiss company J. R. Geigy.
1943-1945	DDT is used on civilians and military troops in Europe for the control of lice and typhus.
1946	The limited use of DDT on crops is permitted by the U.S. Department of Agriculture.
1948	Müller receives the Nobel Prize in Physiology or Medicine for the development of DDT as an insecticide. The first insects to develop resistance to DDT are observed.
1950's	DDT is used widely for agriculture, public health, and domestic pest control. Laboratory and field studies reveal the negative effects of DDT.
1957	The Clear Lake study shows the bioaccumulation of DDT in aquatic life and birds; citizens on Long Island, New York, file a suit in an attempt to halt aerial DDT spraying.
1958	Robert Barker publishes the results of studies that link DDT to declines in robin populations.
1961	Annual production levels of DDT in the United States peak at 160 million pounds.
1962	Rachel Carson publishes *Silent Spring*, which explains the dangers of DDT to a broad audience.
1963	The President's Science Advisory Committee releases a report on pesticide use that becomes the keystone of the drive to ban DDT.
1964	The U.S. Federal Commission on Pest Control is established.
1967	The Environmental Defense Fund (EDF) is formed.
1968	Joseph Hickey and Daniel Anderson publish a report on DDT's impact on declining raptor populations.
1968	The Wisconsin Hearings, the first major legal challenge to the use of DDT, begin.
1969	Malaria is virtually eliminated in China, largely as a result of DDT use.
1969	Michigan and Arizona become the first states to ban DDT use.
1969	The EDF files petitions with U.S. federal agencies seeking the elimination of the use of DDT.
1969	The use of DDT in residential areas is banned in the United States.
1970	The Environmental Protection Agency (EPA) is established.

Year	Event
1972	The EPA bans DDT use in the United States.
1990's	Studies suggest that DDT acts as an endocrine disrupter.
1993	Reports in the *Journal of the National Cancer Institute* claim that DDT may increase the risk of breast cancer.
1998	International negotiations to phase out the production and use of DDT and other persistent organic pollutants begin in Montreal, Canada.

leased by the British government in 1944, the United States Department of Agriculture (USDA) concluded that before DDT could be recommended for use by farmers, more toxicity information was needed. By the 1946 crop year, limited use was permitted even though evidence suggested that DDT might have some acute toxic effects on birds and that it could be stored in animal fat and excreted in milk.

Between 1940 and 1980, at least 4 billion pounds of DDT were used. More than 1,200 different formulations were developed for industrial, agricultural, and public health applications in the United States alone. Annual worldwide usage peaked in 1964 at 440,800 tons. The success of DDT also served as an impetus for chemical companies to begin an intensive search for other organic pesticides.

A New Pollutant

The insecticide properties of DDT are related to its ability to act as a nerve poison and to freely pass through insect cuticles. Besides causing convulsions, paralysis, and death, DDT can also interfere with calcium-dependent processes. However, animal skin prevents absorption of DDT; thus, it initially appeared to be a safe alternative to metal insecticides.

Although effective, DDT does have undesirable pesticide characteristics. As a broad-spectrum insecticide, DDT kills a wide variety of organisms, including beneficial insects, such as bees. Development of resistance to DDT was observed as early as 1948. Because of its chemical composition, DDT is preferentially stored in animal fat and is therefore not readily excreted by animals that ingest it. This fat solubility and DDT's persistence in the environment are responsible for the accumulation of DDT in the food chain.

The first clear evidence of the bioaccumulation of DDT came from a case study in Clear Lake, California. Between 1949 and 1957, DDT was used to control gnats on the lake. By the mid-1950's, the health of fish-eating birds in the area began to decline; several bird species, especially grebes, were dying in large numbers. Since no infectious agent was found, scientists used new analytical methods developed to measure compounds in tissues. High levels of DDT were detected in plankton, fish, and birds in and around Clear Lake. The studies also clearly showed biomagnification: Levels of pesticide residues were found to be sequentially higher at each step in the food chain, with concentrations in grebes and gulls up to 100,000 times greater than in the formulations of DDT that were sprayed.

DDT toxicity was becoming evident throughout the United States. DDT was used in efforts to control Dutch elm fungal disease in the Midwest and New England, since the fungus is spread by the elm bark beetle. Several studies between 1954 and 1958 noted sharp declines in robin populations—in some areas by as much as 70 to 90 percent. Extensive aerial spraying for gypsy moths in the 1950's from Michigan to New England coincided with significant declines in many species of songbirds and bees. Ironically, this impacted populations of some of the natural predators of the intended target pests.

The DDT spraying also had a negative impact on agriculture. In addition to reduced pollination caused by the loss of bees, farmers were discovering that cows' milk and farm produce were contaminated with pesticide residues. In the Pacific Northwest, DDT used to control the spruce budworm devastated salmon populations. Coastal spraying along the Atlantic Ocean to control the salt marsh mosquito and malaria took a heavy toll on migrating birds, marine life, and raptors.

The effect of DDT on raptor populations is well known. In 1968 an article in *Science* magazine by wildlife ecologists Joseph Hickey and Daniel Anderson reported that the decline in raptor population was largely caused by eggshell breakage caused by chlorinated hydrocarbons. Calcium processes were altered in birds containing high DDT levels in their fatty tissues, resulting in the production of eggs with thin shells. The young did not hatch since the eggs were crushed during incubation. The American eagle and the osprey ended up on the brink of extinction largely as a result of DDT use.

Ban on DDT

In the late 1950's the first DDT-related lawsuits were filed over losses to farmers and beekeepers and in attempts to stop further aerial spraying. The most notable case of the time was filed in 1957 by a group of citizens led by the well-known ornithologist, Robert Cushman Murphy, in order to gain an injunction to stop the spraying of DDT over Long Island, New York. The injunction was not granted, but the case went all the way to the U.S. Supreme Court, which declined to hear it.

Perhaps one of the most significant events leading to the ban of DDT was the publication of the book *Silent Spring* in 1962 by Rachel Carson. The author described the negative environmental impact of pesticides such as DDT, and the subsequent public outcry led to a dramatic decline in DDT use. The production of DDT in the United States peaked in 1961, and global production began to decline around 1964. As a result of the controversy spawned by Carson's book, the President's Science Advisory Committee was charged with reviewing pesticide use. The committee report, published in 1963, called for legislative measures to safeguard the health of the land and people against pesticides. The Federal Commission on Pest Control was established in 1964, and four governmental committees studied DDT in-depth between 1963 and 1969. Ultimately, these investigations led to the establishment of the Environmental Protection Agency (EPA) in 1970.

DDT was also a major impetus for the formation of associations whose missions were aimed at protecting the environment and public health. The newly formed Environmental Defense Fund (EDF) initiated a series of court hearings and lawsuits related to DDT in the late 1960's. In October, 1969, it filed petitions with the USDA and the Department Health, Education, and Welfare (HEW) seeking elimination of the use of DDT. When no effective action was taken, the EDF, along with other environmental groups and individuals, took the case to court. On May 28, 1970, the U.S. Court of Appeals for District of Columbia rendered two major rulings on DDT in response to the EDF's litigation. Besides leading to the eventual ban of DDT, these rulings set important environmental law precedents: They provided power to membership associations (environmental groups) and served to protect public interests.

In 1972 the EPA banned the use of DDT and highly restricted the use of other chlorinated hydrocarbons. However, the ban applied only to DDT use within U.S. borders; it still allowed companies to produce DDT for export, which they continued to do for several years.

Persistence of DDT

Even after the ban of DDT in the United States and several other countries, the pesticide is still found in high levels in marine animals and other wildlife. DDT can be detected in the tissues of almost every person on earth, especially indigenous people living in the Arctic and workers from insecticide production plants and agriculture.

New concerns about DDT's toxicity have arisen as a result of studies published beginning in the early 1990's. Data suggests that DDT and its metabolites can act as endocrine disrupters—compounds that mimic sex hormones in animals. Also known as environmental estrogens, these compounds may decrease sperm count and fertility, affect the onset of puberty, alter male and female characteristics in wildlife, and increase the risk of cancer of reproductive organs. It also appears that DDT can be transferred across the placenta, increasing the risk of birth defects and impairing brain development. DDT levels average 1 part per million (ppm) in human breast milk and can be transferred to infants. Concentrations as low as 2 ppm are known to have damaging hormone-disruption effects in birds.

Attempts to find alternatives for eradication of malaria are underway, and countries are being urged to decrease their reliance on DDT for mosquito control. In June, 1998, a round of international treaty negotiations began in Montreal, Canada, to discuss the eventual phase-out of production and use of a group of human-made compounds—including DDT—referred to as persistent organic pollutants (POPs). Most of these compounds are classified as endocrine disrupters, and all are known to bioaccumulate in animals. However, despite the environmental consequences, worldwide use of DDT continues at levels about 10 percent of those in 1964. Large stockpiles of the insecticide are known to exist in Africa and Asia. For many nations in Africa and Latin America, DDT remains the most economical and least acutely toxic weapon against mosquito-borne tropical diseases.

Diane White Husic

SUGGESTED READINGS: Rachel Carson's classic *Silent Spring* (1962) was the first nontechnical portrayal of DDT's effects on the environment. Frank Graham, Jr., discusses the history of DDT and the events that occurred after the publication of Carson's book in *Since Silent Spring* (1970). The link between DDT and the environmental movement is investigated in Thomas Dunlap's *DDT: Scientists, Citizens, and Public Policy* (1981). *Silent Spring Revisited* (1987), edited by Gino Marco, Robert Hollingworth, and William Durham, addresses numerous environmental and regulatory issues related to pesticides from a scientific, but very readable, perspective. The persistent problems with DDT and other estrogen-disrupters are examined in *Our Stolen Future: Are We Threatening Our Fertility, Intelligence, and Survival?* (1996), by Theo Colborn, Dianne Dumanoski, and John P. Myers.

SEE ALSO: Agricultural chemicals; Biomagnification; Carson, Rachel; Pesticides and herbicides.

Dioxin

CATEGORY: Pollutants and toxins

Dioxin, a toxic by-product of particular manufacturing processes, is called a ubiquitous chemical because it is found everywhere. Exposure to even trace amounts can cause severe health problems in humans and other organisms.

There are approximately seventy-five different types of dioxin. However, the term "dioxin" is commonly used to refer to a variety known as 2,3,7,8-tetrachlorodibenzo-p-dioxin (TCDD), a highly toxic chemical that has caused great concern among environmentalists. Dioxin can be destroyed by exposure to direct sunlight in the presence of hydrogen, but the chemical can remain under the surface of the ground for ten years or longer.

One of the earliest documented cases of dioxin exposure occurred in West Germany in 1957 when thirty-one workers at a chemical plant developed chloracne, a skin disease that is one of the characteristic effects of exposure. In 1977 investigators in the Netherlands discovered dioxin in fly ash from a municipal incinerator. By 1980 it had been found that practically all organic substances, when burned, produce dioxin.

At first it was thought that chloracne was the only effect of exposure to dioxin. As time went on, however, it was discovered that dioxin was highly toxic to experimental animals. Researchers found that guinea pigs could be killed with as little as 1 microgram of dioxin per kilogram of body weight. However, hamsters could take a dose of 5,000 micrograms per kilogram. Further experimentation showed that the organs within animals were also affected differently. This kind of difference among animal species and organs was unheard of in any previously tested substance, and it invalidated the usual method of testing animals. Because of this, dioxin became known as the first environmental hormone. That is, it acts like a hormone in animals and plants because it has such a strange effect on various organs.

Scientists also began to suspect that dioxin could be involved in causing various cancers. The most famous exposure incident occurred between January, 1965, and April, 1970, when Agent Orange was used in Vietnam to kill trees and plants that provided hiding places and food for North Vietnamese soldiers. One of the herbicides used was contaminated with dioxin during the manufacturing process. After the war, the National Academy of Sciences reviewed more than six thousand studies of dioxin exposure and came to the conclusion that four kinds of cancer could be linked to the chemical.

After the Environmental Protection Agency detected high levels of dioxin in soil along the banks of the Woonasquatucket River in Rhode Island, signs were posted warning residents not to eat local fish. Many researchers believe dioxin to be a cancer-causing agent. (AP/Wide World Photos)

Other studies of dioxin exposure have produced mixed results. Thirty-seven thousand people were exposed to dioxin in Seveso, Italy, in 1976 after a factory explosion. Although no immediate human fatalities were traceable to the accident, a significant number of male births in the region were interrupted before term during the four months following exposure. Dioxin was therefore suspected of being a factor in the decline of male births worldwide. Two hundred workers were exposed to dioxin in 1949 at Nitro, West Virginia, and several died from cancer. Sweden reported six times the normal cancer rate in people exposed to dioxin, but Finland traced 1,900 cases of exposure and found no harmful effects. A Veteran's Administration (VA) study of eighty-five thousand Vietnam veterans found lower-than-normal rates of cancer. Many researchers have concluded that such disparate findings are attributable the observation that dioxin does not seem to cause cancer itself, but rather acts as an influence on other cancer-causing chemicals.

Another well-publicized dioxin incident occurred in 1983 in Times Beach, Missouri, when a local resident tried to settle the dust on the town's roads by wetting it with a liquid containing dioxin. The federal government bought the entire town and moved everyone out to prevent any further exposure. This incident illustrates the policy of using preventive measures before all the facts are known. In 1983 it was not known that dioxin caused any of the cancers later linked to exposure.

When the Environmental Protection Agency (EPA) issued its report on dioxin in September of 1994, it claimed that perhaps no more than 14 kilograms (30 pounds) of dioxin are released into the U.S. environment annually. However, even this trace amount is unacceptable since dioxin is suspected of being an endocrine disrupter, which means that trace amounts can disrupt the effects of other hormones and cause numerous disorders. An EPA lab worker was quoted as saying, "I don't know a hormone system that dioxin doesn't like to disrupt." To make matters worse, there are about seventy other chemicals in the environment suspected of being disrupters, including dichlorodiphenyl-trichloroethane (DDT) and other insecticides that were used for many years. The 1994 EPA report also stated that dioxin may be responsible for damaging immune systems and creating other hormone-related diseases, such as diabetes. For example, mice treated with dioxin readily die after exposure to viruses that ordinarily would have no effect on them.

Dioxin has expanded the view of environmental chemicals to include environmental hormones and hormone disrupters. These seventy or more substances are like no other chemicals known in the past, raising the critical question of how to regulate them when so little is known about their effects. All previous ways of measuring are invalid because these substances do not have uniform effects. One animal is killed by a tiny exposure and another is seemingly unharmed, while both may have internal effects that remain undiscovered.

Robert B. Bechtel

Suggested Readings: For a thorough review of all chlorinated compounds, see Sub Ramamoorthy and Sita Ramamoorthy, *Chlorinated Organic Compounds in the Environment* (1998). For a general guide on how to avoid exposure to dioxin, read Lois Gibbs's *Dying from Dioxins: A Citizen's Guide to Reclaiming Our Health and Rebuilding Democracy* (1995).

See also: Agent Orange; Italian dioxin release; Times Beach, Missouri, evacuation.

Diquat

Category: Pollutants and toxins

The herbicide diquat, which has been used in the United States since the 1950's, is potentially toxic if swallowed or inhaled. As is the case for many herbicides, diquat may be released into wastewater and subsequently absorbed into soil, potentially contaminating drinking supplies.

Though diquat is not produced in the United States, nearly one million pounds of the compound are imported each year. Approximately two-thirds are utilized as a desiccant or defoliant, with another one-third used for aquatic weed control. Diquat is readily adsorbed into clay particles in the soil, sediment in water, and the surface of weeds. While it is biodegradable if it is adsorbed into plant life and subject to photodegradation by light, diquat bound to sediment remains stable for weeks or even months. Ultimately, however, the herbicide undergoes degradation through the action of soil flora.

The extensive overgrowth of weeds affecting both crops and waterways has created significant problems in management. Regulations pertaining to use of herbicides remain under the auspices of the Environmental Protection Agency (EPA) and its various offices. Cutbacks in funding, however, have hampered research into the short- and long-term effects of using herbicides in the control of weed problems; therefore, the long-term effects of diquat are unclear.

Since diquat is a nonselective herbicide, its use is limited. It is used as a growth regulator in suppressing the flowering of sugarcane and to control aquatic weeds in the absence of endangered plant species. However, the most important use of diquat is the desiccation of potato haulm or seed crops such as clover or alfalfa. Such practices are generally carried out to prepare a crop for harvest. Application results in loss of moisture from the leaf, usually killing the plant. Desiccation may also be utilized in prevention of seed loss resulting from scattering of seed upon opening of the pod. Application of diquat prior to harvest of crops such as alfalfa greatly reduces seed loss.

The primary metabolic effect of diquat seems to be its ability to divert electrons activated during photosynthesis into the production of toxic compounds such as hydrogen peroxide. Peroxide production in turn results in membrane damage to those parts of the plant in contact with the herbicide. Human exposure to diquat is primarily occupational, with agricultural workers who use the chemical at highest risk. Most actual poisonings have been intentional, as a means of suicide. Nevertheless, though diquat is not as toxic as some herbicides, the level of toxicity is mainly a function of degree. Exposure to significant levels of the chemical may cause damage to the central nervous system and kidneys, while cataract formation is the most common effect of chronic exposure.

Richard Adler

SEE ALSO: Agricultural chemicals; Hazardous and toxic substance regulation; Pesticides and herbicides.

Disposable diapers

CATEGORY: Waste and waste mangement

Each year 19 billion soiled diapers end up in landfills. Though heated debates continue about whether reusable cotton diapers are more earth-friendly than disposable diapers, the percentage of parents using disposable diapers rose 15 percent between 1991 and 1997.

Single-use diapers made of wood pulp with a cellulose liner and plastic backing account for 2 percent of landfill space. Though the wood pulp and human waste contained in disposable diapers is biodegradable, it is sealed in the non-biodegradable plastic backing. Environmentalists claim that the number of disposable diapers reaching landfills is too high. Proponents of disposable diapers say garbage in general is the issue, not disposable diapers in particular. They also claim that the energy needed to wash and dry cotton diapers causes more pollution and greater environmental harm than the 12,300 tons of disposable diaper waste per day.

Pampers, the first disposable diapers, were first marketed in 1961 but did not catch on quickly. Their only convenience then was that they could be thrown away. Caregivers still had to fold and pin them. By 1970 they accounted for only 25 percent of diaper sales. As their popularity grew, more brands were marketed. The competition prompted improvements such as greater absorbency, leak guards, resealable closures, different designs for boys and girls, and decorative plastic backings. By 1980 use of disposable diapers had reached 65 percent.

While manufacturers worked on improving the convenience of the product for consumers, they also worked to address environmental issues created by the third-largest single product to enter the waste stream. The major issue was that disposables are not truly disposable. They are thrown away, but they do not decompose. Manufacturers experimented with "green" disposables, which were made from plastic films that eventually broke down in water. Others had backings with high cornstarch content so they would degrade. However, landfills do not contain the oxygen and water needed for such biodegradable disposable diapers to break down.

Proctor & Gamble, the makers of Pampers, conducted two separate pilot programs during the late 1980's and early 1990's. One was intended to create a partially flushable diaper. This was partly in response to complaints that fecal matter should enter the sewage system, not landfills. However, the program was discontinued after only three years because consumers did not want to bother with removing the flushable inner section and discarding the plastic cover. Sewage treatment officials also claimed that diapers would overtax the sewage treatment system. Finally, the diaper was not compatible with water-saving, low-flush toilets.

In another program, Proctor & Gamble set up centers to recycle diapers. Organic materials were separated from the plastics, which were processed and recycled into other products. The company found low marketability of products made from recycled diapers, however. Disposable diaper manufacturers are now looking at the possibility of composting the organic portions of disposable diapers to help ease the burden on landfills.

Proponents of disposable diapers continue to argue that diapers in landfills are not the real issue. The problem is garbage in general and the limited space for landfills. They argue that human urine and waste that travels to landfills with diapers is not really an issue, either. Pet fecal matter also ends up in landfills, as does medical waste. Proponents also point out that the amount of garbage going into landfills during a baby's diapering life is minimal when compared with the total amount of garbage adults send to landfills each day.

In conjunction with disposables as a garbage issue, those opposed to disposable diapers claim cloth diapers are the better alternative. Proponents of disposables disagree. Whether parents launder their own cloth diapers or have a diaper service perform this chore, energy is burned, and pollution is added to the environment. Plenty of hot water and bleach must be used to wash diapers. This requires energy, as does running the machine and drying the diapers afterward. Diaper services consume even more energy in picking up soiled diapers and delivering clean diapers, not to mention adding to traffic congestion.

When considering the energy and resources used to manufacture both types of diapers, neither appears to be an ideal option. Cloth diapers are made from cotton, which is a renewable resource. However, growing cotton consumes huge amounts of resources. It is one of the most water-intensive crops grown. Also, huge amounts of pesticides are used on the plants, and bleach is used in processing cotton, which causes wastewater pollution. More energy is used when the cotton is spun into cloth and manufactured into diapers.

Disposable diapers use wood pulp, a less renewable resource than cotton. They also contain plastics made from a nonrenewable petrochemical resource. Proponents of disposable diapers claim that the lengthy process required to manufacturing cotton diapers, in addition to the laundering of these diapers for the first thirty months of a baby's life, far exceeds the energy output required for manufacturing disposable diapers. Despite the arguments and counterarguments, the environmental impact of both cotton and disposable diapers remains a problem.

Lisa A. Wroble

Suggested readings: Diaper debates on whether "Use of Cloth Diapers Can Reduce Garbage" are covered in articles by Francesca Lyman (yes) and Robert J. Samuelson (no) in *The Environmental Crisis: Opposing Viewpoints* (1991). *In Defense of Garbage* (1993), by Judd H. Alexander, covers the issue of disposable diapers weighed against the total garbage generated by babies as compared to adults that enter landfills each year. In *Garbage! Where It Comes From, Where It Goes* (1990), Evan Hadingham and Janet Hadingham discuss disposable diapers along with plastics in the chapter "Down in the Dumps." Virginia I. Postrel and Lynn Scarlett cover both sides of the diapering issue in their article "Talking Trash," included in *Taking Sides: Clashing Views on Controversial Environmental Issues* (1995), edited by Theodore D. Goldfarb.

See also: Landfills; Plastics.

Dodo birds

Category: Animals and endangered species

Dodo birds, flightless birds native to the island Mauritius in the Indian Ocean, were hunted into extinction during the seventeenth century by European sailors in search of food.

Dodo birds were indigenous to the Mascarene Islands in the Indian Ocean east of Madagascar. The largest island of the group is Mauritius, 765 kilometers (475 miles) east of Madagascar. The other large islands are Réunion, 225 kilometers (140 miles) southwest of Mauritius, and Rodrigues, 644 kilometers (400 miles) east of Mauritius. The Mauritius species *Raphus cucullatus* is, strictly, the dodo bird, but some people consider the closely related "solitaire," *Raphus solitarius*, of Réunion Island and some *Pezophaps solitaria* of Rodrigues Island also to be dodo birds. These species were pigeons that, in isolation, had evolved to large size and odd forms.

The dodo bird of Mauritius was flightless and has been described as having had a rather globular body supported on legs that were, for a bird, short and stout. It had a short neck crowned with a large head ending in a long, heavy beak that was hooked downward at the end. The nostrils were well forward on the beak. The similar birds of Réunion and Rodrigues Islands were somewhat slimmer and had longer legs.

The male dodo is reported to have weighed about 22.7 kilograms (50 pounds), as compared to 6.8 to 9 kilograms (15 to 20 pounds) for an adult male North American wild turkey. The female was reportedly somewhat smaller. The bill was about 22.9 centimeters (9 inches) long. The wings were rudimentary, displaying only three or four feathers, and the tail was equally rudimentary, consisting of a few short, curly feathers. The body feathers of the Mauritius dodo were blue-gray, and the sparse wing and tail feathers were white with a splash of yellow. Overall, the dodo appears to have been a rather unattractive bird.

The dodo evolved in the absence of humans and predators and initially showed no great fear of sailors who hunted and captured them, giving rise to the expression "dumb as a dodo." They were slow moving, and humans could outrun a dodo on open ground. One observer reported that Dutch sailors on one ship caught twenty-four dodos one day and twenty the next. The birds apparently laid a single egg in a nest on the ground.

Some visitors to the islands reported that the meat of dodos was hard and tough, even after prolonged boiling. Others claimed that the breast and belly were good, while the rest of the bird was tough. It may have been a matter of season because one source said that the Rodrigues solitaire was extremely fat and good to eat from March to September.

Seafarers first visited the Mascarene Islands and reported the existence of dodos about 1507. The dodo of Mauritius was extinct by 1681; the solitaires of Réunion had disappeared by 1746, and those of Rodrigues were gone by 1790. The extinctions were not sudden; the Mauritius dodo survived for 174 years beyond first human contact. Macaques, which are fond of eggs, and pigs were introduced to Mauritius in the sixteenth century, and, along with heavy predation by humans, were probably an important factor in the ultimate extinction of the birds.

Robert E. Carver

SEE ALSO: Extinctions and species loss; Hunting.

Dolly the sheep

CATEGORY: Biotechnology and genetic engineering

Dolly the sheep was the first clone of an adult animal. The creation of Dolly and subsequently cloned animals has raised a host of ethical issues and opened the door to possible means of improving human health and the environment.

Ian Wilmut of the Roslin Institute in Edinburgh, Scotland, announced the birth of Dolly the sheep, the first clone of an adult animal, on February 5, 1997. Scientists and the general public were shocked with this announcement, since it was believed that cloning an animal from an adult cell was, at the time, technically impossible.

A clone is an organism developed from a single cell isolated from another organism. The cell donor and the clone are genetically identical. Prior to the creation of Dolly, no attempts at cloning an animal from adult cells had been unequivocally successful. In the early 1980's clones of mammals had been created by using donor cells from young embryos. Adult cells and embryonic cells have identical genetic material, or deoxyribonucleic acid (DNA); however, adult cells produce proteins specific to the type of cell they become. For example, brain cells produce neurotransmitters and do not produce hemoglobin, even though they possess the hemoglobin gene. Scientists believed that the structure of the DNA in an adult cell had been irreversibly altered during the process of maturation to gain this specificity and therefore could not be used to produce a clone.

Wilmut and his colleagues at the Roslin Institute used a novel approach to clone Dolly. The donor cells were sheep udder cells from a six-year-old pregnant ewe. The cells had been frozen for about three years, and the donor was long deceased. The researchers believed that prior attempts at creating clones from adult cells failed because the cells were too active or in the wrong phase of their life. To make the cells quiescent, the cells were starved for several days. Meanwhile, the genetic material of eggs from a different breed of sheep was removed (enucleation). The starved cells were then fused with the enucleated donor eggs and implanted into surrogate mother sheep of a third breed. Of 277 attempts, only one of these experiments went

Dolly the sheep, the first animal to be cloned from adult cells. (AP/Wide World Photos)

full-term, resulting in the birth of Dolly. Dolly looks strikingly different from the breed of the egg donor or the surrogate mother but identical to the breed that donated the adult DNA, an observation that provides suggestive evidence that she developed from the donor DNA.

Initially there were many questions surrounding the validity of the experiment that produced Dolly. Some scientists believed that she could have been the result of a contaminating fetal cell, and the results of the experiment were not easily replicated by other researchers. However, in July of 1998 Japanese scientists announced the birth of two calves cloned from adult cow uterus cells, and researchers in Hawaii successfully produced more than fifty mouse clones from adult mouse ovarian cells. DNA analysis of Dolly has also virtually proven that she was indeed the first clone of an adult animal. In addition, Dolly gave birth to a lamb on April 13, 1998, showing that she was a healthy young adult whose ability to reproduce was not compromised by her unusual origin. In May of 1999, however, researchers found that the telomeres in Dolly's cells were shorter than those in other mammals of similar age. Telomeres become progressively shorter as an organism ages, but it was not clear whether Dolly's lifespan would be shorter than usual.

There are numerous potential applications for this cloning technology. Scientists envision using cloning to create animals that produce valuable pharmaceuticals in their milk or human proteins such as the clotting factor needed by hemophiliacs. In fact, Wilmut's research was sponsored by the Scottish pharmaceutical company PPL Therapeutics, Ltd. In July, 1998, Wilmut announced that his group had produced sheep cloned from fetal cells that had been altered to carry a human gene. The gene was shown to be active in the mammary glands of the sheep, thus paving the way for pharmaceuticals produced in animal milk. However, little is known about how introducing genetically altered animals outside the lab might affect other animals they encounter or the environment in general.

Some researchers envision entire herds of genetically identical cattle. Since it is very difficult to produce prize milk or meat-producing animals consistently with traditional breeding methods, repeated cloning of one prize animal would greatly speed the process. A potentially serious problem with genetically identical herds is that genetic diversity allows species to survive changes in their environment and attacks by disease. Diseases affecting only a few individuals of a genetically diverse species may become rampant in a genetically identical one. If genetic diversity is lost, it could lead to the extinction of that species.

Because of human impact on the environment, the world is witnessing an exceedingly high rate of extinction for many species. While many governments and private organizations are attempting to limit that impact, the extinction rate continues to rise. Zoos and nature conservation organizations are working with captive breeding programs, and in some cases have successfully reintroducing animals once on the brink of extinction. Many others, though, reproduce poorly outside their native habitat. Cloning may provide a means to save some species. For example, giant pandas are exceedingly rare and rarely reproduce successfully in captivity. While cloned pandas would be genetically identical and would face the problems already outlined, they would not be lost forever. Cloning may provide a temporary solution until endangered species can be restored to the wild.

Karen E. Kalumuck

SUGGESTED READINGS: An excellent account of the technical background that lead to Dolly's birth, as well as the history of cloning and its social conflicts, can be found in Gina Kolata's *Clone: The Road to Dolly, and the Path Ahead* (1998). In "A Sheep in Sheep's Clothing?" *Discover* (January 22, 1998), Christopher Wills has written a concise overview of the Dolly phenomenon and a synopsis of the scientific controversy surrounding her origin. The original announcement of Dolly to the scientific community, accessible to those with some science background, is contained in Ian Wilmut's "Viable Offspring Derived from Fetal and Adult Mammalian Cells," *Nature* 385 (February 27, 1997).

SEE ALSO: Biotechnology and genetic engineering; Cloning; Genetically engineered organisms; Wilmut, Ian.

Dolphin-safe tuna

CATEGORY: Agriculture and food

Dolphin-safe tuna has been captured using methods that minimize dolphin mortality. The International Conservation Act of 1992 prohibits the sale, purchase, transport, or shipment into the United States of any tuna that does not meet the dolphin-safe standards established by the Dolphin Protection Consumer Information Act.

The eastern Pacific tuna fishery covers the ocean from California to Hawaii and south to the equator and Chile. During the late 1950's, tuna fisherman discovered that dolphins and yellowfin tuna frequently fed in the same areas within this region. Since dolphin are mammals, they must remain near the surface and are therefore easier to locate. Fisherman exploited this knowledge by encircling the dolphin schools with 1-mile-long purse-seine nets in order to capture the yellowfin tuna swimming below the surface. Using helicopters and speedboats, fisherman herded dolphins and tuna into the large nets. Dolphins, of no value to commercial fishermen, became entangled in the nets and drowned. Estimates place the decline in the dolphin population since the 1950's at between 20 and 45 percent. By the early 1970's, the number of dolphins killed annually in the eastern Pacific Ocean had risen to more than 300,000.

Public awareness of the plight of the dolphins began as early as 1972 with the passage of the Marine Mammal Protection Act, which was intended to prevent exploitation of dolphins and related aquatic animals. The act limited the number of dolphins that could be killed annually to 20,500. Unsatisfied, conservationists and animal rights activists pushed for a tuna boycott, a move that succeeded largely as a result of the widespread popularity of dolphins. By 1977 annual dolphin deaths in purse-seine nets had de-

Cans of tuna with the dolphin-safe seal of approval. (Yasmine Cordoba)

clined to about 25,450 animals. The respite was brief; as the United States fishing fleets declined during the early 1980's, foreign vessels began to enter eastern Pacific fishery. This led to an increase in dolphin deaths, bringing the numbers back up to more than 100,000 deaths in 1986.

In 1990 declining sales and concern about public opinion led the three major canners of tuna sold in the United States to stop buying tuna caught in purse-seine nets. At the same time, much of the United States tuna fishing fleet moved into the Western Tropical Pacific Ocean, where tuna and dolphin do not habitually swim in the same waters.

The Dolphin Protection Consumer Information Act set standards for labeling tuna as "dolphin safe," but the regulations included few provisions for enforcement. Critics claimed that the canneries were simply using the phrase as a marketing tool rather than a true guarantee of dolphin-friendly fishing practices. In July, 1997, the Marine Mammal Protection Act was amended to redefine the dolphin-safe label. Under the new guidelines, dolphins can be caught in nets as long as they are not harmed. In early 1998 twelve nations reached an agreement that lifted the embargo on yellowfin tuna from Mexico and other tuna-fishing nations while promoting the recovery of individual dolphin stocks, especially those that were depleted.

P. S. Ramsey

See also: Commercial fishing; Drift nets and gill nets; Marine Mammal Protection Act; Ocean pollution.

Donora, Pennsylvania, temperature inversion

Dates: October 26-October 29, 1948
Category: Atmosphere and air pollution

The first outdoor air pollution disaster in U.S. history occurred in Donora, Pennsylvania, in October of 1948. The incident was prolonged because of the presence of a five-day surface temperature inversion over the city.

Donora is an industrial town of 14,000 located 40 kilometers (25 miles) southeast of Pittsburgh, Pennsylvania, in the Monongahela River valley.

It is surrounded by hills that are 107 meters (350 feet) in elevation. In 1948 four industries with 6-meter (20-foot) smokestacks were located along a 4.8-kilometer (3-mile) stretch of the Monongahela's banks: a steel mill, which used high-sulfur coke; a zinc smelting plant, which used high-sulfur ores; a wire manufacturing plant; and a sulfuric acid plant. In addition, high-sulfur coal was used to generate the area's electricity and heat most homes and businesses. Therefore, sulfur dioxide fumes and hydrocarbon particulates were common in the air of Donora.

On the morning of October 26, 1948, a surface high-pressure weather cell with light winds settled over the eastern United States. A blocking ridge in the upper air and the westward position of the jet stream prevented any other weather systems from moving into the area. On Tuesday night the cool, heavy air from the mountain slopes drained into Donora's valley, which, combined with the sinking air of the high-pressure cell, created a strong surface temperature inversion.

A temperature inversion occurs whenever the temperature of the atmosphere increases with height. Since cool air is heavier than warm air, a temperature inversion discourages surface air from rising and dissipating surface air pollutants into the upper air. The air within Donora's valley, which was filled with the pollutants being emitted by industries, shops, and homes, became trapped at the surface with no way of dispersing either vertically, because of the temperature inversion, or horizontally, because of the surrounding mountains. To complicate matters, radiation fog, which had begun to form during the afternoon, blocked sunlight from reaching the surface of the earth, preventing surface heating from breaking the inversion.

This meteorological situation remained over Donora for five days. The result was an unprecedented air pollution episode. By Wednesday morning, visibility was so poor that local residents were becoming lost. On Friday, respiratory-related health complaints from residents began to pour in; four thousand people became ill from the polluted air between 6:00 A.M. and midnight. The first death occurred early Saturday morning, with sixteen more deaths and two thousand illnesses by midnight. Three more deaths occurred on Sunday even though relief came as an approaching front pushed the high-pressure cell out of the area and broke the temperature inversion. Subsequent rain washed most of the pollutants out of the air, the wind swept the smoke away, and the disaster was over.

The lessons learned from this event were slow to be realized, but it did eventually prompt an increase in air pollution research. It caused many people to question the acceptance of billowing smokestacks as a necessary part of economic progress and to recognize the importance of the association between certain meteorological conditions and air pollution episodes.

Kay R. S. Williams

SEE ALSO: Air pollution; Black Wednesday; London smog disaster; Smog.

Dredging

CATEGORY: Water and water pollution

The removal of sediment from streams, lakes, and oceans is referred to as dredging. The sediment is sometimes removed for the purpose of extracting valuable ore minerals. Many waterways must be continuously dredged to remove sediment so that ships can continue to move in and out of ports.

Sediment that contains ore minerals is formed by the weathering of nonore minerals (gangue) and chemically stable, hard, and high density ore minerals out of their sources. The minerals are carried downstream, where the ore minerals are concentrated by density differences from the gangue minerals. Diamonds, cassiterite (a major tin mineral), ilmenite (a titanium ore mineral), gold, sand, and gravel have been mined by dredging. The sand and gravel have been used for construction and road metal. Material dredged from harbors along coastlines is often taken out to sea for dumping.

In stream dredging for ore minerals, the dredge sits in a small lake that it has formed in a river and scoops up sediment. The ore mineral is separated from the gangue minerals inside of the dredge; the ore is collected, and the gangue is tossed out the rear. In this fashion, the dredge gradually expands the lake in front of it and fills in the lake behind it. Large quantities of sediment may eventually be disrupted during this process. For example, large dredges are used to mine tin from sediment in southeast Asia to more than 50 meters (164 feet) below the surface of the water. This results in more than 5 million cubic meters (1.77 million cubic feet) of sediment being processed per year. Large quantities of gold have also been mined in this fashion in the western United States. In areas where land is expensive and a local source of sand and gravel is required, the sands and gravels in the river may be removed by dredges. For example, this has been done along the Kansas River in the Kansas City, Missouri, metropolitan area for many years.

There is little or no direct pollution produced as a result of dredging, but the unsightly piles of sediment dumped behind the dredge in streams may cover organisms and produce turbid waters that may hurt or kill organisms that need clean water to survive. For instance, oysters are especially susceptible to harm in turbid waters. Also, erosion patterns, river currents, and ocean currents may be affected by dredging. For example, dredging in the Kansas River has removed far more sediment than can be naturally replenished. This has resulted in increased erosion of river banks, a wider and deeper river channel, a lowered water surface, a steeper bed gradient, and loss of vegetation and farmland along the river. A final concern of dredging results from dredging sediment in areas where the sediment is already polluted, such as in the San Francisco Bay area in California, as removal of this material may spread the pollutant even further.

Robert L. Cullers

See also: Sedimentation; Water pollution.

Drift nets and gill nets

Category: Animals and endangered species

Drift nets and gill nets kill not only the target fish but also other marine species, such as stingrays, dolphins, and fish with no marketable value. The use of such nets in commercial fishing is therefore restricted under international treaty.

Gill nets are made from thin nylon twine that is nearly invisible in water. The rectangular nets range from 15 to 366 meters (50 to 1,200 feet) in

Fishing boat with nets. Several international treaties regulate the use of drift nets and gill nets, both of which entangle not only target fish but also any other fish that swims into them. (Ben Klaffke)

length. Floats on the top and weights on the bottom allow them to hang in the path of fish like floating walls. The fish swim into them and get caught in the netting at their gills. Gill nets are used from fishing boats, as well as near shore. They can also be used across rivers to catch migrating fish such as salmon.

Drift nets are similar to gill nets. They are made of the same material and also trap fish by the gills. Drift nets are much longer than gill nets, measuring up to 3 nautical miles (55.56 kilometers) and are used mostly in open waters. Like lobster traps, they are set out at night and pulled in during the day. Like gill nets, they trap marine life other than the target fish. Dolphins, seals, sea turtles, whales, sea lions, manta and stingrays, and marine birds are often trapped in drift nets and die.

When used in tropical waters, the harvested fish in drift nets sometimes rot before being gathered by the fishing boats. This is compounded when one fishing boat sets out eight to ten nets, covering up to 30 nautical miles (555.6 kilometers). Because the nets are nearly invisible in the open ocean, they create hazards for other boats in the area. Propellers can easily get tangled in the nets or slice a net into several pieces. When this happens, the portions that break free to float through the ocean become "walls of death." The filmy nylon netting catches marine life that will never be harvested onto a fishing boat.

Both types of nets create another problem that upsets the balance of the ocean ecosystem. Nets with very small openings also capture juvenile fish, which are either thrown back into the ocean, dead or dying, or become part of the by-catch. By-catch is all the harvested fish and marine life other than the target fish. For example, if a fishing vessel is seeking tuna, only tuna will be kept when the nets are pulled in. Everything else is by-catch. Whether the juvenile fish are thrown back or become by-catch, they do not live to breed more fish. Regulations now require all fishing nets to have openings large enough to allow juvenile fish to swim through. They then live long enough to breed.

The United Nations passed a resolution in 1989 banning the use of drift nets in the South Pacific after 1991. A temporary halt, or moratorium, took effect in November, 1991, on drift net use in international waters, which are all waters more 200 (3,704 kilometers) nautical miles from shore. On January 1, 1993, all drift nets longer than 1.5 nautical miles (27.78 kilometers) were banned from use in international waters. Most nations have agreed to uphold this treaty.

Lisa A. Wroble

SEE ALSO: Commercial fishing; Dolphin-safe tuna; Turtle excluder devices.

Drinking water

CATEGORY: Water and water pollution

In some parts of the world, inadequate amounts of drinking water threaten the lives of humans and animals. In other parts of the world, water exists in adequate quantities, but contaminants limit its usefulness.

Drinking water in most developed countries is protected from contaminants by national regulations. However, in many parts of the developing world, even minimal water quality standards that protect human health are not in place. The lack of sufficient quantity and quality of this basic natural resource can lead to widespread cases of waterborne diseases and suffering. The World Health Organization has estimated that 80 percent of all sickness and disease in developing countries can be attributed to waterborne infectious agents, such as viruses, bacteria, protozoans, and parasites. For example, during 1995 there were one billion cases of diarrheal diseases resulting in more than three million deaths worldwide. In the United States during 1993, a protozoan named *Cryptosporidium* in the public water supply of Milwaukee caused more than one hundred deaths and 400,000 cases of illness.

In the United States the primary legislation protecting drinking water supplied by public water systems is the Safe Drinking Water Act of 1974 and its amendments. The law mandated that the U.S. Environmental Protection Agency (EPA) set national water quality standards for

Maximum Concentrations of Certain Constituents Allowed in U.S. Drinking Water

Constituent	Concentration (milligrams per liter)
Arsenic	0.05
Cadmium	0.005
Chromium	0.1
Selenium	0.05
Lead	0.015
Mercury	0.002
Cyanide	0.2
Benzene (volatile organic)	0.005
Atrazine (pesticide)	0.003
Malathion (pesticide)	0.014
Phenol	0.030

drinking water. The states are responsible for enforcing the standards. The EPA has set two types of national standards. Primary standards protect human health through the establishment of maximum contaminant levels for nearly one hundred chemical and biological agents. Secondary standards, which are advisory only, protect taste, odor, and appearance of the drinking water.

The protection of source water for human consumption is the most critical issue to consider. In the United States, about 50 percent of the people, including 95 percent of those in rural areas, depend on underground aquifers for drinking water. However, underground aquifers are being threatened by various sources: agricultural activities, leaking septic systems, and direct injection of wastes. In one Iowa study, fertilizers and pesticides were found to have contaminated one-half of all aquifers and wells. High nitrates from these sources are dangerous to the health of infants and can be transformed into cancer-causing nitrosamines. In Florida more than one thousand wells have been closed because of excessive levels of chemicals, mostly a pesticide called ethylene dibromide.

Roy Darville

See also: Clean Water Act and amendments; Water pollution; Water quality; Watersheds and watershed management.

Dubos, René

Born: February 20, 1901; Saint-Brice, France
Died: February 20, 1982; New York, New York
Category: Ecology and ecosystems

French-born American bacteriologist and environmental writer René Dubos discovered tyrothricin, the first antibiotic to be made available on the commercial market. Dubos was more widely known, however, as a writer who explored the manner in which humans interacted with their environment.

René Dubos was born into a family of poor French shopkeepers. He excelled as a student in the College Chaptal in Paris and decided to pursue a career as a scientist. Later, inspired by the work of Louis Pasteur, about whom he wrote two books, Dubos determined to follow bacteriology. In 1924 he settled in the United States and eventually became a U.S. citizen. From 1924 to 1927 he undertook study and research at Rutgers University in New Jersey and then moved to Rockefeller University in New York City. Except for a brief period at Harvard (from 1942 to 1944), he remained associated with Rockefeller University until he became director of Environmental Studies at the State University of New York College at Purchase in 1971.

Dubos first achieved international recognition through his discovery of the first commercially marketed antibiotic, tyrothricin, in 1939. Its discovery awakened medical researchers to the enormous possibilities for soil-produced antibacterial agents. During the 1950's Dubos developed an interest in ecology; over the next twenty-five years, he became one of the world's most outspoken and heralded environmentalists.

Aided by his second wife, Letha Jean Porter, a

former research associate, Dubos wrote prolifically about environmental issues and the way humans ought to relate to the natural world. He emphasized that the quality of human life (mental, physical, and spiritual) primarily depended upon people establishing a symbiotic relationship with nature. He rejected the claims of more extreme environmentalists that human life would become extinct in a planet filled with chemical waste.

Dubos's book *The Unseen World* (1962) gained him international recognition as an influential commentator on the environment. There followed a series of books over the next thirty years, including *So Human an Animal* (1968), *A God Within* (1972), and *The Wooing of the Earth* (1980). In 1969 he was cowinner of the Pulitzer Prize in general nonfiction for *So Human an Animal.* In the 1960's and 1970's Dubos achieved near-celebrity status and was often featured in major news magazines and newspapers.

Dubos's most important contribution to the environmental dialogue was a report published in 1972 as *Only One Earth.* Maurice Strong, general secretary for the United Nations (U.N.) Conference on the Human Environment, commissioned Dubos to chair an international committee of experts that would prepare a commentary on the condition of the earth for the United Nations Environmental Conference held in Stockholm, Sweden, in 1972. Dubos then chose well-known British environmentalist Barbara Ward to assist in preparing and writing the final report. Much of the Dubos-Ward report focused on the need to maintain a balance in the earth's ecosystems—a balance that was in jeopardy. Although many of the experts consulted were displeased that *Only One Earth* did not predict doom, the report inspired the United Nations to focus attention to worldwide environmental problems. As a consequence, the Stockholm Conference established the United Nations Environment Programme to observe and coordinate information about environmental issues on a worldwide basis.

Ronald K. Huch

SEE ALSO: Spaceship Earth metaphor; United Nations Environmental Conference; United Nations Environment Programme.

Ducktown, Tennessee

CATEGORY: Land and land use

Copper mining operations that began in Ducktown, Tennessee, during the late nineteenth century led to the eradication of plant life and subsequent soil erosion in the region. The transformation of forestland into an artificial desert demonstrated the hazards of extracting natural resources without consideration for its effect on the surrounding area.

In 1843 a gold rush took prospectors to the southeastern corner of Tennessee. However, copper was found instead, which led to the founding of several mining companies in 1850. The Copper Basin, an area of approximately 48,000 acres surrounding the area where the city of Ducktown is located, was the only place in the eastern United States to produce significant amounts of copper, and by 1902 Tennessee was the sixth-largest copper-producing state in the nation.

After the copper ore was mined, it had to be roasted in order to separate the copper from the rest of the ore, which included zinc, iron, and sulfur. To do this, firewood was harvested from the surrounding forests and put into heaps in roofed sheds with open sides. The ore was piled onto the heaps of firewood, which were then lit. These heaps, which covered acres of ground, were then allowed to roast for three months.

The smoke from the ores was highly sulfuric. Not only did it hinder vision, but it also allowed sulfuric gas, sulfuric dust, and sulfuric acid to descend on the earth. The extent of the pollution was first acknowledged around 1895, when individual citizens filed lawsuits for damage to vegetation. The state of Georgia filed similar lawsuits in 1904 and 1905. The heap method was soon abandoned. Instead, raw ore was smelted in furnaces without being roasted first.

Companies also built taller smokestacks to take advantage of higher air currents that would disperse the smoke, but that only spread the sulfuric vapors over a larger area. It was not until 1911 that the problem of sulfuric pollution could be controlled to some extent by capturing the smoke for the valuable sulfuric acid that could be derived from it. By then the years of sulfuric pollution and systematic deforestation had turned the area into the only desert east of the Mississippi River. Erosion washed away the topsoil, leaving a bare, rocky terrain that, at its greatest extent, covered about 130 square kilometers (50 square miles). Copper mining continued until 1986, when it became impossible to compete with the price of imported copper.

During the 1930's copper mining companies worked with the Tennessee Valley Authority (TVA) to reforest the area. More than fourteen million trees were planted over the next several decades, and grasses were also planted. Little resulted from the plantings until the 1970's, when improved methods of fertilization and soil churning increased the plants' chance of survival. By the 1990's only about 1,000 acres of land appeared to be bare, with large gullies and less than three hundred trees per acre. The restoration of the forest was not totally accepted by local residents, however, who believed that a small portion of the desert should be saved as an example of severe environmental degradation.

Rose Secrest

See also: Desertification; Erosion and erosion control; Reclamation.

Dust Bowl

Date: 1930's
Category: Land and land use

The Dust Bowl was an environmental disaster marked by huge dust storms in the southern region of the Great Plains of the United States during the 1930's. The Dust Bowl revealed the damage that mechanized agriculture could cause if not accompanied by a program of soil management.

Droughts periodically occur in the Great Plains of the United States. During such periods, winds pick up loose soil and create dust storms, especially during the spring months. Settlers reported numerous examples of this natural phenomenon during the nineteenth century. During the twentieth century, new agricultural practices and overgrazing by cattle speeded soil erosion. Tractors and other machines allowed farmers to plow larger areas for planting wheat. In the process, they destroyed the natural grasses whose root systems stabilized the soil. Because the wheat replaced the grasses, most farmers remained unaware that they were contributing to a coming catastrophe.

In 1931 a severe drought struck the Great Plains; it centered on the Texas and Oklahoma panhandles, northeastern New Mexico, eastern Colorado, and southwestern Kansas. The wheat crop withered in the fields, and its root systems were no longer able to support the soil. As the drought continued, soil particles that normally clustered together separated into a fine dust. When the winds blew in early 1932, they lifted the dust into the air, marking the beginning of an environmental disaster that a newspaper reporter later dubbed the Dust Bowl.

Although their number and severity increased, dust storms remained an issue of local and regional concern for the first two years. However, as the drought continued into 1934, the storms grew so immense that they caused damage in areas far from the plains. A storm that emanated from Montana and Wyoming in May, 1934, deposited an estimated twelve million tons of dust on Chicago, Illinois. Ships 483 kilometers (300 miles) offshore in the Atlantic Ocean reported that dust from the same storm landed on their decks. Incidents such as these provoked national concern over the growing crisis on the plains.

Scientists identified two types of dust storms: those caused by winds from the southwest and those resulting from air masses moving from the north. While no less damaging, the more frequent southwest storms tended to be milder than the terrifying northern storms, which came to be known as "black blizzards." Huge walls of dust, sometimes more than 1.6 kilometers (1 mile) high, rolled across the plains at 100 kilometers per hour (62 miles per hour) or faster, driving frightened birds before them. The sun would disappear, it would become as dark as night, and frightened people would huddle in

A farmer's son sits on a sand dune near his home in Liberal, Kansas, during the 1930's. The Dust Bowl was caused by a combination of drought and poor soil conservation practices. (Library of Congress)

their homes, their windows often taped shut. On occasion people stranded outside during these severe storms would suffocate. Some black blizzards lasted less than one hour, while a few reportedly continued for longer than three days.

Most historians argue that the Dust Bowl was one of the worst ecological disasters in the United States, one that could have been mitigated had farmers practiced soil conservation in the years before drought struck. Instead, farms were ruined, causing some 3.5 million people to abandon the land. Many of them moved into small towns on the plains, while others journeyed to California in search of opportunity. Cattle and wildlife choked to death. Human respiratory illnesses increased markedly during the Dust Bowl era, and a number of people died from an ailment known as "dust pneumonia." Anecdotal evidence indicates that many people grew depressed as the dust storms continued year after year.

The mid-1930's marked the peak of the Dust Bowl, with seventy-two storms that reduced visibility to less than 1.6 kilometers (1 mile) reported in 1937. The return of the rain in the late 1930's eased the crisis, and by 1941 the disaster was over. However, by that time ecologists and farmers had begun soil conservation measures in response to the crisis. The United States government provided expertise and financial support for many of these efforts. Farmers practiced listing, a plowing process that made deep furrows to capture the soil and prevent it from blowing. Alternating strips of planted wheat with dense, drought-resistant feed crops such as sorghum slowed erosion by blocking wind and retaining moisture, which prevented the soil from separating into dust. To stop erosion on lands not farmed, natural grasses were planted. The government also sponsored the Shelterbelt Project, a program that used rows of trees to form windbreaks. Millions of trees were planted throughout the Great Plains, with more than 4,828 kilometers (3,000 miles) of shelterbelts created in Kansas alone.

Despite the experiences of the 1930's, once the drought ended many farmers returned to the farming practices that had damaged their fields. Soil conservation experts worried that the region would suffer a return of the Dust Bowl when the rains stopped. Their predictions came to pass in 1952, when another drought led to a series of dust storms, including several storms with wind gusts clocked at 129 kilometers (80 miles) per hour. That drought ended in 1957, but in accord with a twenty-year cycle, the region again faced a shortage of rainfall in the early 1970's. At that time some analysts confidently predicted that the Dust Bowl was a thing of the past. They claimed that irrigation with aquifer water from deep wells would prevent soil erosion. However, shrewd observers pointed out that the fate of the region was now tied to a resource, aquifer water, that would become increasingly precious in the coming years. The possibility that the Great Plains could again witness a devastating ecological catastrophe like that of the 1930's remains.

Thomas Clarkin

SUGGESTED READINGS: Donald Worster's *Dust Bowl: The Southern Plains in the 1930's* (1979) identifies capitalism as the primary cause of the ecological crisis. R. Douglas Hurt's *The Dust Bowl: An Agricultural and Social History* (1981) offers a thorough discussion of agricultural practices and soil conservation measures. Paul Bonnifield's *The Dust Bowl: Men, Dirt, and Depression* (1979) is an easy-to-read overview of the Dust Bowl era.

SEE ALSO: Erosion and erosion control; Soil conservation; Strip farming.

E

Earth Day

Date: inaugurated April 22, 1970
Category: Ecology and ecosystems

Earth Day is an annual observance held on April 22 to promote public concern for environmental problems. In many ways the first Earth Day in 1970 marked the birth of the environmental movement, as twenty million Americans either engaged in demonstrations or gathered to hear speeches. Although participation was disappointing during the 1970's, a growing number of grassroots organizations celebrated Earth Day throughout the 1980's. An estimated 200 million people around the world participated in Earth Day 1990.

In June, 1969, Democratic senator Gaylord Nelson, having observed how anti-Vietnam War "teach-ins" had influenced public opinion, conceived of the idea of a large teach-in to educate the general public about the importance of environmental issues. He suggested that such an event should be planned for April 22, a day when many states commemorated Arbor Day and a day that would not conflict with final exams on college campuses. Recognizing the potential of the idea, a small group of concerned citizens founded Environmental Action to sponsor the event. The organization was able to raise the modest sum of $125,000, and a dynamic young law student, Denis Hayes, was put in charge of publicizing and coordinating activities. Senator Nelson and Republican representative Paul McCloskey were named official cochairpersons.

Numerous historical factors contributed to the great success of the first Earth Day. By the late 1960's a growing number of organizations were helping to sensitize the public to environmental problems, and an unprecedented number of publications on environmental themes were being produced by prominent writers, such as Rachel Carson, Walter Udall, Lynn White, and Paul Ehrlich. Of even greater significance, Americans in locations throughout the country were witnessing the harmful effects of environmental damage. In 1968 and 1969 members of Congress reflected public opinion as they considered nearly 140 environmental bills. Three months before Earth Day, President Richard Nixon signed the National Environmental Policy Act, which required analysis and review of public projects. For many people, especially the young, the impressive achievements of the Civil Rights movement represented a model for reform based upon a moral appeal. In addition, the spirit of youthful rebellion embodied in the antiwar movement was inspiring parallel movements throughout society. In short, Senator Nelson could not have chosen a more auspicious context for the launching of his idea.

Conservative business and political leaders were not enthusiastic about the idea of Earth Day. Although President Nixon's press secretary announced the administration's support for the day, the president took no active role in any of the events. While some members of the administration suspected that the observance was a means for advancing the agenda of liberal Democrats, Nixon's secretary of the interior, Walter Hickel, urged Nixon to proclaim a national holiday and become an active participant. Hickel later wrote that he gave "marching orders" to department personnel to visit college campuses and that fifteen hundred employees of the department did so. The White House was embarrassed when the press reported that Controller General James Bentley had spent $1,600 of public funds to telegram warnings of a possible left wing plot after he observed that Earth

Day fell on former Soviet leader Vladimir Lenin's birthday. Bentley was forced to apologize and pay for the telegrams with his own money.

APRIL 22, 1970

In all measurable ways, the first Earth Day was a huge success. In New York City, Chicago, and Philadelphia large crowds gathered to hear speeches by politicians, poets, ecologists, and other concerned citizens. Some fifteen hundred college campuses, as well as ten thousand elementary and secondary schools, scheduled programs of one kind or another. The National Education Association estimated that about ten million schoolchildren participated in some kind of environmental activity for the day. Also, about two thousand communities planned environmental ceremonies of one kind or another.

There was an impressive diversity of activities throughout the United States, and the atmosphere was euphoric and theatrical. In Washington, D.C., about ten thousand young people attended a rock concert in front of the Washington Monument. The University of Wisconsin held fifty-eight separate programs. To dramatize air pollution problems caused by internal-combustion engines, several universities held enthusiastic automobile-wrecking events called "wreck-ins." Some localities also held "bike-ins." In New York City, traffic was closed to automobiles along Fifth Avenue for two hours. Many idealistic people helped proenvironmental efforts. At the University of Washington, four hundred people planted trees and shrubs during a "plant-in" in an abandoned area near the campus. In Ohio one thousand students from Cleveland State University gathered litter and loaded it into garbage trucks. In hundreds of communities groups of Boy Scouts and Girl Scouts held cleanup campaigns and picked up litter.

Earth Day was an occasion for numerous speeches, including many by the best-known spokespersons of the environmental movement. Barry Commoner, Paul Ehrlich, Ralph Nader, and the aging René Dubos were among the speakers in greatest demand. Many politicians could also be seen at various rallies, and both houses of Congress were adjourned for the day. Senator Nelson spoke on nine university campuses in Wisconsin, California, and Colorado. Senator Thomas McIntyre, who delivered fourteen speeches in his home state of New Hampshire, set the record for the greatest number of speeches given on Earth Day.

No major acts of violence marred the celebrations of Earth Day, but there were scattered incidents of militancy. At Boston's Logan Airport,

An Earth Day gathering in New York City in 1974. (Bernard Gotfryd/ Archive Photos)

thirteen demonstrators were arrested for blocking traffic during a demonstration to protest a proposed expansion of the airport. In Washington, D.C., about twenty-five hundred demonstrators assembled before the Department of the Interior to protest the approval of oil leases. Students at the University of California at Berkeley conducted a sit-in to register their disapproval of recruiters from Ford Motor Company, while at the University of Texas, twenty-six students were arrested for perching in trees to try to prevent the trees' destruction.

Earth Day 1990

In 1971 Earth Day was expanded into Earth Week, but the expanded observance was not successful. For a few years Earth Day attracted limited interest. By the mid-1980's, however, the celebration of Earth Day was regaining its original popularity, and environmentalists viewed the celebrations as a repudiation of President Ronald Reagan's conservative environmental polices. Leaders of the environmental movement wisely decided to concentrate their efforts on commemorating the twentieth anniversary of Earth Day, and Denis Hayes was chosen as the chairperson of the occasion.

On April 22, 1990, an estimated 200 million people in 140 countries participated in Earth Day. Organizers claimed that this was the largest grassroots demonstration in history. For this occasion, Hayes had the assistance of a large coalition of environmentalist and other socially conscious groups, and together they raised and spent about $4 million. The day was celebrated by marches, rallies, parades, concerts, and a large assortment of activities in all continents. Although the largest demonstrations took place in the developed industrial countries, there were scattered events within the poorer and less developed regions of the world.

In Boston, a crowd of 200,000 people turned out; in New York City's Central Park, the various rallies attracted an estimated 750,000 participants; in Washington, D.C., some 125,000 people participated in a demonstration at the Capital Mall; in St. Louis, an estimated ten thousand people planted ten thousand trees on the banks of the Mississippi River. Throughout the day, speeches were given by Hayes, Gaylord Nelson, Morris Udall, Barry Commoner, Bruce Babbitt, Senator Edward Muskie, and countless others. In a special American Broadcasting Companies (ABC) television broadcast called *The Earth Day Special*, Bette Midler played an abused Mother Earth who collapsed as a result of global warming, deforestation, and toxic poisoning. Although the events of the day were almost uniformly peaceful, ecoguerrilla groups attracted headlines by destroying oil-exploration gear and pouring sand in fuel tanks of logging machinery.

At a time of controversy over issues such as the northern spotted owl, President George Bush and his administration appeared distrustful of Earth Day 1990. However, President Bush, who had earlier referred to himself as "the environmental president," addressed several crowds via a telephone hookup. Although the Bush administration's policies were often criticized, there was a tendency for speakers to spend more time attacking the earlier policies of former president Ronald Reagan. In contrast to twenty years before, the speeches of Earth Day 1990 were more somber and realistic in recognition of the fact that environmental problems would not be quickly solved in a painless manner.

Earth Day 1990 was truly a global festival. In Brazil, a concert by Paul McCartney paid special attention to the environment. In West Germany, Green organizations sponsored the ceremonial planting of trees. Thailand's top rock band, the Carabano, held a We Love the Forest concert. In Hong Kong a day-long educational entertainment featured singers, mimes, and exhibits of "green" consumer goods. Other activities included a roadway "lie-down" by five thousand protesters against car fumes in Italy, an 800-kilometer (500-mile) human chain across France, a trash-cleanup campaign on Mount Everest in Nepal, and a "flyby" of three thousand kites made by schoolchildren in Tours, France.

Impact of Earth Day

By 1990 it appeared clear that yearly celebrations of Earth Day had become a mainstream institution. The day's silver anniversary in 1995 attracted considerable interest, but organizers decided to concentrate greater efforts on the thirtieth anniversary in the year 2000.

The 1970 celebration of Earth Day tended to be a predominantly white, middle-class affair. Many African Americans were suspicious that the day would detract from the issues of racial and economic justice. While such views did not completely disappear, they tended to decline with time. In 1969 polls indicated that only 33 percent of African Americans wanted the government to pay more attention to environmental issues; by 1976, approximately 58 percent of African Americans expressed the same viewpoint. By the time of Earth Day 1990, African American leaders had more evidence that pollution tends to be especially severe in areas where poor and marginalized people live, and thus they were able to use the day to publicize the issue of environmental racism.

Although celebrations of Earth Day are primarily important as a reflection of public opinion at a particular time, there is evidence that such celebrations have some impact on public attitudes and that they can solidify the commitment of environmental organizations. In a 1965 Gallup Poll, only 17 percent of the responding Americans said they considered the reduction of air and water pollution to be one of the most pressing problems demanding governmental action. Immediately after Earth Day 1970, the figure was 53 percent, but by 1980 it had fallen to 24 percent. Ironically, the environmental policies of President Ronald Reagan appeared to encourage the popularity of Earth Day during the 1980's, and the success of Earth Day 1990 is partially explained by a survey of the time in which 80 percent of Americans said they would support more strenuous environmental efforts, regardless of costs.

Without doubt, there are always faddish, trendy elements in the celebrations and speeches of Earth Day. At times the rhetoric seems rather excessive, and environmentalists are often offended to observe that some of the nation's worst polluters attempt to exploit the day as a form of commercial advertising. The important point, however, is that Earth Day promotes education and reflection about serious problems that are experienced by people in their daily lives.

Thomas T. Lewis

SUGGESTED READINGS: Many of the 1970 speeches are in *Earth Day—The Beginning: A Guide for Survival* (1970), compiled and edited by the national staff of Environmental Action. Earth Day is analyzed within the context of the environmental movement in Victor Scheffer's *The Shaping of Environmentalism in America* (1991). A useful sociological study of the topic is in Riley Dunlap and Angela Mertig, *American Environmentalism: The U.S. Environmental Movement, 1970-1990.* Philip Shabecoff presents an optimistic view of the observance in *A Fierce Green Fire* (1993). Jacqueline Switzer concentrates on conservative critics in *Green Backlash: The History and Politics of Environmental Opposition in the U.S.* (1997). To get the flavor of Earth Day each year, consult magazines such as *Conservationist* and *Sierra* in April. For young people, an excellent work is Sylvia Whitman's *This Land Is Your Land: The American Conservationist Movement* (1994).

SEE ALSO: Animal rights movement; Earth Summit; Environmental education; Green movement and Green parties.

Earth First!

DATE: founded 1980
CATEGORY: Preservation and wilderness issues

Earth First! is a radical environmental movement that gained notoriety during the 1980's for its use of ecological sabotage to protest the abuse of wilderness areas.

Earth First! was founded in the United States in 1980 by several people who were concerned about the old-line conservation organizations' lack of passion and commitment. They were also inspired by Edward Abbey's novel *The Monkey Wrench Gang* (1975), which recounted the adventures of a small group of people who sabotaged construction equipment in the American Southwest to stop environmental destruction. The founders included Howie Wolke, Mike Roselle, and Bart Koehler, but the most notable was Dave Foreman, a one-time Barry Goldwater Republi-

Earth First!ers lead a demonstration during the Forest Summit in Oregon in 1993. Members of the radical environmental movement have criticized the U.S. Forest Service for allowing logging to occur in national forests. (Reuters/Steve Dipaola/Archive Photos)

can and later an official of the Wilderness Society. Many of the early Earth First!ers had some preservation experience and expertise.

Always more a movement than a formal organization, Earth First! has become the most notorious and controversial environmental group in the United States. They are committed to naturalist John Muir's wilderness preservation ideals, as well as the principles of deep ecology. Their philosophy is expressed in the slogan "No compromise in defense of Mother Earth." Enemies of Earth First! include both private and public developers, as well as the U.S. Forest Service, which is seen as being too willing to capitulate to developers. The movement is also highly critical of mainstream conservation organizations, which are considered to have become passive and bureaucratized.

Earth First!ers use various strategies to inhibit wilderness destruction, including ecological sabotage, also called ecotage and monkeywrenching. Their first success came in 1981 when they unfurled a 300-foot-long black polyethylene "crack" on the face of the Glen Canyon Dam, located just south of the Utah-Arizona border, in symbolic opposition to the recently created Lake Powell. Black plastic quickly gave way to other types of monkeywrenching, such as spiking trees (a controversial tactic in which nails are pounded into trees in old-growth forests to stop logging), disabling bulldozers and other equipment, pulling up survey stakes, and blocking logging roads. Such actions have been widely condemned and criticized. Earth First! also runs the Biodiversity Project, which seeks to protect natural ecosystems, and has organized several task forces to educate the public on such issues as overgrazing and grizzly bear habitat.

In 1985 Foreman published *Ecodefense: A Field Guide to Monkeywrenching*, a detailed how-to book on ecotage that claimed to neither condemn nor condone the use of such tactics. Foreman and three others were arrested in 1989, accused of planning to sabotage power lines in Arizona. Many of the early Earth First!ers subsequently left the movement, including Roselle, who joined Greenpeace, and Foreman, who became active in the Sierra Club.

The original Earth First!ers were anarchistic individuals and outlaws, as Foreman indicated in his comment that "there was a certain gonzo, coyote flavor to it all." Change was suggested during the 1990 Redwood Summer—a conscious attempt to emulate the 1964 civil rights Freedom Summer—when a number of Earth First!ers protested lumbering in California's redwood forests. The protesters demonstrated a more communal, countercultural, and leftist impulse, with greater involvement by women and attempts to appeal to the loggers as a class against the corporations. One of the organizers, Judi Bari, was seriously injured when a bomb was placed in her automobile.

During the 1990's Earth First! declined in public prominence. However, its confrontational campaign to save the redwoods continued, epitomized by Julia "Butterfly" Hill's year-long occupation of a redwood tree in 1997-1998, which gained international media attention. Earth First!, always a small movement in numbers, had changed, but there was a direct connection between Glen Canyon's 1981 black plastic "crack" and Hill's sit-in, both of which were highly visible attempts to preserve the environment by unorthodox and controversial means.

Eugene Larson

SEE ALSO: Abbey, Edward; Deep ecology; Ecotage; Ecoterrorism; Foreman, Dave; Glen Canyon Dam; Monkeywrenching.

Earth resources satellite

Category: Land and land use

Since their inception in the early 1970's, earth resources satellites have become an essential tool for observing and assessing the earth's resources and environment, particularly where large-scale or long-term processes are involved.

The earliest efforts to collect images of earth from orbiting satellites focused on military and meteorological applications. During the 1960's, however, as weather satellites and manned space missions provided more and more photographs and television images of the earth, scientists became increasingly aware of how much information about the planet's surface could be obtained from space. In 1972 the United States launched the first earth resources technology satellite (ERTS), which was dedicated to observing the planet's resources. With the launch of the second satellite in 1975, the ERTS program was renamed Landsat. As of mid-1998, a total of five Landsat platforms had been successfully sent into orbit, although Landsat 5 was the only satellite remaining in operation. A sixth Landsat was launched in 1993 but never achieved orbit. Landsat 7 was scheduled for launch in 1999.

Other nations have developed similar programs devoted largely or entirely to resource monitoring from orbit. These programs include such satellite series as the French Space Agency's Système Probatoire d'Observation de la Terre (SPOT), the European Space Agency's Earth Remote Sensing Satellites (ERS), Russia's Resurs (Resource) satellites, Japan's Marine Observation Satellites (MOS) and Japanese Earth Resources Satellite (JERS), and India's Indian Remote Sensing Satellites (IRS).

Earth resources satellites generally have an orbit such that, as the earth rotates, the satellite's sensors are provided with an ever-changing view of the planet below. Every few weeks, the satellite completes a full survey of the earth's surface and begins a new one. As the satellite collects data, it transmits them to ground stations. From there, the data are processed into color images, maps, and other useful forms.

Sensors aboard the satellite obtain data by scanning the planet's surface for electromagnetic radiation. Active sensing systems, such as the radar instruments aboard ERS 1 and JERS 1, emit microwave energy, then measure how the surface below reflects it. Passive sensors, such as the multispectral scanner and thematic mapper carried aboard Landsat 5, detect visible, near-infrared, and thermal-infrared wavelengths reflected or emitted from the planet's surface. Various surface features emit, reflect, or absorb energy in a distinctive way, making the interpretation of satellite images possible.

Satellite data facilitate the study of earth's land, oceans, atmosphere, and life as an integrated system. From satellite imagery, researchers derive information on soil and rock types, water quality, fires, floods, plant health, human and animal populations, and urban development. Ongoing worldwide imaging facilitates observation of complex, large-scale global phenomena such as El Niño events and climate change. Through regular, repeated imaging of areas, researchers are able to observe environmental change over time, such as urban expansion, deforestation, and desertification. Remote sensing data can maximize the effectiveness of resource exploration and exploitation efforts while minimizing their environmental impact. Data from earth resources satellites are invaluable to scientists, planners, resource managers, and anyone who must observe or make informed decisions regarding natural resources, land use, and the environment.

Karen N. Kähler

SEE ALSO: Climate change and global warming; Forest management; Range management; Urban planning.

Earth Summit

DATES: June 3-June 14, 1992
CATEGORY: Ecology and ecosystems

The Earth Summit was convened in Rio de Janeiro, Brazil, in 1992 as a follow-up to the Stockholm Conference of 1972. Considerable changes had taken place over those twenty years: Many countries had negotiated treaties and agreements to manage the earth's resources and preserve the environment. Yet serious degradation of the environment was still occurring, and international debates continued, particularly between the countries of the Northern and Southern Hemispheres.

The United Nations (U.N.) General Assembly passed a resolution on December 20, 1988, that the U.N. Conference on the Environment and Development (UNCED), or Earth Summit, would convene in 1992. The first preparatory meeting for the Earth Summit was held in 1990. The conference was based on the belief that there were environmental issues that could only be addressed at the global level; sustainable development was the core concept on which their deliberations would take place.

It was evident at the first meeting that the same issues between the Northern and Southern Hemispheres that had been prevalent at Stockholm twenty years before still remained. The Group of 77, representing the developing nations in the South, repeated demands for preferential, noncommercial, and concessional technology transfer from the advanced states in the North. They also wanted additional financial resources made available to the less-developed states. The Northern nations were divided on these issues. Some, particularly the Nordic countries, felt that there would never be agreement on environmental issues unless the North made substantial economic concessions to the South. They wanted a working group established to consider the South's demands for technology transfer and to discuss resource issues.

The United States and the European Community were opposed to formalizing the demands of the South. In fact, the U.S. delegation was working under orders from the White House to keep such issues off the agenda. The United States had its way when a working group was established and given a mandate to discuss only legal and institutional issues. The only forum in which the South could bring up its issues was the plenary session at Rio. In response, the South changed its strategy, making their participation in any agreements to be reached at Rio conditional on recognition of their demands.

At the second meeting of the Preparatory Committee, Malaysia presented a declaration, supported by the other developing countries, that they would only negotiate the issue of forests if the North paid for the South's participation in any agreement that was reached. The declaration also demanded that the advanced economies reduce their energy consumption and sign a climate change agreement.

THE CONFERENCE

Delegations representing 178 countries, with more than 110 heads of state, met in Rio de Janeiro from June 3 to June 14, 1992, for the Earth Summit. With the nongovernmental organizations, which met in a parallel Global Forum, and other observers, the Rio meeting involved twenty-five thousand people. The conference had been convened to address urgent problems related to environmental protection and socioeconomic development. The U.N. General Assembly had added "development" to the official conference name to include the concerns of developing countries that still, as they had at Stockholm, believed that economic growth was of greater importance to them than protection of the environment.

The developed and developing countries each brought different agendas to Rio. The developed nations were concerned with specific environmental problems such as ozone depletion, global warming, deforestation, and acid rain. The developing countries focused on the negative impact of the economic policies of the North on poorer countries whose economies were slow to improve or, in some cases, were experiencing negative growth patterns.

One of the major areas of disagreement between developed and developing countries was over a possible treaty on forests. At the March, 1991, Preparatory Committee meeting, the United States tried to get approval for negotiations on a forest treaty outside of UNCED. Instead, the committee banned any outside negotiations and created an ad hoc group within the conference to study the topic of forests. Although all the world's forests have suffered from deforestation, the major depletion of forests has taken place in Africa, with Asia also experiencing significant losses. However, the countries of the South insisted that the issue not focus on their forests alone.

Agreements Reached at Rio

Notwithstanding the problems that occurred in getting the participants to consider issues that they considered intrusive into their own internal affairs, the results of the conference were significant. The Convention on Climate Change and the Convention on Biological Diversity were signed at Rio. In addition, the Rio Declaration and the Forest Principles were endorsed, and Agenda 21 was adopted. The Commission on Sustainable Development (CSD) was created to monitor the results of the summit.

Agenda 21 is a three-hundred-page plan for achieving sustainable development in the twenty-first century. Since 1992 more than 1,800 cities and towns worldwide have created their own "local Agenda 21," and 150 countries have established national advisory councils to promote dialogue among government officials, environmentalists, and businesspeople on sustain-

Representatives from more than 170 nations stand for a moment of silence during a 1997 meeting at the United Nations. The special session was called to review progress on environmental issues since the 1992 Earth Summit in Brazil. (AP/Wide World Photos)

able development policies for their own countries. In addition, countries have been able to reach agreement on new treaties on desertification and high-seas fishing.

A range of environmental topics were included in Agenda 21. These topics came under the classifications of agriculture and rural development, biotechnology, business and industry issues, capacity building in developed countries, changing consumption patterns, children's issues, combating poverty, resources, deforestation, desertification, hazardous wastes, human health, human habitat, and nongovernmental organization (NGO) participation. Agenda 21 entered into force on June 13, 1992, when the participating governments signed it. It proposes comprehensive strategies and programs to counteract environmental degradation and promote sustainable development. Its strategies are intended to create a balance between the environment and development, and emphasize the importance of participation by all countries.

The U.N. Framework Convention on Climate Change entered into force on March 21, 1994. The main objective of the convention is to achieve stabilization of the production of greenhouse gases. It sets out principles through which countries can acquire an understanding of global warming, including such issues of importance to the South as the sharing of research and development and technology transfer.

The U.N. Convention for Biological Diversity entered into force on December 29, 1993. The convention addresses conservation of biological species; use of genetic resources; technology transfer; identification and monitoring of problems; research, training, and education; and impact assessment.

The Rio Declaration on Environment and Development entered into force on June 13, 1992, at the time of adoption by the participating states. Its purpose is to establish cooperation among member states in order to reach agreements on laws and principles promoting sustainable development. It creates a balance between the sovereign right of states and their responsibility to the rest of the international community. The declaration reaffirms and updates Principle 21 of the Stockholm Declaration and itself contains twenty-seven principles. It addresses the environmental impact of development, poverty, ecosystem protection, sharing of scientific ideas, public participation and access to information, implementation of legislation, economic policies of states, internationalization of environmental costs and the "polluter pays principle," notification of pollution incidents and environmental impact statements, and indigenous cultures.

The Statement of Forest Principles was a compromise reached by the environmental ministers representing each country. Principles, considered "soft law," are not binding on the states and allow them to find some areas of agreement without committing themselves. While acknowledging the right of states to exploit their own forest resources, nations are encouraged to manage such resources without causing long-term damage to them.

The Commission on Sustainable Development is a high-level group created to monitor and report on implementation of the agreements reached at the summit. It has been referred to as a "watchdog" over the actions of the states in carrying out their obligations under the agreements. The commission reports to the U.N. General Assembly through the U.N. Economic and Social Council (ECOSOC). It is also intended to serve as an ongoing forum for negotiating further global policies on the environment and development.

The conference delegates agreed to a proposal that called for a treaty to be negotiated on the topic of desertification. They also agreed on a mechanism for financing programs discussed at Rio, including the projects included in Agenda 21. Developed countries were urged to meet a target of providing 0.7 percent of their gross national product to help developing countries implement sustainable development programs. This funding is to be distributed by the Global Environment Facility (GEF), which is an agreement among the U.N. Environment Programme, the U.N. Development Programme, and the World Bank.

Follow-up to the Earth Summit

The governmental leaders participating in the Earth Summit agreed to a five-year review of progress, to be held in 1997. The review took place in a special session of the U.N. General Assembly, held in New York from June 23 to June 27, 1997. Its formal name was A Special Session of the General Assembly to Review and Appraise the Implementation of Agenda 21. The session studied how well countries, international organizations, and various sectors of civil society responded to the challenges given them by the Earth Summit.

The objectives of the 1997 session were to revitalize and energize commitments to sustainable development, recognize failures and determine the reasons for them, recognize achievements and identify actions to build on them, define priorities beyond 1997, and raise the profile of issues that had not been sufficiently dealt with at Rio. While recognizing the progress that had been made, the state participants concluded that the global environment is continuing to deteriorate. The more than fifty heads of state did agree to further action on freshwater, energy, and transportation issues. There was little, however, in the way of concrete commitments, although the five-year review process will continue.

Colleen M. Driscoll

SUGGESTED READINGS: For an understanding of the issues and the concept of sustainable development, see *Our Common Future* (1987), World Commission on Environment and Development. *Time for Change: A New Approach to Environment and Development* (1992), by Hal Kane and Linda Starke, is a guide to the Earth Summit. *The Global Partnership for Environment and Development: A Guide to Agenda 21* (1992), by UNCED, discusses the twenty-seven principles discussed at the summit. A. Adede, *International Environmental Law from Stockholm 1972 to Rio de Janeiro 1992: An Overview of Past Lessons and Future Challenges* (1992), provides a historical perspective on the work of states to protect the environment through global law.

SEE ALSO: Global Environmental Facility; Sustainable development; United Nations Environment Programme; United Nations Environmental Conference.

Eaubonne, Françoise Marie-Thérèse d'

BORN: March 12, 1920; Paris, France
CATEGORY: Philosophy and ethics

A French novelist, poet, essayist, and journalist, Françoise d'Eaubonne coined the term "ecofeminism" to designate the relationship between ecology and feminist theory, both of which are built on the concept of the interconnectedness of living and nonliving beings.

Although she is well known in the French-speaking world as the author of more than fifty volumes of fiction, nonfiction, essays, and verse, Françoise d'Eaubonne is little familiar to English-speaking readers. Of her dozens of books, only a handful have been translated into English. Her significance to English readers stems from her early advocacy of ecofeminism, which would become a powerful segment of the environmental movement.

The daughter of an insurance executive and a professor, Eaubonne began publishing fiction and poetry in the 1950's. She had published more than a dozen works by the time of her 1974 book *Le Feminisme ou la mort* (feminism or death), in which she coined the term "ecofeminism" to denote the connection between ecology and feminism, as well as the potential for women to effect reforms promoting the well-being of the human species and the earth. In 1979, she elaborated on her ideas in *Ecologie-Féminisme* (ecofeminism). According to Eaubonne and her followers, ecofeminism is both a theory and a sociopolitical movement. As a sociopolitical movement, ecofeminism recognizes the historical connection between women and nature. Ecofeminists believe that the human condition can be transformed in an environment in which people live at peace with one another, nonhuman species, and the natural environment, and they maintain that such a transformation is necessary if life on earth is to survive.

Ecofeminists believe that feminism and ecology need each other. Feminism cannot liberate women from oppression if it does not work to liberate the environment from human exploitation, because an ecology-based science that fails to recognize the oppression of peoples becomes irrelevant and useless. According to ecofeminists, women especially have an important role in saving the environment.

Ecofeminism has forced a re-examination of the idea of development, particularly in Third World countries, where women have traditionally been the managers of resources and providers of food for their people. Ecofeminism has effected a re-evaluation of development in the Third World, based on sustainability rather than on resource extraction and commodity production for profit maximization.

Critics claim that the woman-nature connection claimed by ecofeminists defines women primarily as biological beings. Feminism is largely an attempt to overcome the view that women are primarily childbearers and nurturers. While ecofeminists do not deny this, they tend to ignore the new roles women have assumed in traditional male professions and occupations.

According to critics, the validity of ecofeminism as a theory is questionable because of its references to mysticism, the Goddess tradition, magic, and witchcraft. Ecofeminists especially romanticize prehistoric cultures and the simple lifestyles they depict. Essentially, critics claim that ecofeminism cannot really be called a theory, but is more of an ideology built on a belief in the universal ideal of the life-giving, caring, nurturing woman.

Ecofeminists represent a wide range of disciplines, including the natural, social, and behavioral sciences, as well as philosophy, theology, and literature.

Alexander Scott

SUGGESTED READINGS: Ecofeminism.

Echo Park Dam proposal

DATE: 1956

CATEGORY: Preservation and wilderness issues

The Sierra Club's successful attempt to stop construction of the proposed Echo Park Dam in Dinosaur National Monument in Utah during the 1950's signaled the emergence of the environmental movement as a powerful political force in the United States. The proposed dam would have been located 3.2 kilometers (2 miles) below the confluence of the Green and Yampa Rivers in northwestern Colorado.

The United States Bureau of Reclamation first suggested building a high dam at the site during the 1930's but delayed formally asking Congress for authorization for the project until 1950. The terms of the Organic Act of 1916, passed following the controversy surrounding construction of the Hetch Hetchy Dam in Yosemite National Park in California, prohibited such a project, but administrators within the bureau believed legislators would be willing to make an exception for Echo Park. Subsequent events proved them wrong.

At the time that the Bureau of Reclamation asked Congress for permission to build a high dam within the boundaries of Dinosaur National Monument, few people expected any significant opposition. In the years following the Great Depression, both the federal government and the general public saw dam development as good for the economy and thus good for the country. True, noted photographer Ansel Adams had helped mobilize opposition to hydroelectric development on the Kings River in California a decade earlier, leading to the creation of Kings Canyon National Park, but the Kings River area was home to giant sequoia trees. Its scenic wilderness value was obvious. Dinosaur National Monument, in contrast, appeared barren. As long as the Echo Park Dam would not inundate the dinosaur fossil quarries, advocates of wilderness preservation initially voiced few objections. According to David Brower, executive director of the Sierra Club at the time, even members of his organization described the monument as being nothing but sagebrush.

This changed in 1952 following a Sierra Club member's trip through Dinosaur National Monument. The home movie he took of the canyons within the monument persuaded Brower and others to take a closer look. In 1953 the Sierra Club began organizing rafting trips along the Green River through Dinosaur National Monument. As more people traveled through the spectacular river canyons, opposition to dam construction within the boundaries of the monument grew. Other wilderness preservation and conservation groups, such as the Wilderness Society and the Izaak Walton League, along with prominent writers and politicians, joined with the Sierra Club in fighting the Echo Park Dam proposal.

In 1956 this coalition of preservationists and conservationists won. The Bureau of Reclamation dropped its plans for the Echo Park Dam. The victory for the environmental preservationists proved bittersweet, however. In exchange for the cancellation of plans for Echo Park, environmentalists agreed not to fight the Bureau of Reclamation's plan to build Glen Canyon Dam on the Colorado River, a decision Brower later regretted. Still, by preventing construction of the Echo Park Dam, environmentalists reaffirmed the important principle that no industrial development should ever take place within a national park.

Nancy Farm Männikkö

See also: Brower, David; Dams and reservoirs; Glen Canyon Dam; Sierra Club.

Ecofeminism

Category: Philosophy and ethics

Ecofeminism is an environmental ethic that asserts that a strong link exists between the male domination of both women and nature. As such, ecofeminism is derived from a blending of feminism and a concern for the environment.

There are four broad categories of ecofeminism. Liberal ecofeminism has its roots in the movement to gain political and economic rights for women. By the 1960's some feminists were becoming concerned with environmental problems and began connecting masculine domination of women with what they saw as masculine domination of the environment. Liberal ecofeminists maintained that environmental problems came from the rapid development of natural resources and the failure of government to regulate environmental pollution. Liberal ecofeminists advocate civic activism and governmental action to improve the environment, making it a better place for the nurturing of families.

Marxist feminism provides the philosophical background for the second variety of ecofeminists, who believe that the major problem with the environment is derived from capitalist exploitation of resources and people. Marxist ecofeminists maintain that reproduction, rather than production, should be the central concept for the achievement of a sustainable, just world. If resources are controlled by the workers, then the reckless accumulation of profits, with the concomitant exploitation of the environment and people, will cease. Marxist ecofeminism is a call for women to work toward achieving a just society that will be less exploitive of nature. Neither Marxist ecofeminism nor liberal ecofeminism has a negative view of the impact of science and technology on nature.

Cultural ecofeminism celebrates the relationship between women and nature. Some cultural ecofeminists maintain that there was a period in the past during which women played a dominant political and social role. During this period, humankind and nature existed in harmony. The development of masculine domination of women and nature, at times associated with the development of science and technology, led to the present exploitation of women and the environment. Cultural ecofeminists tend to see a very personal relationship between women and nature derived from the child-bearing role of women. Some call for direct action by women to improve the position of women and the environment, and they may be suspicious of the traditional institutions of society as well as science and technology. Cultural ecofeminism has a variety of critics, including some feminists and deep ecologists.

Social ecofeminism is an evolving movement that traces its roots to feminism and the social ecology of Murray Bookchin. Social ecofeminists indicate that the development of humane, decentralized communities would improve human life and lead to a diminished impact on nature.

They advocate a rethinking of patriarchal social hierarchies. They accept a positive role for science and technology, although they are critical of how society presently uses technology. As social feminism continues to evolve, some of its advocates seem to incline in the direction of cultural ecofeminism, while others incline in the direction of socialist ecofeminism.

The blending of feminist and environmental thought into ecofeminism is multifaceted in philosophical orientation and its call to action. All strands of ecofeminism are tied together by an emphasis on the reproductive role of women as an aspect of the reproductive role of nature.

John M. Theilmann

See also: Deep ecology; Eaubonne, Françoise Marie-Thérèse d'; Environmental ethics; Social ecology.

Ecology

Category: Ecology and ecosystems

Ecology is the science that studies the relationships among organisms and their biotic and abiotic environments. The term "ecology" is commonly, but mistakenly, used by people to refer to the environment or to the environmental movement.

The study of ecological topics arose in ancient Greece, but these studies were part of a catch-all science called natural history. The earliest attempt to organize an ecological science separate from natural history was made by Carl Linnaeus in his essay *Oeconomia Naturae* (1749; *The Economy of Nature*, 1749), which focused on the balance of nature and the environments in which various natural communities existed. Although the essay was well known, the eighteenth century was dominated by biological exploration of the world, and his science did not develop.

Early Ecological Studies

The study of fossils led some naturalists to conclude that many species known only as fossils must have become extinct. However, Jean-Baptiste Lamarck argued in his *Philosophie zoologique* (1809; *Zoological Philosophy*, 1914) that fossils represented the early stages of species that evolved into different species that were still living. In order to refute this claim, geologist Charles Lyell mastered the science of biogeography and used it to argue that species do become extinct and that competition from other species seemed to be the main cause. English naturalist Charles Darwin's book *On the Origin of Species* (1859) blends his own researches with the influence of Linnaeus and Lyell in order to argue that some species do become extinct, but existing species have evolved from earlier ones. Lamarck had underrated and Lyell had overrated the importance of competition in nature.

Although Darwin's book was an important step toward ecological science, he and his colleagues mainly studied evolution rather than ecology. However, German evolutionist Ernst Haeckel realized the need for an ecological science and coined the name *oecologie* in 1866. Yet it was only in the 1890's that steps were actually taken to organize this science. Virtually all of the early ecologists were specialists in the study of particular groups of organisms, and it was only in the late 1930's that some efforts were made to write textbooks covering all aspects of ecology. Since the 1890's, most ecologists have viewed themselves as plant ecologists, animal ecologists, marine biologists, or limnologists.

Nevertheless, general ecological societies were established. The first was the British Ecological Society, which was founded in 1913 and began publishing the *Journal of Ecology* in the same year. Two years later, ecologists in the United States and Canada founded the Ecological Society of America, which began publishing *Ecology* as a quarterly journal in 1920; in 1965 *Ecology* began appearing bimonthly. These two societies have been the leading organizations in ecology ever since, though other national societies have also been established. More specialized societies and journals also began appearing; for example, the Limnological Society of America was established in 1936 and expanded in 1948 into the American Society of Limnology and Oceanography. It publishes the journal *Limnology and Oceanography*.

Grasslands were one of the five major biomes identified by the International Biological Program during its effort to study large-scale ecological environments. (McCrea Adams)

Although Great Britain and Western Europe were active in establishing ecological sciences, it was difficult for their trained ecologists to obtain full-time employment that utilized their expertise. European universities were mostly venerable institutions with fixed budgets; they already had as many faculty positions as they could afford, and these were all allocated to the older arts and sciences. Governments employed few, if any, ecologists. The situation was more favorable in the United States, Canada, and Australia, where universities were still growing. In the United States, the universities that became important for ecological research and the training of new ecologists were mostly in the Midwest. The reason was that eastern universities were similar to European ones in being well established with scientists in traditional fields. An exception in the East is Duke University, which was established in 1924.

Ecology After 1950

Ecological research in the United States was not well funded until after World War II. With the advent of the Cold War, science was suddenly considered important for national welfare. In 1950 the U.S. Congress established the National Science Foundation, and ecologists were able to make the case for their research along with the other sciences. The Atomic Energy Commission had already begun to fund ecological researches by 1947, and under its patronage the Oak Ridge Laboratory and the University of Georgia gradually became important centers for radiation ecology research. Another important source of research funds was the International Biological Program (IBP), which, though international in scope, depended upon national research funds. It got underway in the United States in 1968 and was still producing publications in the 1980's. Even though no new funding sources were created for the IBP, its existence meant that more research money flowed to ecologists than otherwise would have.

Ecologists learned to think big. Computers became available for ecological research shortly before the IBP got underway, and so computers and the IBP became linked in ecologists' imaginations. Earth Day in 1970 helped awaken Americans to the environmental crisis, and they expected ecologists to advise on environmental policy. The IBP encouraged a variety of studies, but in the United States, biome (large-scale environments) and ecosystem studies were most prominent. The biome studies were grouped under the headings of desert, eastern deciduous forest, western coniferous forest, grassland, and tundra (a proposed tropical forest program was never funded). Although the IBP has ended, a number of the biome studies have continued at a reduced level.

Ecosystem studies are also large scale, at least in comparison with many previous ecological studies, though smaller in size than a biome. The goal of ecosystem studies was to gain a total understanding of how an ecosystem—such as a lake, a river valley, or a forest—works. IBP funds enabled research students to collect data, while computers processed the data. However, ecologists could not agree on what data to collect, how to compute outcomes, and how to interpret the results. Therefore, thinking big did not always produce impressive results.

Plant and Animal Ecologies

Because ecology is enormous in scope, it was bound to have growing pains. It arose at the same time as the science of genetics, but since genetics is a cohesive science, it reached maturity much sooner than ecology. Ecology can be subdivided in a wide variety of ways, and any

Early studies of marine ecology focused on coastal ecosystems. (McCrea Adams)

collection of ecology textbooks will show how diversely it is organized by different ecologists. Nevertheless, self-identified professional subgroups tend to produce their own coherent findings.

Plant ecology progressed more rapidly than other subgroups and has retained its prominence. In the early nineteenth century, German naturalist Alexander von Humboldt's many publications on plant geography in relation to climate and topography were a powerful stimulus to other botanists. By the early twentieth century, however, the idea of plant communities was the main focus for plant ecologists. Henry C. Cowles began his studies at the University of Chicago in geology but switched to botany and studied plant communities on the Indiana dunes of Lake Michigan. He received his doctorate in 1898 and stayed at that university as a plant ecologist. He trained others in the study of community succession.

Frederic E. Clements received his doctorate in botany in the same year from the University of Nebraska. He carried the concept of plant community succession to an extreme by taking literally the analogy between the growth and maturation of an organism and that of a plant community. His numerous studies were funded by the Carnegie Institute in Washington, D.C., and even ecologists who disagreed with his theoretical extremes found his data useful. Henry A. Gleason was skeptical; his studies indicated that plant species that have similar environmental needs compete with each other and do not form cohesive communities. Although Gleason first expressed his views in 1917, Clements and his disciples held the day until 1947, when Gleason's individualistic concept received the support of three leading ecologists. Debates over plant succession and the reality of communities helped increase the sophistication of plant ecologists and prepared them for later studies on biomes, ecosystems, and the degradation of vegetation by pollution, logging, and agriculture.

Animal ecology emerged from zoology. A good illustration of the transition is the career of Stephen A. Forbes, professor of zoology and entomology at the University of Illinois and head of the State Laboratory of Natural History. His responsibilities focused his attention on the practical uses of zoology for agriculture and for fish and wildlife management; he also had a theoretical interest in both evolution and ecology. He brought together these various interests in his 1887 essay "The Lake as a Microcosm."

One important aspect of the early history of animal ecology was the attempt to mathematically understand and describe the growth or decline of animal populations. Mathematics is a universal language, and the fluctuation of animal populations is a universal problem. Therefore, this aspect of ecology developed globally rather than regionally. It was also possible to use the same mathematical methods to study population changes from the standpoints of ecology, evolution, and genetics. This situation promoted a lively exchange and rapid progress in the development of population ecology in the United States, Great Britain, Australia, Italy, and the Soviet Union. The great challenge was to develop equations that could help predict the pattern of population fluctuations. It turned out to be easier to develop mathematical models than to understand or predict the fluctuations of real populations. Nevertheless, these efforts eventually paid off in the ability of fish and wildlife biologists to gauge the level of harvesting that could maintain stable populations versus the level that would cause a population to decline.

Marine Ecology and Limnology

Marine ecology is viewed as a branch of either ecology or oceanography. Early studies were made either from shore or close to shore because of the great expense of committing oceangoing vessels to research. The first important research institute was the Statione Zoologica at Naples, Italy, founded in 1874. Its successes soon inspired the founding of others in Europe, the United States, and other countries. Karl Möbius, a German zoologist who studied oyster beds, was an important pioneer of the community concept in ecology. Great Britain dominated the seas during the nineteenth century and made the first substantial commitment to deep-sea research by equipping the HMS *Challenger* as an oceangoing laboratory that sailed the world's seas from 1872 to 1876. Its scientists collected so many specimens and so much data that they called upon marine scientists in other countries to help them write the fifty large volumes of reports (1885-1895). The development of new technologies and the funding of new institutions and ships in the nineteenth century enabled marine ecologists to monitor the world's marine fisheries and other resources and provide advice on harvesting marine species.

Limnology is the scientific study of bodies of fresh water. The Swiss zoologist François A. Forel coined the term and also published the first textbook on the subject in 1901. He taught zoology at the Académie de Lausanne and devoted his life's researches to understanding Lake Geneva's characteristics and its plants and animals. In the United States in the early twentieth century, the University of Wisconsin became the leading center for limnological research and the training of limnologists, and it has retained that preeminence. Limnology is important for managing freshwater fisheries and water quality.

Frank N. Egerton

SUGGESTED READINGS: For a theoretical perspective on ecology, see Robert P. McIntosh, *The Background of Ecology* (1985), and Timothy F. H. Allen and Thomas W. Hoekstra, *Toward a Unified Ecology* (1992). For insight into the state of ecology worldwide, see Edward J. Kormondy and J. Frank McCormick, editors, *Handbook of Contemporary Developments in World Ecology* (1981), and Robert W. Hiatt, editor, *World Directory of Hydrobiological and Fisheries Institutions* (1963). For information on American ecology, see Victor E. Shelford, *The Ecology of North America* (1963), and David G. Frey, editor, *Limnology in North America* (1963). On the history of ecology, see Frank N. Egerton, "The History of Ecology," *Journal of the History of Biology* (Summer 1983-Spring 1985), and Eric L. Mills, *Biological Oceanography: An Early History, 1870-1960* (1989). For ecology in human affairs, see Mark J. McDonnell and S. T. A. Pickett, editors, *Humans as Components of Ecosystems* (1993).

SEE ALSO: Balance of nature; Biodiversity; Ecosystems; Food chains; Restoration ecology; Social ecology.

Ecology, concept of

CATEGORY: Philosophy and ethics

Ecology is the scientific study of ecosystems, which are generally defined as local units of nature; examples are ponds, prairies, and coral reefs. Ecosystems consist of both biotic (living) and abiotic (nonliving) components. Included among the biotic components are plants, animals, and microorganisms. The abiotic components are the physical factors of the ecosystem.

The roots of ecology can be traced to the writings of early Greek philosophers such as Aristotle and Theophrastus, who were keen observers of plants and animals in their natural habitats. During the nineteenth century, German biogeographer Alexander von Humboldt and English naturalist Charles Darwin wrote detailed descriptions of their travels. They recognized that the distribution of living things is determined by such factors as rainfall, temperature, and soil.

The word "ecology" was first proposed in 1869 by the German biologist Ernst Haeckel. It soon came to be defined as "environmental biology," or the effect of environmental factors on living things. At the beginning of the twentieth century, American botanist Henry Cowles established plant succession as a major concept of ecology. During the next few decades, F. E. Clements helped establish plant ecology as a recognized branch of biology. Animal ecology developed separately and slightly later. Perhaps best known was Victor Shelford, who wrote *Animal Communities in Temperate America* in 1913. Reflecting the independent development of plant and animal ecology, most ecological studies in the early twentieth century were concerned with either plant or animal communities, but not both. Furthermore, most were descriptive rather than being involved with explanations of fundamental ecological processes.

However, other lines of research during this time increasingly emphasized interrelationships among all life forms, especially those within lakes. From such beginnings emerged the concept of the ecosystem. The term, first used by British ecologist Arthur G. Tansley in 1935, is now considered the foundation stone of ecology.

Ecologists place all the organisms of an ecosystem into three categories: producers, consumers, and decomposers. Producers include algae and green plants that, because of their photosynthetic ability, produce all the food for the ecosystem. Consumers are animals that feed directly on the producers. Decomposers are bacteria and fungi that break the large organic molecules of dead plants and organisms down into simpler substances. Ecosystems are dynamic. Each day, matter (nutrients) cycles through ecosystems as consumers eat producers, then moves to decomposers, which release nutrients into the soil, air, and water. Producers absorb nutrients as they photosynthesize, thus completing the cycle. Energy flows from the sun to producers, then consumers, and finally to decomposers.

Other changes in ecosystems occur over longer periods of time. Large-scale disturbances such as fire, logging, and storms initiate gradual, long-term changes known as ecological succession. Following a major disturbance, an ecosystem of pioneer species exists for a while but is soon replaced by a series of other temporary ecosystems. Eventually a permanent, or climax, ecosystem is formed, the nature of which is primarily determined by climate. Although generally considered to be stable, climax communities are subject to gradual changes caused by climatic fluctuations or subsequent disturbances.

Ecologists attempt to name and classify ecosystems in a manner similar to the way that taxonomists name and classify species. Ecosystems may be named according to their dominant plants, such as deciduous forests, prairies, and evergreen forests. Others, such as coral reefs, are named according to their dominant animals. Physical factors are used to name deserts, ponds, tidal pools, and other ecosystems.

The science of ecology has developed a few major branches as well as several areas of specialization. Some would consider plant and animal ecology as major branches, but a more meaningful distinction can be made between theoretical, or academic, ecology on one hand and applied, or practical, ecology on the other. Among the specialties that have developed from theoretical ecology are autecology (study at the level of individuals or species), synecology (study at the community level), and the ecosystems approach, which is largely concerned with the flow of energy and the cycling of nutrients within ecosystems. As one might expect, theoretical ecologists are generally associated with universities where their basic research contributes to the understanding of a great diversity of ecosystems.

Applied ecologists are employed by a variety of governmental and environmental agencies, as well as by universities. Their primary objective is to apply fundamental principles of ecology and related disciplines to the solution of specific problems. Forestry, although it developed independently from ecology, may be considered a specialty within the field of applied ecology. Foresters must be knowledgeable of a wide range of factors that influence the accumulation of tree biomass. Wildlife management, once concerned with only game species, has been extended to include a wide range of nongame animal species as well. Another branch of applied ecology is conservation biology. Concerned with biodiversity in all its aspects, conservation biologists attempt to prevent the extinction of threatened species around the globe. Disturbed ecosystems are rehabilitated by scientists working in a related field called restoration ecology.

The efforts of ecologists, whether theoretical or applied, represent the best attempts to understand and solve the many environmental challenges that humankind faces. Climate change, pollution, and other global and local problems contribute to a loss of biodiversity; all are made worse by increasing human populations.

Thomas E. Hemmerly

SUGGESTED READINGS: A wide range of college-level textbooks of general ecology is available; especially recommended is *Ecology and Field Biology* (1996), by Robert L. Smith. For an overview of the practical aspects of ecology, see Edward I. Newman, *Applied Ecology* (1993). Perhaps the most readable of the conservation biology texts is *Fundamentals of Conservation Biology* (1996), by Malcolm L. Hunter, Jr.

SEE ALSO: Balance of nature; Biodiversity; Ecosystems; Extinctions and species loss; Food chains; Tansley, Arthur G.

Ecosystems

CATEGORY: Ecology and ecosystems

The word "ecosystem" was coined by English ecologist Arthur G. Tansley in 1935 to define a specific area of the earth and the attendant interactions among organisms and the physical-chemical environment present at the site.

Ecosystems are viewed by ecologists as basic units of the biosphere, much like cells are considered by biologists as the basic units of an organism, self-organized and self-regulating entities within which energy flows and resources are cycled in a coordinated, interdependent manner to sustain life. Disruptions and perturbations to, or within, the unit's organization or processes may reduce the quality of life or cause its demise. Ecosystem boundaries are arbitrary and are usually defined by the research or management questions being asked. An entire ocean can be viewed as an ecosystem, as can a single tree, a rotting log, or a drop of pond water. Those systems with tangible boundaries—such as forests, grasslands, ponds, lakes, watersheds, seas, or oceans—that separate them from contiguous ecosystems are especially useful to ecosystem research.

ECOSYSTEM RESEARCH

The ecosystem concept was first put to use by American limnologist Raymond L. Lindeman in the classic study he conducted on Cedar Bog Lake, Minnesota, which resulted in his article, "The Trophic Dynamic Aspect of Ecology" (1942). Lindeman's study, along with the publication of Eugene P. Odum's *Fundamentals of Ecology* (1953), converted the ecosystem notion into a guiding paradigm for ecological studies, thus making it a concept of theoretical and applied significance.

Ecologists study ecosystems as integrated components through which energy flows and resources cycle. Although ecosystems can be divided into many components, the four fundamental ones are abiotic (nonliving) resources, producers, consumers, and decomposers. The ultimate sources of energy come from outside the boundaries of the ecosystem (solar energy or chemothermo energy from deep-ocean hydrothermal vent systems). Since this energy is captured and transformed into chemical energy by producers and translocated through all biological systems via consumers and decomposers, all organisms are considered as potential sources of energy. Since the first law of thermodynamics states that energy cannot be created or destroyed, it follows the tenets of the second law of thermodynamics upon reaching the end of its flow and enters the state of entropy. In this state energy is completely randomized and divested of its ability to do work.

Abiotic resources (water, carbon dioxide, nitrogen, oxygen, and other inorganic nutrients and minerals) primarily come from within the boundaries of the ecosystem. From these, producers utilizing energy synthesize the biomolecules, which are transformed, upgraded, and degraded as they cycle through the living systems that comprise the various components. The destiny of these bioresources is to be degraded to their original abiotic forms and be recycled.

The ecosystem approach to environmental research is a major endeavor. It requires amassing large amounts of qualitative and quantitative data relevant to the structure and function of each component. These data are then integrated among and between the components in an attempt to determine linkages and relationships. This is attractive because it utilizes the holistic approach to the examination of complex systems. The ecosystem approach to research involves the use of systems information theory, predictive models, and computer application and simulations. As ecosystem ecologist Frank B. Golly stated in his book *A History of the Ecosystem Concept in Ecology* (1993), the ecosystem approach to the study of ecosystems is "machine theory applied to nature."

ECOSYSTEM RESEARCH PROJECTS

Initially ecosystem ecologists used the principles of Tansley, Lindeman, and Odum to determine and describe the flow of energy and resources through organisms and their environment. The objectives of this research were to answer the fundamental academic questions that plagued ecologists, such as those concerning controls on ecosystem productivity: What are the connections between animal and plant productivity? How are energy and nutrients transformed and cycled in ecosystems?

Once fundamental insights were obtained, computer-model-driven theories were constructed that linked and integrated the components to provide an understanding of the biochemophysical dynamics that govern ecosystems. Responses of ecosystem components could then be examined by altering or manipulating parameters within the simulation model. Early development of the ecosystem concept culminated, during the 1960's, in defining the approach of ecosystem studies. Ecosystem projects were primarily funded under the umbrella of the International Biological Program (IBP). Other funding came from the Atomic Energy Commission and the National Science Foundation. The intention of the IBP was to integrate data collected by teams of scientist at research sites that were considered typical of wide regions. Although the IBP was international in scope, studies in the United States received the greatest portion of the funds—approximately $45 million during the life of IBP (1964-1974).

Five major IBP ecosystem studies involving grasslands, tundra, deserts, coniferous forests, and deciduous forests were undertaken. The Grasslands Project, directed by George Van Dyne, received the largest portion of IBP funding, primarily because of the dynamism of the director. The Grasslands Project set the research stage for, and gave direction to, the other four endeavors. However, since the research effort was so extensive in scope, the objectives of the IBP were not totally realized. Because of the large number of scientists involved and their cultural, visionary, and motivational differences, little coherence in results was obtained even within the same project. Other problems plagued the early development of the concept. Evolutionary biologists-ecologists faulted the concept's ability to link with the evolutionary theory. This link could not be established because of the lack of fundamental knowledge of individuals, species, populations, and behavior. A more pervasive concern, voiced by environmentalists and scientists alike, was that little of the information obtained from the ecosystem simulation models could be applied to the solution of existing environmental problems.

An unconventional project partially funded by the IBP called the Hubbard Brook Watershed Ecosystem—located in New Hampshire and studied by F. Herbert Bormann and Gene E. Likens—redirected the research approach for studying ecosystems from the IBP computer-model-driven theory to more conventional scientific methods of study. Under the Hubbard Brook approach, an ecosystem phenomenon is observed and noted. A pattern for the phenomenon's behavior is then established for observation, and questions are posed about the behavior. Hypotheses are developed to allow experimentation and manipulation in an attempt to explain the observed behavior. This approach requires detailed scrutiny of the ecosystem's subsystems and their linkages. Since each ecosystem functions as a unique entity, this approach has more utility. The end results provide insights specific to the activities observed within particular ecosystems. Explanations for these observed behaviors can then be made in terms of biological, chemical, or physical principles.

Utility of Concept

Publicity from the massive ecosystem projects and the publication of Rachel Carson's *Silent Spring* (1962) helped stimulate the environmental movement of the 1960's. The public began to realize that human activity was destroying the bioecological matrices that sustained life. By the end of the 1960's, the applicability of the IBP approach to ecosystem research was proving to be purely academic and provided few solutions to the problems that plagued the environment. Scientists realized that, because of the lack of fundamental knowledge about many of the systems and their links and because of the technological shortcomings that existed, ecosystems could not be divided into three to five components and analyzed by computer simulation.

The more applied approach taken in the Hubbard Brook project, however, showed that the ecosystem approach to environmental stud-

Ecosystems with tangible boundaries, such as this rain forest in Olympic National Park in Washington State, are particularly useful to ecologists who are studying the complex interrelationships among organisms and their environments. (Jim West)

ies could be successful if the principles of the scientific method were used. The Hubbard Brook study area and the protocols used to study it were clearly defined. This ecosystem allowed hypotheses to be generated and experimentally tested. Applying the scientific method to the study of ecosystems had practical utility for the management of natural resources and for testing some possible solutions to environmental problems. When perturbations such as diseases, parasites, fire, deforestation, and urban and rural centers disrupt ecosystems from within, this approach helps answer problems and defines potential mitigation and management plans. Similarly, external causative agents within airsheds, drainage flows, or watersheds can also be considered.

The principles and research approach of the ripening ecosystem concept are being used to define and attack the impact of environmental changes caused by humans. Such problems as human population growth, apportioning of resources, toxification of biosphere, loss of biodiversity, global warming, acid rain, atmospheric ozone depletion, land-use changes, and eutrophication are being holistically examined. Management programs related to woodlands (the New Forestry program) and urban and rural centers (the Urban to Rural Gradient Ecology, or URGE, program), as well as other governmental agencies that are investigating water and land use, fisheries, endangered species, and exotic species introductions, have found the ecosystem perspective useful.

Ecosystems are also viewed as systems that provide the services necessary to sustain life on earth. Most people either take these services for granted or do not realize that such natural processes exist. Ecosystem research has identified seventeen naturally occurring services, including water purification, regulation, and supply, as well as atmospheric gas regulation and pollination. An article by Robert Costanza and others, "The Value of the World's Ecosystem Service and Natural Capital," placed a monetary cost to humanity should the service, for some disastrous reason, need to be maintained by human technology. The amount is staggering, ranging from $16-54 trillion per year and averaging $33 trillion. Humanity could not afford this; the global gross national product is only about $18-20 trillion.

Academically, ecosystem science has been shown to be a tool to dissect environmental problems, but this has not been effectively demonstrated to the public and private sectors, especially decision makers and policymakers at governmental levels. Additionally, the idea that healthy ecosystems provide socioeconomic benefits and services is controversial. In order to bridge this gap between academia and the public, Scott Collins of the National Science Foundation suggested to the Association of Ecosystem Research Centers that ecosystem scientists be "bilingual"; that is, they should be able speak their scientific language and translate it so that the nonscientist can understand.

Richard F. Modlin

SUGGESTED READINGS: An outstanding basic text that covers ecosystems and ecological concepts is Stanley I. Dodson et al., *Ecology* (1998). A complete treatment of the ecosystem concept and ecosystem management is made by Kristiina A. Vogt et al. in *Ecosystems: Balancing Science with Management* (1997). Gene E. Likens provides critical analyses of ecosystem functions and structures in *The Ecosystem Approach: Its Use and Abuse* (1992). A history of the ecosystem concept and a discussion of its impact on understanding problems and their potential solutions is contained in Frank B. Golly, *A History of the Ecosystem Concept in Ecology* (1993). Gretchen C. Daily, editor, *Nature's Services: Societal Dependence on Natural Ecosystems* (1997), contains several essays in which scientists examine how ongoing processes in ecosystems provide the necessary biochemophysical pathways that serve the needs that allow biological life to exist on earth.

SEE ALSO: Balance of nature; Biodiversity; Biosphere concept; Biosphere reserves; Ecology; Tansley, Arthur G.

Ecotage

CATEGORY: Philosophy and ethics

Ecotage refers to sabotage tactics used by radical environmentalists to stop developers from destroying wilderness areas.

In 1972 the group Environmental Action published the handbook *Ecotage!* in which they compiled ideas about how to sabotage environmentally destructive projects. Edward Abbey's novel *The Monkey Wrench Gang* (1975), featuring a small group of ecoguerillas who destroyed construction equipment to stop development in the southwestern United States, inspired Dave Foreman and others to start the radical environmental movement Earth First! In 1985 Foreman published *Ecodefense: A Field Guide to Monkeywrenching* (1985), a manual of ecotage methods and related issues such as safety and security. In the early 1990's Foreman wrote that "those willing to commit ecotage are needed today as never before." Several other environmental advocates, such as Greenwar International, also promoted ecotage.

These proponents of ecotage, known as ecoteurs, were dismayed by industrial development of wilderness areas that the government refused to protect. Frustrated that civil disobedience did not achieve their goals, ecoteurs decided to preserve the environment by illegally damaging machinery used to degrade wilderness areas. Such militant acts of destruction often impeded future development efforts and reduced industrialists' profits.

Many ecoteurs, including Foreman, had belonged to mainstream environmental groups during the 1960's and 1970's but became disillusioned by the dominance of conservative political leaders during the 1980's. Ecoteurs were critical of environmentalists in such organizations as the Sierra Club for being passive, ignoring opportunities to preserve the wilderness, and appeasing industrialists and governmental agencies by sacrificing the environment. They denounced environmentalists with anthropocentric attitudes who viewed the environment as a source of production and resources to fulfill human needs.

Most ecoteurs considered themselves a symbiotic part of the environment and justified their bold, destructive conduct as acts of self-defense. "It is time to act heroically and admittedly illegally in defense of the wild," Foreman stated, "to put a monkeywrench into the gears of the machinery destroying natural diversity." He stressed that ecotage "is sabotage, not terrorism, because it's about property destruction. It's saying, I'm operating as part of the wilderness, defending myself."

Sometimes calling themselves "ecowarriors," ecoteurs strategically selected their targets for ecotage in an effort to disrupt environmentally harmful development. They focused on obstructing logging and strip mining operations. Common monkeywrenching procedures included tree spiking, in which nails were driven into trees in old-growth forests to deter loggers. They also poured sand into the gas tanks of trucks and bulldozers and sometimes slashed tires and blew up machinery. Other measures included pulling up survey stakes, cutting power lines, blockading roads, stealing machinery parts, and ruining tools. Some ecoteurs damaged offices belonging to the U.S. Forest Service to protest logging in national forests, while others sprayed baby seals with dye so their fur would be unusable for coats. The Federal Bureau of Investigation (FBI) and industrial leaders offered rewards for the arrest and conviction of ecoteurs.

Many of the people in the Earth First! movement came to believe that ecotage often did more to ostracize people than protect the environment; therefore, they reassessed their protest tactics during the 1990's. Columns in the *Earth First! Journal* discussed individuals' ecotage efforts, but Earth First!ers disagreed about the role of ecotage in the environmental movement. While some people believed all ecotage should be ceased, others specified curtailing certain acts of sabotage that might result in injuries to loggers or miners.

Elizabeth D. Schafer

SEE ALSO: Abbey, Edward; Earth First!; Ecoterrorism; Environmental ethics; Foreman, Dave; Monkeywrenching; Preservation.

Ecoterrorism

CATEGORY: Philosophy and ethics

Ecoterrorism is a term coined by critics of radical environmental activists to characterize the environmentalists' clandestine attempts to disrupt environmental damage or prevent cruelty to animals as undemocratic acts of terrorism.

Ecoterrorism began in the United States in the 1970's, peaked in the early 1990's, and waned later in that decade. Other analogous terms, such as "ecological sabotage," "monkeywrenching," "ecotage," and "decommissioning," roughly connote the same concept without pejorative implications. Monkeywrenching was coined by Edward Abbey in his novel *The Monkey Wrench Gang* (1976). Ecotage refers to acts of sabotage for environmental ends. The radical environmentalist group Earth First! probably engaged in more ecoterrorist activities during that period than any other group, including People for the Ethical Treatment of Animals (PETA), the Sea Shepherd Conservation Society, and the Animal Liberation Front (ALF). A variety of radical activities and ways to carry them out were described in *Ecodefense: A Field Guide to Monkeywrenching* (1985), written by David Foreman, cofounder of Earth First!

Lorenz Otto Lutherer and Margaret Sheffield Simon chronicle the history of ecoterrorism conducted by animal rights groups in *Targeted: The Anatomy of Animal Rights Attack* (1992). They describe numerous cases of break-ins, vandalism, theft of animals and equipment, and threats of violence against researchers and businesspeople. Animal rights activists who engage in these activities, these authors maintain, are a menace to society, scientific and economic progress, and democracy itself.

The types of activities that led critics to characterize the activists as ecoterrorists can be divided into three categories. The first involves legal protests that are still clear cases of monkeywrenching. Examples are sit-ins in front of offices, laboratories, factories, bulldozers and even locomotives, sometimes accentuated by activists chaining themselves to gates or trees. Activists often engaged in a process called tree spiking, in which metal spikes are driven into trees to prevent logging with chain saws.

The second group of activities is characterized by their illegal nature, such as decommissioning machinery by pouring sand, sugar, or water into gas tanks and damaging oil exploration equipment by smashing distributors and spark plugs. Most break-ins at animal research facilities are included in this category. The third group consists of more daring but potentially hazardous operations such as the ramming of whaling ships by Sea Shepherd activists.

Although some environmentalists oppose violence, others consider monkeywrenching as "the conscience of the environmental movement." However, most agree that the term "ecoterrorism" more aptly applies to those who plunder the earth and its atmosphere in the name of capitalism and progress. Some environmental activists compare themselves to resistance fighters of World War II, seeing environmental damage as equivalent to the Holocaust and other war crimes meriting drastic reprisals. In general, however, activists believe action ought to be nonviolent. They are aggressive only for preventative measures. For example, to avoid injuries, antilogging activists inform the loggers and mill workers about trees that contain spikes by sending letters, telephoning, or marking trees with paint.

Opinions regarding the moral implications of ecoterrorism vary. Many environmental activists, whether or not they approve, agree that it works. However, critics and even some activists feel that some radical activists have damaged the reputations of many hard-working, peace-loving environmentalists around the world.

Chogollah Maroufi

SEE ALSO: Earth First!; Ecotage; Foreman, Dave; Monkeywrenching.

Ecotourism

Category: Preservation and wilderness issues

Ecotourism is the promotion of tourism in wilderness areas to fund preservation and conservation efforts. While supporters claim that ecotourist dollars help save endangered wilderness areas, critics note that ecotourism can contribute to the continued destruction of fragile ecosystems.

Improvements in travel after World War II, especially the development of jet aircraft, dramatically increased the number of tourists in all areas of the globe. With this increase came an interest in visiting exotic locations to enjoy unspoiled landscapes, view unusual wildlife, or participate in recreational adventures. The rise in tourism as a leisure activity brought economic benefits to some regions but was often accompanied by negative social and environmental impacts. At times, local communities were displaced, and ecosystems were altered in order to provide hotels and other services for guests. The growing number of tourists threatened the very vistas and animals that had lured visitors in the first place.

Despite these problems, environmentalists have recognized the potential for tourism to benefit preservation efforts. While many developing countries perceive the exploitation of natural resources as a potential source of revenue—an attitude that threatens ecosystems such as rain forests—environmentalists believe that these resources can provide an alternative source of income—tourist dollars—that will give governments an incentive to preserve wilderness areas. Coupled with a "no-impact" ethic, ecotourism is seen as a method of saving ecosystems that are quickly disappearing.

Ecotourism proponents point to the regions that have successfully used ecotourism to preserve environments and support local communities. Tourism in a rain forest in Ecuador has prevented oil exploration and provides income to native peoples in the area. A former director of a mountain gorilla project in Africa credits ecotourism with the survival of mountain gorillas and their habitats. An estimated 40 percent of the tourists visiting Costa Rica participate in ecotourist activities. Kenya, one of the most popular ecotourist destinations, receives more than 650,000 visitors per year, generating about $350 million in income. Estimates indicate that each elephant in Kenya's wildlife preserve has brought more than $14,000 into Kenya, close to $1 million over its lifetime. The profits provide the Kenyan government with a powerful incentive to ensure the elephants' survival.

Ecotourism is not without its drawbacks. Observers in Costa Rica, for example, have noted that although some national parks are large, visitors want to see specific sites, which leads to overcrowding, trail erosion, and pollution. Scientists have noted changes in the behavioral patterns of local wildlife. Growth in ecotourism also promotes development outside protected areas, with attendant environmental degradation. In addition, ecotourism requires an appreciation of the no-impact philosophy, which not all ecotourists necessarily possess. Studies have revealed that the typical ecotourist is highly educated and wealthier than the average tourist. These individuals are more likely to have an interest in the environment. However, as ecotourism becomes more popular, some critics fear that individuals with little interest in preservation might come to dominate the ecotours.

In addition to the behavior of tourists, the actions of some tour operators also threaten the attractiveness of ecotourism as an environmentally friendly activity. As ecotourism grows in popularity, the number of companies that offer ecotours also grows. Lacking any regulation or even consensus on what constitutes ecotourism, operators can sell their product as an ecotour even if it does not meet the standards of the term as it is usually understood. One Costa Rican project touted as an ecodevelopment included a shopping center and a golf course, facilities that most people agreed had little to do with ecotourism.

Studies indicate that local communities usually do not benefit from ecotourist activities. In many countries, foreign interests often owned facilities and tourist sites, thus ensuring that profits will flow out of the local area. In Nepal local families make little money while serving as porters for tourists. In areas where locals do profit, there are still problems. Some communities in Costa Rica have moved from a subsistence to a market economy, a transition that belies the ethic of maintaining the integrity of local cultures.

Critics maintain that the concept of ecotourism is flawed. They argue that ecotourists merely pave the way for mass tourists, people who demand the comforts of home while they visit remote areas. Moreover, the developing nations that offer ecotourist attractions are often the least able to invest the funds necessary to counter the impact of tourism. Only about 2 percent of the tourist dollars spent in Kenya go into park management. The Ecuadoran government appears unable to manage the deluge of tourists visiting the Galápagos Islands. Although government regulations allow only 25,000 tourists per year, almost 60,000 people visited the islands in 1994, promoting a frenzy of economic development that caused appreciable environmental damage. Opponents claim that ecotourism is merely a variation of tourism that will inevitably despoil the very areas it is intended to protect. Environmental advocates recommend that tourists carefully review the literature of any organization that offered ecotours to ensure that it meets the standards of environmental and cultural preservation.

Thomas Clarkin

SUGGESTED READINGS: Deborah McLaren's *Rethinking Tourism and Ecotravel: The Paving of Paradise and What You Can Do to Stop It* (1998) advocates ecotourism while acknowledging its potential downsides. Lesley France edited *The Earthscan Reader in Sustainable Tourism* (1997), a collection of brief essays that includes material on ecotourism. Erlet Cater and Gwen Lowman edited *Ecotourism: A Sustainable Option?* (1994), which includes case studies examining ecotourism in several different regions. Elizabeth Boo's *Ecotourism: The Potentials and Pitfalls* (1990) also offers several case studies.

See also: Environmental economics; Green marketing; National parks; Nature reserves.

Ehrlich, Paul

Born: May 29, 1932; Philadelphia, Pennsylvania
Category: Population issues

Environmental philosopher Paul Ehrlich has published several influential books that deal with such topics as the dangers of overpopulation and the possible effects of nuclear war.

Paul Ehrlich, the son of William and Ruth Ehrlich, displayed an early interest in nature study. Following high school, he enrolled at the University of Pennsylvania, graduating in 1953 with a zoology degree. Ehrlich conducted his graduate work at the University of Kansas, earning his M.A. in 1955 and his Ph.D. in 1957. His doctoral research was in the field of entomology, and his first published book dealt with identification of butterflies. He married Anne Howland in 1954. They had one child, a daughter named Lisa Marie.

Following his graduation in 1957, Ehrlich worked as a research associate on various studies. He joined the faculty of Stanford University's Biology Department in 1959 and became a full professor in 1966. During this time his interest in ecology and conservation developed more fully. He was promoted to Bing Professor of Population Studies at Stanford in 1976.

Ehrlich is best known to the general public for his vigorous support of conservation, including what some consider radical ideas for preserving resources of the earth. Foremost among the topics he addressed was the potentially devastating effect that increased human consumption could have upon the environment. His book *The Population Bomb* (1968) was widely circulated and caused much discussion regarding worldwide population growth. Ehrlich maintained that increased population, coupled with decreased food production, would, in the following few decades, result in billions of deaths from starvation. This prediction, however, did not come true—although population growth is increasing, food production, with modern agricultural technology, is increasing at an even faster rate. Ehrlich suggested that the government place limitations upon the number of children a couple could have (he limited his family to only one offspring) and went so far as to suggest forced vasectomies for males in overpopulated countries that exhibited rapid birth rates.

Ehrlich's suggestions were not limited to population control. He criticized developed countries, especially the United States, for unwise and excessive consumption of natural resources. He predicted that as resources became depleted, inflation would follow. Developing countries, unable to afford even basic necessities, would suffer the most. He urged the U.S. government to pass legislation mandating limited consumption of natural fuels, proper treatment and disposal of wastes, and extensive conservation of fish and wildlife areas.

Ehrlich also predicted increases in air pollution, ozone depletion, the number of extinct species, and the number of deaths caused by acquired immunodeficiency syndrome (AIDS); decreases in food production because of poor farming practices; and growing disparity between rich and poor unless corrective measures were taken. His predictions increased public awareness of environmental problems, and efforts have been made by government officials and grassroots organizations to address these concerns.

Following his first book on population, Ehrlich wrote several other books—often coauthored with his wife—detailing his concerns for conservation and ecological restraint. Among them are *The Population Explosion* (1990), *Healing the Planet* (1991), and *Betrayal of Science and Reason* (1996).

Gordon A. Parker

See also: Antinuclear movement; Population-control movement; Population growth.

El Niño

CATEGORY: Weather and climate

El Niño refers to the sudden appearance of warm water off the Peruvian coast around Christmas. The condition is usually localized and lasts only a short time, but the warm-water pattern occasionally establishes itself more strongly, leading to the development of a quasiperiodic El Niño.

The term "El Niño" (Spanish for "the Christ child") has now been expanded to include large warm-water anomalies covering extensive portions of the tropical Pacific Ocean off the coast of Latin America that persist for many months. It has been known for many years that when warm water appears off the coast of Peru, atmospheric pressure drops over the eastern Pacific and rises over Australia and the Indian Ocean, a pattern known as the Southern Oscillation. Because of this relationship, major El Niño events are usually associated with other global weather phenomena, including drought in Africa and Australia, and the failure of the Indian monsoon. Many meteorologists refer to the global pattern as El Niño-Southern Oscillation (ENSO).

CAUSES AND PREDICTION

When an El Niño develops in the eastern Pacific, the sea surface temperature and rainfall in the eastern tropical Pacific are at their seasonal peaks. Major El Niño occurrences are closely tied to global weather patterns and the circulation of currents in the Pacific Ocean. Variations in Indian monsoon circulation sometimes precede variations in the Southern Oscillation, indicating that there is a possible feedback mechanism linking the atmospheric phenomenon to the oceanic phenomenon. The period of the Southern Oscillation is irregular, with a return period of about three to four years, so about two quasiperiodic El Niños occur per decade. The amplitude of the Southern Oscillation is highly irregular. If some global atmospheric perturbation contributes to the amplitude of the Southern Oscillation as a quasi-periodic El Niño is developing, a major El Niño might be expected

Demolished homes in Pacifica, California, teeter on the edge of a cliff that was washed away by 1997-1998 El Niño storms. (AP/Wide World Photos)

Years in Which El Niño and La Niña Were Observed, 1900-1998

El Niño Years (warm water in eastern Pacific)	La Niña Years (cold water in eastern Pacific)
1902, 1905, 1911, 1914, 1918, 1923, 1925, 1930, 1932, 1939, 1941, 1951, 1953, 1957, 1965, 1969, 1972, 1976, 1982, 1986, 1991, 1994, 1997	1904, 1908, 1910, 1916, 1924, 1928, 1938, 1950, 1955, 1964, 1970, 1973, 1975, 1988, 1995, 1998

Note: Many El Niños begin in one calendar year and end during the following calendar year. Only the beginning year is listed.

to occur. However, if a global perturbation subtracts from the amplitude of the Southern Oscillation, even the quasiperiodic El Niño might be weak.

The point at which scientists have decided that a major El Niño was occurring has been historically contentious. A network of buoys has been established to help augment satellites in determining sea surface temperatures in the Pacific Ocean. When sea surface temperatures reach 3 to 5 degrees Celsius above normal for the season in the eastern equatorial Pacific, scientists can be fairly certain that an El Niño is occurring. Some years an El Niño begins off the coast of Peru, slowly propagating westward. Sometimes an El Niño begins in the western Pacific and propagates slowly eastward, as the 1982-1983 El Niño did. During a major El Niño, the normal westerly trade winds subside, and the height of the eastern Pacific sea surface rises. This is coupled with a decline in the height of the western Pacific sea surface, which sometimes causes submerged coral reefs in the western Pacific to appear above the ocean surface.

Since historical records of El Niño began, long-term variations in its strength have been observed. The 1920's and 1930's experienced only weak El Niño events. In contrast, the El Niños of the 1980's and 1990's were strong events. El Niño is a result of a complex interplay of atmospheric and oceanic forces, and the reasons for the waxing and waning in strength of El Niño over time periods of decades are not understood. Some scientists have noted that unusual El Niños have followed volcanic eruptions. Major volcanic eruptions cause the formation of a stratospheric aerosol layer, which may lead to more solar radiation being reflected back into space. The 1951 El Niño followed the eruption of Mount Lamington in Papua New Guinea; the 1982 El Niño followed the eruption of El Chichon in Mexico and Galunggung in Indonesia; the 1991 El Niño followed the eruption of Mount Pinatubo in the Philippines.

Just as El Niños start at the same time during a calendar year (December-January) in the eastern Pacific, they end at the same time during the calendar year in the western Pacific. The 1972 El Niño ended when easterly winds replaced westerly winds over the western tropical Pacific. Although it is tempting to focus on a single parameter such as sea surface temperatures or the period of the Southern Oscillation as a predictor of El Niño, history shows that there are many factors, both atmospheric and oceanic, that must all contribute to the development of a strong El Niño event.

Consequences

When unusually warm water appears off the coast of Peru, the local anchovy fishing industry falters. Sport fishing off Baja California, California, and Oregon enjoys a boom, as marlin and other highly prized fish usually found in more tropical southern waters move north.

During El Niño episodes, the Intertropical Convergence Zone, a band of major tropical convection circling the globe, moves southward.

This southward shift in precipitation patterns causes torrential rains in some places that are normally dry and dry conditions in places that are usually wet. In the Galápagos Islands, El Niño brings much higher than normal precipitation in March, April, and May. During major El Niños, Peru and Ecuador experience torrential rains and flooding. In Guayaquil, Ecuador, El Niño was credited for causing more than 3 meters (9.8 feet) of rain between October, 1982, and January, 1983. During the 1997-1998 El Niño, severe flooding occurred in Ecuador along rivers where rain forests had been cleared to establish shrimp farms.

During many major El Niño events, countries bordering the western Pacific and Indian Oceans experience droughts. During nine of the seventeen El Niño years from 1901 to 1995, India received 10 percent or less of its normal rainfall, while in eight of the ENSO years, rainfall was normal. However, monsoon rainfall has never exceeded 110 percent of the normal rainfall during an El Niño year. In El Niño years, Sri Lanka tends to receive higher than normal rainfall. During the 1982-1983 El Niño, Indonesia and Australia were drought stricken. Early in the decade of the 1990's, southern Africa experienced its worst drought of the century, probably worsened by the quasiperiodic El Niño that began in late 1991. El Niño years in Japan are associated with mild winters, cool summers, and lengthy rainy seasons.

Many diverse ecological, environmental, and economic events throughout the world are often attributed to El Niño occurrences. Sometimes these events may indeed be related to El Niño, but some occurrences can be attributed to other factors. Sometimes conflicting claims are made about the effects of El Niño. Just as there is no consistent relationship between the failure of the Indian monsoon and El Niño years, other weather claims may hold up during a statistically significant number of years but not in all years.

Although the 1997 El Niño was credited with causing the unusually mild winter of 1997-1998 in the eastern United States, the 1976 El Niño was credited with causing extreme cold in the same region in December, 1976, and January, 1977. The Sonoran Desert of Arizona and California tends to be wet in El Niño years. The Florida drought of 1998 was attributed to El Niño, although the warm-water anomaly off Peru had virtually disappeared by July, when forest fires were plaguing Florida. Because Texas often experiences more precipitation during the growing season after an El Niño, farmers planted crops requiring more moisture than usual in 1998. The summer of 1998 was unusually dry in Texas and Oklahoma, and many of the affected farmers realized that El Niño-based forecasts might be less than reliable. The failure of the Soviet harvest in 1972 was attributed by some to El Niño. In El Niño years, Moscow frequently experiences very cold winters; December, 1997, fit into this pattern.

La Niña

As the warm-water anomalies in the eastern Pacific fade, scientists know that El Niño is ending. When a large body of colder-than-average water establishes itself off the coast of Peru, along with a strong ridge of high pressure, meteorologists announce that a "La Niña" is occurring. During these periods, the Pacific sea surface height in the eastern Pacific is measurably lower than when an El Niño is occurring. The westerly trade winds are also strong. During a La Niña, the average temperature of the tropical troposphere may be 1 degree Celsius lower than during an El Niño. La Niñas are less common than El Niños. La Niñas tend to be much more variable in strength than El Niños.

During periods when the waters of the eastern Pacific have been observed to be anomalously cold, the Pacific Northwest tends to be wetter and cooler than normal, especially during winter. During a La Niña year, winter temperatures in the southeastern United States are often warmer than normal. In the United States, some link La Niña years to very hot summers; the summer of 1988, which was very hot and dry, is the prototype summer for this weather phenomenon. Tropical cyclones are more common on the northern Australian coast during La Niña events. Widespread flooding in eastern Australia tends to occur early in the calendar year (late summer) during La Niñas.

Anita Baker-Blocker

Suggested readings: For an understandable introduction to the atmospheric and oceanic aspects of El Niño, read *El Niño, La Niña and the Southern Oscillation* (1990), by S. George Philander. Thomas Y. Canby, "El Niño's Ill Wind," *National Geographic* (February, 1984), portrays the devastation wrought by the 1982-1983 El Niño. For a look at the historical effects of El Niños, see Michael E. Moseley and James B. Richardson, "Doomed by Natural Disaster," *Archeology* 45 (November-December, 1992). The agricultural implications of El Niño are portrayed in "El Niño and U.S. Corn Belt Rainfall," *Weekly Weather and Crop Bulletin* (May 12, 1992). Worldwide ecological effects of El Niño are discussed in Robert C. Brock, "El Niño and World Climate: Piecing Together the Puzzle," *Environment* 26 (April, 1984). *El Niño, Historical and Paleoclimatic Aspects of the Southern Oscillation* (1992), edited by Henry F. Diaz and Vera Markgraf, provides an in-depth treatment of most aspects of the phenomenon.

See also: Climate change and global warming.

Enclosure movement

Category: Land and land use

The European enclosure movement involved a conversion from the medieval agricultural system of open fields and lands held in common to fenced or hedged fields and pastures. Enclosures, along with improvements in agricultural techniques, dramatically changed land use.

Under the European, feudal system of agriculture, typically three large, open fields without fences or hedges surrounded each village. Each field was left fallow for one year out of three. The fields were subdivided into long, narrow strips. Each farmer's allotted strips were scattered over the open fields rather than adjoining one another. Farmers were allowed to graze a set number of animals on pastureland held in common. During certain seasons—especially after harvest—the arable and hay lands were opened up to grazing by the livestock of the whole community. Woodlands and other uncultivated lands were also held in common.

In England during the thirteenth century, farmers' allotments of arable land began to be consolidated into fields enclosed by hedgerows or fences. The ancient common lands were divided among the farmers and enclosed. In the process, large areas of forestland and uncultivated land were converted into plowland. English enclosure reached its height during the late eighteenth and early nineteenth centuries and was largely complete by the mid-nineteenth century. The process occurred somewhat later in Continental Europe. Enclosure drove many farmers off the land, leading to social unrest.

Agricultural efficiency increased within enclosed fields. Farmers could rotate crops and pasture scientifically, without regard to what their neighbors did. The resulting increase in soil fertility helped remove the need for fallow fields. Farmers could raise livestock more easily with herds in enclosures and could grow fodder without having it eaten by livestock belonging to others. With enclosed land to grow fodder, farmers could maintain livestock through the winter. The manure from the growing herds also improved soil fertility.

Over time, as the hedgerows naturally accumulated plant species, they became important wildlife habitat, compensating somewhat for the conversion of woodland into fields and pastureland during enclosure. More than eight hundred kinds of plants have been found in British hedgerows, including such woody perennials as blackthorn, hawthorn, oak, beech, ash, hazel, roses, crabapple, and holly. Most of Great Britain's woodland birds and small mammals use hedgerows at some time during their lives. For many species, hedgerows are the only remaining habitat.

The landscape of rectangular, hedged fields largely persisted in Britain until the 1950's, as fields remained small and were regularly rotated between crops and pasture. However, many farmers began selling off their livestock and turning to crops cultivated with large equipment that required a broad expanse of open land, such as wheat. In the process, many of the hedgerows, some of great antiquity, were destroyed. This loss transformed the appearance of the countryside and was detrimental to wildlife.

Another major threat to hedgerows has been neglect. The strict maintenance that hedgerows require costs more than many farmers are willing to pay. In 1989 the British government began paying subsidies for planting and maintaining hedgerows.

Jane F. Hill

SEE ALSO: Grazing; Land-use policy; Range management.

Endangered species

CATEGORY: Animals and endangered species

A plant or animal species is considered endangered when its numbers are so reduced that it is in danger of becoming extinct. Extinction, which occurs when the last member of a species dies, is the ultimate catastrophe for a species.

Extinction of a species does not occur in a vacuum. Causes, typically environmental, are many and often complex. Likewise, because of the many intricate, interconnected relationships existing within ecosystems, the loss of any member may have a ripple effect, eventually having profound negative results. For example, the extinction of a single insect, bird, or bat species may result in the extinction of one of more plant species dependent on the animal species for pollination. If the plant is a critical item in the diet of certain animals, they too may be adversely affected.

Most nonbiologists are not likely to set a high priority on the conservation of plant or animal parasites. Yet a large percentage of all known species are parasitic on other forms of life known as hosts. Furthermore, all species are host to one or more parasitic species. These host-parasite relationships can be considered a part of the "glue" that holds together species of an ecosystem. Thus extinction or even endangerment of a host species may affect an entire ecosystem as well as cause the loss of its parasitic species.

Paul Ehrlich and Anne Ehrlich introduce their book *Extinction: The Causes and Consequences of the Disappearance of Species* (1981) by referring to fictitious "rivet poppers"—workmen whose job it is to remove rivets from the wings of airplanes. The expectation is that many could be removed without the wings falling off. By analogy, the Ehrlichs consider many world leaders—politicians, bureaucrats, industrialists, engineers, churchmen, and even some scientists—to be rivet poppers. By their policies and practices, they espouse programs that will, by design or neglect, result in the loss of endangered species. Ecosystems, by their nature, are somewhat redundant: They are likely to continue to function even after the loss of several species. Ecologists refer to this capacity as "resistance." Ecosystems also possess resilience, or the ability to recover after disturbances, including those in which species are lost. However, just as one would not wish to fly in an airplane from which even a few rivets have been removed, it seems only prudent to take reasonable steps to prevent endangered species from becoming extinct.

Extinction is the conclusion of a long, gradual process typically involving a considerable span of time. When a species undergoes a drastic reduction in the extent of the range, accompanied by a reduction in the number of individuals, it may be designated as a rare species. As this trend continues, it is likely to be considered threatened prior to being recognized as endangered.

NATURAL AND HUMAN-CAUSED EXTINCTIONS

Of the ten to fifty million species (1.5 million described and named) that inhabit the earth, a large percentage are of special concern—rare, threatened, or endangered. It is accepted that some of these special concern species achieved that status because of natural processes. New species came into existence when segments of a preexisting species became separated into populations isolated from one another. Such geographic isolation permits the development of genetic, behavioral, or structural differences between the groups. When new species emerge by this process, it is known as "speciation."

New species are subjected to many natural forces, some of which are potentially harmful.

Included are various geological events and climatic shifts that may drastically alter the environment. Other forces of nature are biological, such as competition with closely related species or diseases caused by microorganisms. As a result of a combination of various negative factors, a species may have its geographic range and numbers gradually but effectively restricted. This shrinkage is generally accompanied by a reduction in the genetic diversity of the species. Because of this loss of genetic diversity, the species gradually loses its ability to adapt to its changing environment, which causes a further reduction in its numbers. At this point, the species is on a downward spiral leading toward extinction. According to this model, the death of a species is just as normal as was its birth many millennia earlier.

The rate of the natural extinction process varies greatly in various species existing under different conditions, but it is generally quite slow. One estimate is that only one species every one thousand years becomes extinct by natural processes. This may be a low estimate and certainly does not include catastrophic events in geological history. Perhaps the most notable event of this kind was the collision of the earth with asteroids sixty to seventy million years ago believed to have caused the demise of the dinosaurs.

Biologists are convinced that the current rate of species loss far exceeds these low predictions from natural causes. Evidence that the current acceleration is caused by human activities is overwhelming: Humans are polluting the air and water, destroying habitats, overconsuming goods, and making the earth a less desirable place for the welfare of its biota.

FACTORS CONTRIBUTING TO SPECIES LOSS

Whether because of their intrinsic nature or environmental conditions, some species are naturally more predisposed to becoming endangered or extinct than others. As one would expect, species with a smaller number of individuals are more vulnerable than those with more. Each species has a critical population size. Once

Numbers of Endangered and Threatened Species, 1996
As Listed by U.S. Government

	Mammals	Birds	Reptiles	Amphibians	Fishes	Snails	Clams	Crustaceans	Insects	Arachnids	Plants
Total listings	335	274	112	21	116	23	59	17	33	5	496
Endangered species, total	307	252	79	15	76	16	53	14	24	5	406
United States	55	74	14	7	65	15	51	14	20	5	405
Foreign	252	178	65	8	11	1	2	—	4	—	1
Threatened species, total	28	22	33	6	40	7	6	3	9	—	90
United States	9	16	19	5	40	7	6	3	9	—	90
Foreign	19	6	14	1	—	—	—	—	—	—	—

Source: U.S. Department of Commerce, *Statistical Abstract of the United States, 1996*, 1996. Primary source, U.S. Fish and Wildlife Service.

Note: Numbers reflect species officially listed by U.S. government; actual worldwide totals of species that could be considered threatened or endangered are unknown but are much higher.

the numbers fall below that size, it is especially subject to extinction. Natural populations undergo year-to-year fluctuations in numbers; therefore, a small population will "crash" more readily than a large one.

Several categories of animal and plant species are at high risk of becoming endangered or extinct. Among these are species restricted to special habitats. Most such animal or plant species, by becoming tolerant of an unusual situation, lose their ability to compete in a more general one. An example would be a plant endemic to the rocky soils of the cedar glades of middle Tennessee that is unable to inhabit adjacent forests. Another category is island species: If threatened by humans, predators, or diseases, they cannot easily escape. A disproportionate number of animals known to live on islands have become extinct. Large species with a low reproductive rate are also at risk. Large species require more space; therefore, the number of large specimens occupying a given area is lower than the number of smaller ones. Also, most large species, whether whales or trees, are likely to reproduce less often than smaller ones. Even when given protection, it is difficult for them to increase their numbers.

Neotropical migratory birds such as warblers, orioles, and tanagers winter in tropical Central or South America or the Caribbean, and breed in eastern North America. Their migratory pattern is advantageous in that they can take advantage of the availability of summer food in the north while escaping the harsh condition in winter. Migration, however, is a process that is fraught with danger. As the tropical forests in which they spend the winter are destroyed and the temperate forests in which they breed are fragmented, neotropical migrants may be threatened; thus, they are subject to double jeopardy. Since 1966 the United States Fish and Wildlife Service has conducted surveys on changing numbers of neotropical migrants. Between 1978 and 1987, 61 percent of the species declined in number. During the same time, changes in the number of nonmigratory neotropicals were much less pronounced, underscoring the greater vulnerability of migratory birds to habitat destruction.

Among the other at-risk species are those at the end of long food chains. Animals such as hawks, owls, and various cat species suffer when any of the links in their food chain are affected. Also, they may be more subject to damage by toxic materials because of chemical amplification along the food chain. Finally, species of economic value are also in a precarious situation. Many examples exist of animals that were hunted to extinction; the often-cited example is the passenger pigeon. Plants used medicinally have been subjected to overcollecting. Special regulations are now in effect to protect ginseng, which has been dug in eastern North America for more than two centuries.

Protecting Endangered Species

Once a species has been identified as endangered, all aspects of its biology become of great value: its numbers, distribution, reproductive capacity, and other information about its total life cycle and all of its environmental relationships. If much data is not already available, it must be obtained from field observations, experimentation, or by whatever means possible. Only then is it possible to make plans leading to its recovery.

One approach to protect endangered plants and animals is to focus on the individual species. Federal legislation has afforded special protection to endangered species, including the Endangered Species Preservation Act of 1966 and the Endangered Species Act of 1973, which has been amended several times. These laws are considered the most stringent and comprehensive in the world. Critics view the Endangered Species Act as a major stumbling block to economic progress. A classic example is the delay, in 1977, of the construction of Tellico Dam in the mountains of eastern Tennessee because of the presence of the snail darter, a small endangered fish. The dam was later completed, and other populations of the fish were unexpectedly found elsewhere, resulting in the snail darter being removed from the endangered list. Some people, especially those of the private sector, believe that business and public interests should take precedence over wildlife, especially plants and animals of no obvious value. Of the many conflicts that have arisen over enforcement of the laws, almost all have been settled by some reasonable compromise.

Debate over the Endangered Species Act continued into the late 1990's. Whereas some would like to see it weakened, environmentalists would like to see it made more comprehensive. Of the six hundred species that have been listed as endangered, most of the success stories have involved "charismatic megafauna," such as bison, wolves, and the bald eagle. Less glamorous endangered species, such as fungi, wildflowers, liverworts, mosses, and insects, receive much less attention even though their role in ecosystems may be more important.

Once a plant or animal species has been listed as endangered, a recovery plan is designed to prevent its extinction. If successful, the species will increase in abundance so that it can be taken off the list. Some of the measures included in recovery plans of animals include translocation (moving individuals from one location to another) and hacking (release of birds after being raised in captivity). The distribution of a plant species can be extended by seeding or transplanting. In all cases, careful monitoring is important.

Protecting Critical Habitats

As necessary as it is to sometimes single out particular species in need of immediate attention, it is now recognized that, in the long run, the best way to conserve species is to preserve their habitats. U.S. secretary of the interior Bruce Babbitt wrote,

> We need a new approach: one that encourages us to think ahead and plan for the future; one that encourages us too look at whole ecosystems and not just tiny parcels of land; one that stresses compromise and balance between people and nature.

In practice, all natural areas are not viewed as being equally worthy of protection. Certain areas contain large numbers of rare and endemic species. Large areas, especially when roadless, are of more ecological value than several smaller fragmented areas with the same total area.

National parks were established in the United States beginning in the nineteenth century. Also included in restricted-use lands are roadless units of the National Wilderness Preservation System administered by the Forest Service and the Fish and Wildlife Service. A number of national forests are managed according to a multiple-use principle that includes timber production but also affords some degree of protection to habitats and biota.

In addition to the government, private conservation organizations have been of tremendous value in acquiring and protecting sensitive landscapes. The National Audubon Society, begun in the late nineteenth century to protect nongame birds from slaughter, soon began to acquire land for wildlife sanctuaries. The Sierra Club, founded by nineteenth century conservationist John Muir, has been concerned from its beginning with the establishment and protection of preserves. The Nature Conservancy, a newer organization, identifies critical areas, which are acquired, protected, and managed until they can be transferred to a governmental unit for preservation.

It has been realized for some time now that battered ecosystems, even if they cannot be converted into their former pristine state, can be greatly improved or restored. The success of restoration efforts depends on being aware of, and cooperating with, natural long-term processes, especially ecological succession.

As parks and preserves were being established during the nineteenth and early twentieth centuries, sites were chosen primarily for their scenic value or their unsuitability for agricultural or other "practical" uses. Other areas, often those of more ecological importance, were left unprotected. Also, many of these preserved areas are now known to be too small and too fragmented for the processes of adaptation and evolution to occur naturally. Consequently, humans are faced with the problem of how best to enhance existing preserves and establish new ones in order to accomplish the overall goal of protecting endangered species.

Thomas E. Hemmerly

Suggested readings: Textbooks for courses in conservation biology are concerned with endangered species and the extinction process. Among the more useful are *Environmental Sci-*

ence: Action for a Sustainable Future (1997), by Daniel D. Chiras, and *Principles of Conservation Biology* (1997), by Gary K. Meffe et al. *Extinction: The Causes and Consequences of the Disappearance of Species* (1981), by Paul Ehrlich and Anne Ehrlich, is a classic in the field. Among the books concerned specifically with endangered plants are *The Conservation of Medicinal Plants* (1991), edited by Olayiwola Akerele et al., and *Principles and Practice of Plant Conservation* (1994), by David R. Given. *Environmental Restoration* (1990), edited by J. J. Berger, also deals effectively with this topic.

SEE ALSO: Biodiversity; Endangered Species Act; Endangered species and animal protection policy; Extinctions and species loss.

Endangered Species Act

DATE: passed 1973
CATEGORY: Animals and endangered species

The Endangered Species Act (ESA) was designed by the United States government to protect species threatened with extinction. The ESA outlines a process for the listing of protected species, authorizes appropriate regulations, and provides state subsidies and funding for habitat acquisition. Despite many amendments and regulations added since 1973, the fundamental purpose of the legislation has remained the same.

The U.S. Congress first demonstrated concern for the conservation of species in the Lacey Act of 1900, which prohibited the transportation in interstate commerce of any fish or wildlife taken in violation of national, state, or foreign laws. Following the extinction of passenger pigeons, the Migratory Bird Treaty Act of 1918 authorized the secretary of the interior to adopt regulations for the protection of migratory birds.

In the Endangered Species Preservation Act of 1966, Congress declared that the preservation of species was a national policy. The statute authorized the secretary to identify native fish and wildlife threatened with extinction and to purchase land for the protection and restoration of such species. The Endangered Species Conservation Act of 1969 further empowered the secretary to list species threatened with "worldwide extinction" and prohibited the importation of any listed species into the United States. The only species eligible for the list were those threatened with complete extinction. Although the 1966 and 1969 statutes did not include any penalties for destroying species on the list, the legislation was the most comprehensive of its kind enacted by any nation.

In legislative hearings of 1973, it was reported that species were being lost at the rate of about one per year and that the pace of disappearance seemed to be accelerating, with potential damage to the total ecosystem. The majority of Congress concluded that it was necessary to stop a further decline in biodiversity, and President Richard Nixon signed the ESA into law on December 28, 1973.

The ESA provides that any species of wild animals or plants may receive federal protection whenever the species has been listed as "endangered" or "threatened." The statute defines "endangered" to mean that the species is currently in danger of becoming extinct within a significant geographical region. The term "threatened" means that the species probably will become endangered within the foreseeable future. The definition of a "species" includes any subspecies or any distinct population that interbreeds within a specific region. Species found only in other parts of the world are eligible for inclusion on the U.S. list. The only creatures not eligible for inclusion are those insects that are determined to pose an extreme risk to human welfare.

The act makes it a federal offense to take, buy, sell, or transport any portion of a threatened or endangered species. Listed animals, however, may be taken in defense of human life, and Alaskan natives are allowed to use listed animals for subsistence purposes. Additional exemptions may be granted for special cases involving economic hardship, scientific research, or projects aimed at the propagation of a species. Individuals may be fined $10,000 for each violation of the law committed knowingly and $1,000 for a violation committed unknowingly. Harsher

A female Canadian Lynx is released into the Colorado wilderness during the winter of 1999 as part of the effort to revive the species from the brink of extinction. Lynxes were added to the endangered species list in 1998. (AP/Wide World Photos)

criminal penalties are available in extreme cases.

The ESA assigned most enforcement and regulatory powers to the heads of two executive departments. The secretary of commerce, through the National Marine Fisheries Service (NMFS), has responsibility over threatened and endangered marine species. The secretary of the interior exercises formal responsibility for the protection of other species, but the secretary delegates most of the work to the U.S. Fish and Wildlife Service (FWS), which is assisted by the Office of Endangered Species (OES). As of 1994, the regulations under the act took up 350 pages in the Code of Federal Regulations.

In order to benefit from the ESA, the species must be officially designated as either endangered or threatened. The courts have consistently ruled that the act cannot be used to protect an unlisted species. Species may be proposed for listing by the NMFS, the FWS, private organizations, or citizens. Species are listed only after comprehensive investigations, open hearings, and opportunities for public involvement in the decision.

The first list of endangered species, published in 1967, included 72 species. By 1976 the list had grown to 634 species. As of 1995, 1,526 species of plants and animals were listed, including more than 500 that were foreign, and there were almost 4,000 candidate species awaiting a listing determination. Although the FWS is required to prepare a recovery plan for each listed species, only a few have recovered sufficiently to be taken off the list.

The act requires that critical habitat for threatened or endangered species be designated whenever possible. All federal agencies have special obligations to determine whether their projects or actions jeopardize the continued existence of a species. Following the Supreme Court's controversial ruling in *Tennessee Valley Authority v. Hill,* Congress passed the amendments of 1978, which allow consideration for economic factors in the designation of critical habitat. Especially controversial is the section of the act requiring the FWS to formulate and enforce regulations on private lands that provide habitat for listed species. The government must compensate owners in those rare cases when regulations eliminate almost all productive and economic uses of their property, but not when landowners continue to have partial productive use of their land.

Many people in western and rural states are highly critical of the ESA, and they charge that it causes a significant loss of jobs to protect minor subspecies, such as the northern spotted owl. In 1995-1996, a conservative coalition of Republican congressmen tried to pass the Young-Plombo bill, which would have weakened the ESA. The controversy demonstrated, however, that the existing law enjoyed considerable support, and the proposed bill was never passed. Most experts argue that the economic impact of the ESA is minimal on the national economy but that it does cause hardship for small landowners in some instances. Many environmentalists would support revisions of the law that would give less emphasis to particular species and place more concern on the need for sufficient habitat to support a healthy biodiversity, but others fear that such complexity would make the law ineffective.

Thomas T. Lewis

SUGGESTED READINGS: For a detailed legal analysis, see Daniel Rohlf's *The Endangered Species Act: Protection and Implementation* (1989). Charles Mann and Mark Plummer give a historical account of the ESA from a moderate environmentalist perspective and suggest changes in *Noah's Choice: The Future of Endangered Species* (1995). Lewis Regenstein presents an interesting historical account of the 1973 statute in *The Politics of Extinction* (1975). For a detailed and fascinating study of habitat policy and its many controversies, consult *Habitat Conservation Under the Endangered Species Act* (1997), by Reed Noss, Michael O'Connell, and Dennis Murphy. There are many useful articles in *Balancing on the Brink of Extinction: The Endangered Species Act and Lessons for the Future* (1991), edited by Kathryn Kohm.

SEE ALSO: Biodiversity; Convention on International Trade in Endangered Species; Endangered species; Endangered species and animal protection policy; Extinctions and species loss.

Endangered species and animal protection policy

CATEGORY: Animals and endangered species

An endangered species is one that is considered to be in danger of becoming extinct if protection is not provided. Extinction has been a natural process throughout geological history, with some species disappearing and some evolving to take their place. However, the changes caused by human population growth and technology have been too rapid to allow species to adapt, and animals and plants are facing rapid extinction rates.

As human populations clear more forests and other lands for farms, housing projects, or shopping malls, they cause habitat degradation, fragmentation, and destruction. Deforestation, particularly of tropical rain forests, is the greatest single cause of decline in biological diversity in the world. This is followed closely by human-induced biological losses resulting from the destruction of coral reefs and wetlands, and the plowing of grasslands.

Pollution can directly kill many kinds of plants and animals, but it can also alter or destroy habitats. Forests may experience a decline in growth because of acid deposition and air pollution. Heavy sedimentation and surplus nutrients in waterways kill many living species in lakes, rivers, and bays. Slowly degradable pesticides, particularly dichloro-diphenyl-trichloroethane (DDT), have caused large declines in bird species.

The introduction of exotic species by humans is responsible for an estimated 40 percent of all animal extinctions since 1600. When a species is introduced to an area for control of pests, the initial intentions often go awry, as it did after the introduction of the mongoose to Puerto Rico for rat control. The mongoose primarily hunts its prey during the day, whereas rats are nocturnal. The mongoose therefore turned to preying on amphibians, reptiles, and ground-nesting birds. As a result, many of the bird species in Puerto Rico became extinct.

Legal and illegal commercial hunting has led

to the extinction or near extinction of many animal species. Although policies have been in effect to regulate hunting, poaching remains a lucrative business, particularly in underdeveloped countries. Besides poaching for skins and horns of animals, a large number of threatened and endangered species are smuggled into countries for sale as pets and decorative plants.

CONSERVATION AND MANAGEMENT

In order to preserve biodiversity and not lose species that are important to the health and existence of an ecosystem, wildlife conservation and management practices must be put into effect. There are three basic approaches to wildlife conservation and management. The species approach involves giving endangered species legal protection, protecting and managing their habitat, propagating species in captivity, and reintroducing species into safe habitats. In 1903 President Theodore Roosevelt established the first wildlife refuge in the United States. The refuge, located on Pelican Island on the east coast of Florida, was developed to protect the brown pelican, which was endangered. Since then the National Wildlife Refuge System has grown to 456 refuges, over three-fourths of these being wetland refuges for the protection of migratory waterfowl.

Other forms of the species approach to saving diversity include gene banks, botanical gardens, and zoos. Endangered plant species are preserved by storing their seeds in climatically controlled environments. There are approximately 1,500 botanical gardens in the world, which contain about ninety thousand plant species. Many botanical gardens, such as the Kew Gardens in England, are repositories for plant species that are endangered or have even ceased to exist in the wild. Some of these plants are reintroduced into native habitats after being cultivated for decades in these gardens or in seed banks.

Egg pulling and captive breeding are two methods that zoos and animal research centers use for preserving endangered species. Egg pulling involves collecting eggs from endangered species in the wild and hatching the eggs in zoos or research centers, as occurred with California Condors in 1983. Endangered species still in the wild are captured and put into research centers to breed in a controlled environment. When the captive populations become large enough, some of the individuals will be reintroduced into protected habitats. The Arabian oryx, a large antelope species that originated in the Middle East, was saved from extinction by captive breeding programs in San Diego, Los Angeles, and Phoenix zoos.

The second approach to saving biodiversity is the ecosystem approach, which emphasizes preserving balanced populations of species within their native habitats. It involves establishing legally protected wilderness areas and wildlife reserves. An important part of making sure that the habitat is safe is to eliminate all alien species from the area. There were about seven thousand natural reserves, parks, and protected areas throughout the world in 1992. These reserves

Recovery Status of the 711 Species Listed in the Endangered Species Act, 1992 (percentages)

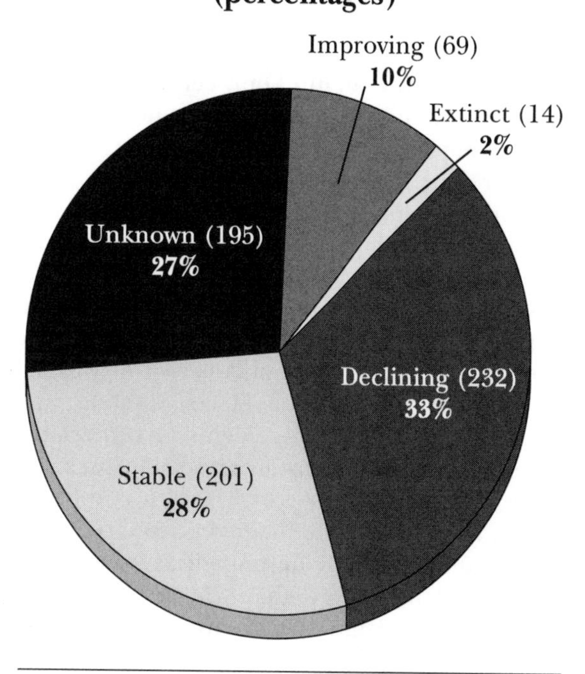

Source: Report to Congress on Endangered and Threatened Species Recovery Program, December, 1992. U.S. Department of the Interior and U.S. Fish and Wildlife Service.

occupy 4.9 percent of the earth's land surface, which is about one-third of the amount that conservationists feel is adequate to give protection to a wide variety of endangered species. The Minnesota Zoo has formed a partnership with the Ujung Kulon National Park in Indonesia to help protect the world's few remaining Java rhinos. Instead of moving rhinos to Minnesota, the Minnesota Zoo helps protect the animals in their native habitat by supplying patrol boats, housing, training, and salaries for Indonesian guards.

The third approach to preserving biodiversity is the wildlife management approach. When it is decided which species or group of species will be managed in a given area, a management plan is put into effect. This plan will look into and decide what kind of cover, food, water, and space a species requires. Action is then taken to grow plants that provide the needed cover and food for the species.

Hunting and International Cooperation

Species, primarily game species, are managed by establishing laws to regulate hunting and hunting quotas. Hunters are required to have licenses and use only certain types of hunting equipment, and can hunt only certain months of the year. Limits are set on the size, number, and sex of animals that can be hunted in a given game refuge.

Management plans and international treaties have been developed to protect migrating game species, such as waterfowl. In North America, waterfowl such as ducks, geese, and swans nest in Canada during the summer and migrate to the United States and Central America in the fall and winter. The United States, Canada, and Mexico have signed agreements to protect the waterfowl from overhunting and habitat destruction.

Some refuges in the United States are set up by building artificial nesting sites, ponds, and nesting islands. More than 75 percent of the refuges in the United States are wetlands established for the protection of migratory birds. In 1986 the United States and Canada entered into an agreement to attempt to double the continental duck population within sixteen years at a cost of US$1.5 billion. They plan to do this by purchasing or improving wildlife habitats in five key areas.

In 1975 a wide-reaching treaty was drafted at the Convention on International Trade in Endangered Species (CITES) and signed by 119 countries. The treaty lists 675 species of endangered and threatened species that cannot be commercially traded, either alive or as a product. Another two hundred species are listed that cannot be exported without an export license. One of the best-known acts of CITES was to ban the international trade in ivory in 1990. This act was created in order to halt the decline of the African elephant from 2.5 million animals in 1950 to approximately 350,000 at the treaty's inception.

In 1980 the International Union for the Conservation of Nature and Natural Resources (IUCN), the World Wildlife Fund (WWF), and the United Nations Environment Programme (UNEP) developed a world conservation strategy. The plan, which was expanded in 1991, attempts to preserve biological diversity, combine wildlife conservation and sustainable development, encourage rehabilitation of degraded ecosystems, and monitor sustainability. Forty countries have planned or established national conservation programs in response to this plan.

United States Laws

The United States has two important laws that control imports and exports of endangered wildlife and wildlife products. The first is the Lacey Act of 1900, in response to declining numbers of egrets because of the commercial value of their feathers as decoration. The Lacey Act prohibited transporting live or dead wild animals or their parts across state borders without a federal permit.

The other law is the Endangered Species Act (ESA), which was established in 1973. The ESA was unique in that while previous wildlife regulations had primarily focused on game animals, the ESA program focused on identification of all endangered species and populations in order to save biodiversity, regardless of the species' usefulness to humans. The act classifies endangered species as being in immediate danger of extinction and threatened species as those that are likely to become endangered in a given habitat in the future. Brown bears, gray wolves, bald eagles, sea otters, orchids, and some other rare plants are classified as being locally threatened even though they can be found in fairly large numbers in some parts of their former habitats.

In spite of ESA protection, the status of a number of species remains critical because of their precarious standing at the time of listing. Although 28 percent are considered stable as a result of the recovery efforts, 2 percent are believed to have gone extinct regardless of their ESA listing. The 27 percent that are listed as "unknown" reflect a lack of research and survey work because of budgetary and staffing constraints.

The ESA provides that a listed species cannot be harassed, harmed, pursued, hunted, shot, trapped, killed, captured, or collected, either on purpose or by accident. It further prohibits importing or exporting endangered species, as well as possessing, selling, transporting, or offering to sell any endangered species. Fines of up to $100,000 and one year imprisonment are imposed on a violator of the act. In 1995 the Supreme Court ruled to extend further protection for endangered species by ruling that habitat essential for species survival must be protected, whether on public or private land.

The United States Fish and Wildlife Service is required to prepare a recovery plan for each species listed by the ESA as officially endangered. Approximately $150 million is spent each year on endangered species in the United States. There are over 1,530 species on the endangered and threatened lists, but more than one-half of the allocated money is spent on ten species or subspecies: bald eagles, northern spotted owls, Florida scrub jays, West Indian manatees, red-cockaded woodpeckers, Florida panthers, grizzly bears, least Bell's vireos, American peregrine falcons, and whoping cranes.

Some of the ESA's recovery plans have proved successful. Bald eagles, which numbered only eight hundred birds in 1970, were able to rebound to eight thousand birds during a twenty-year period, largely because of a ban on DDT. The American alligator was listed as an endangered species in 1967. Its population declined because of habitat destruction and the demand for alligator meat and leather. Because of ESA protection, the alligator is now reestablished in its southern range. The Aleutian Canada goose was once widespread throughout the Aleutian chain in Alaska and the Bering Sea. Its population suffered a drastic decline when commercial fox farmers introduced nonnative foxes to the islands between 1836 and 1930. The geese were devastated by the foxes, the effects of hunters, and the loss of winter habitat. Only two hundred to three hundred geese were thought to exist in 1967 when they were classified as endangered. The recovery plan called for the elimination of the alien foxes from the islands and the relocation of wild family groups of geese to the Aleutians. Measures were taken to protect the geese from hunting and disruptions of their roosting and feeding habitat. Wintering habitats in Oregon and California were protected through easements and inclusion within the National Wildlife Refuge System. The population of the Aleutian Canada goose subsequently increased from fewer than eight hundred in 1975 to approximately 7,900 in the winter of 1991-1992. The goose was upgraded from endangered to threatened.

Aquatic Losses

The loss of aquatic species has attracted less attention than the extinction of land species. With the realization of the importance of healthy freshwater and marine species, however, governments have begun establishing marine preserves. Fishing, construction, tourism, pollution, and other human disturbances are closely regulated and restricted in these areas. The National Marine Sanctuary Program, which was developed in 1972 in the United States, has established twelve sanctuaries, the largest of which is located in Florida. It runs 354 kilometers (220 miles) along the Florida Keys. Fishing and coral collecting are restricted to preserve the delicate ecosystem.

International measures to protect species from destruction or exploitation include the International Convention for the Regulation of Whaling, begun in 1946. Overwhaling worldwide caused a huge decline in whales, from an estimated 4.4 million in 1900 to approximately 1 million by the end of the twentieth century. Overharvesting of whales affected almost every whale species of commercial value. In 1946 the International Whaling Commission (IWC) was established to set annual whaling quotas to prevent overfishing and commercial extinction of whales. However, many whaling countries ignored the suggested quotas. In 1970 the United States stopped all commercial whaling and banned imports of all whale products. In 1974 the IWC began to regulate whaling according to the principle of maximum sustainable yield. When a species fell below the optimal population for such a yield, the IWC issued a ban on hunting that species. The right whale, bowhead whale, and blue whale, all of which were at low levels in 1974, were protected by the IWC.

CLIMATE CHANGE

Curbing contamination of the biosphere by pollutants is probably the most expensive conservation measure. Massive cleanups of contaminated lands and waters can cost millions of dollars. On the international front, organizations have come together to address pollution problems. The Convention on the Prevention of Marine Pollution by Dumping of Wastes and Other Matter met in London, England, in 1972 to address pollution problems. Depletion of the ozone was discussed at the Convention for the Protection of the Ozone Layer in Vienna, Austria, in 1985, and further solutions were addressed at the Montreal Protocol on Substances that Deplete the Ozone Layer in 1987.

Nations at the Earth Summit, held in Rio de Janeiro in 1992 with the support of the UNEP, devised an international treaty based on concerns about the significance of biodiversity for future generations, the sovereignty of a nation over its resources, and a nation's need to conserve and protect its own biodiversity. The treaty detailed a plan that directed industrialized countries to help fund projects for the protection of biodiversity within developing countries. Resources in developing countries formerly viewed a being subject to free access by other nations were given to the national governments of the countries that held the resources. These national governments were given the right to decide who would have access to their resources. The treaty provided for a sharing of technologies, particularly biotechnologies that had been developed from plants originating in developing countries, thereby giving the developing countries substantial benefits from any technology based on their genetic resources. President George Bush refused to sign the treaty in 1992. President Bill Clinton signed the treaty in June of 1993, but the treaty was not binding and did not received a two-thirds majority vote in the U.S. Senate.

Toby Stewart and Dion Stewart

SUGGESTED READINGS: An interesting look at differing views on endangered species is provided in *Endangered Species: Opposing Viewpoints* (1996), edited by Brenda Stalcup. *Noah's Choice: The Future of Endangered Species* (1995), by Charles C. Mann and Mark L. Plummer, gives a unique look at endangered species and policies, as well as their effects on society. *The Endangered and Threatened Species Recovery Program* (1992), prepared by the U.S. Department of the Interior and the U.S. Fish and Wildlife Service, provides eye-opening information on endangered species in the United States and also discusses recovery programs. *Biodiversity* (1988), edited by E. O. Wilson, provides an excellent overview of biodiversity. A worldview of endangered species and programs is contained in *Conserving the World's Biological Diversity* (1990), by Jeffrey McNeely, Kenton Miller, Walter Reid, Russell Mittermeir, and Timothy Werner.

SEE ALSO: Biodiversity; Endangered species; Endangered Species Act; Extinctions and species loss.

Energy conservation

CATEGORY: Resources and resource management

During much of the twentieth century, few people considered the possibility that the energy sources upon which they depended could eventually be depleted. However, the energy crisis of the 1970's shocked people into the realization that the fossil fuels they had taken for granted may indeed be finite, prompting concerns about the efficient use of existing energy sources and the development of renewable energy technologies.

One measure of a society's sophistication is how well it conserves energy. For instance, ancient peoples often lacked weapons powerful enough to kill large or swift game. Consequently, they stampeded herds over cliffs or used fire drives. Even with better weapons, nomadic hunters still used open fires, sending most of the energy into the sky. Hunting is itself wasteful. The hunters must track and kill the game. Successful hunting in one area thins the game, thus requiring the energy cost of moving the entire group to better hunting grounds.

Agriculture is many times more efficient than hunting. That efficiency allowed agricultural societies to support people who did not produce food. Nonproducers were able to advance civilization through such activities as building pyramids, smelting metals, and writing books. Larger populations of civilized people often exhausted firewood supplies. Many, such as the ancient Greeks, developed building styles with south-facing courtyards to collect low winter sunlight, overhangs to block high summer sun, and thick walls to hold heat, thus buffering inhabitants from outdoor temperature swings. Glass windows were a major innovation that allowed light to enter a building while blocking wind. Benjamin Franklin invented a stove consisting of a metallic cylinder that radiated heat in all directions. More important, the metal conducted heat faster than brick, thus collecting more heat from the hot gases before they went out the chimney.

Energy can be conserved in two ways: by doing without or by doing more with less. In a few cases, doing without would increase general well-being. Reducing meat in the average American diet would reduce the energy used to produce grain to feed animals and would also raise health levels. More bicycling and walking would also save energy and increase health. In most cases, however, doing without is unpopular and economically unsound. Energy efficiency (doing more with less), on the other hand, can be increased in virtually every aspect of society. The amount of energy expended to heat and cool buildings, which accounts for about 25 percent of all energy use, could be vastly reduced by improved insulation. Superinsulated houses can function with almost no space conditioning.

Transportation, another 25 percent of energy use, can double or triple in efficiency. The most important technologies are engines, aerodynamics, and lighter materials. The use of aerodynamics and lighter materials for cars and trucks increased markedly after the onset of the oil crises of the 1970's. Researchers are working on developing fuel-cell power and composites for lighter materials. Meanwhile, developments in fiber optics and computers are making it possible to replace business trips with communication.

Manufacturing, another 25 percent of energy use, can be made much more efficient by implementing a number of small fixes that add up to major improvements: more insulation on steam lines, better control of processes, more efficient motors, and better space conditioning. In addition, recycling of wastes into usable resources could decrease energy use. Beyond those conventional technologies, researchers are experimenting with biochemical production, nanotechnology (extremely tiny machines that perform tasks efficiently and even repair structures without disassembly), superconducting motors, and better protective coatings and lubricants.

Appliances represent another 10-15 percent of energy use. Compact fluorescent bulbs, a prime example of increased efficiency, produce several times the power of incandescent bulbs and waste less heat. More efficient electrical motors can cut energy use and waste heat in major energy users, such as refrigerators and air conditioners.

Existing technologies with higher energy efficiency usually have a higher initial cost. Compact fluorescent lights save money over their lifetime, but they cost several times the price of incandescent bulbs. Likewise, high-efficiency cars cost more to manufacture and thus more to buy. Buyers hesitate to spend additional money

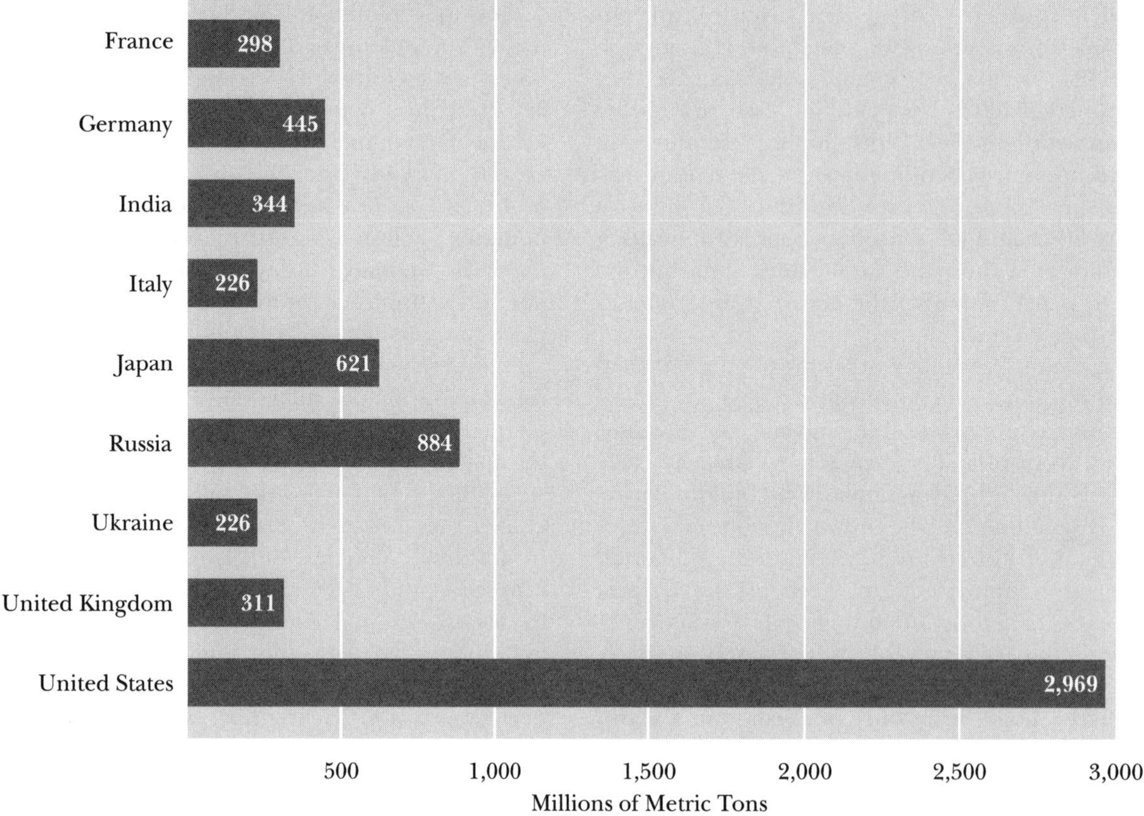

Source: U.S. Department of Commerce, *Statistical Abstract of the United States, 1996*, 1996.
Note: World total energy consumption for 1994 was approximately 11.3 billion metric tons. Figures are in what are called coal equivalents and are based on apparent consumption of coal, lignite, petroleum products, and natural gas as well as hydro, nuclear, and geothermal electricity.

on energy-efficient technology when energy prices are low. Many of the energy efficiency gains of the 1970's were lost after oil prices collapsed during the 1980's. Government support for higher efficiency can include tax supports or loans for more efficient items, requirements for efficiency labeling, and removal of incentives for nonrecycled materials.

Roger V. Carlson

SUGGESTED READINGS: Daniel Yergin and Robert Stobaugh, editors, *Energy Future: The Report of the Energy Project at the Harvard Business School* (1979), details energy issues and argues that energy efficiency technologies provide the best investment opportunities. Marc Reisner, "The Rise and Fall and Rise of Energy Conservation," *Americus* 9 (Spring, 1987), connects energy prices and efforts to improve energy efficiency. *The Energy Controversy: Soft Path Questions and Answers by Amory Lovins and His Critics*, edited by Hugh Nash, highlights the often vitriolic but always brilliant debates on Amory Lovins's radical ideas about energy efficiency. *Scientific American* 263 (September, 1995) is a special issue called "Energy for Planet Earth." Arthur H. Rosenfeld and David Hafemeister, "Energy-Efficient Buildings," *Scientific American* 258 (April, 1988), notes that such buildings could save $50 billion per year in energy costs. Ken Butti and John Perlin, *A Golden Thread: 2500 Years of Solar Architecture and Technology* (1980), traces the history of solar and efficient building design.

See also: Energy-efficiency labeling; Fossil fuels; Lovins, Amory; Recycling; Resource recovery.

Energy-efficiency labeling

Category: Energy

Energy-efficiency labeling is a rating system that measures the amount of electrical energy an appliance uses, the amount of heat that escapes from a house, and the amount of pollution generated by certain power sources.

Energy rating systems were created in an effort to help conserve energy and protect the environment from pollution created when fuel is burned for power. Primary energy sources include coal, petroleum and petroleum products, gas, water, uranium, wind, sunlight, and geothermal energy. A majority of these sources, such as petroleum, are fossil fuels, which are also called nonrenewable energy sources because what is used takes millions of years to replace. Fossil fuels are burned to power cars, heat homes, and generate electrical power. The Environmental Protection Agency (EPA) and the U.S. Department of Energy (DOE) created energy-efficiency labeling programs to lower the amount of electrical power consumed and the amount of pollution generated by electricity plants that burn fossil fuels.

Energy-efficiency labels on major appliances, often called "energy guides," show how much it costs consumers to use the appliance for one year. The estimated cost is based on an average cost per kilowatt hour in the area. Electrical power is measured and billed by kilowatt hours. A watthour is a unit of energy supplied steadily through an electric circuit for one hour. A kilowatt hour is equal to one thousand watthours.

Manufacturers are encouraged to make products that use less electricity or use it more efficiently. Sometimes the higher cost of creating these products is passed along to the consumer, so consumers want to know what their added cost will be to power these products. The energy guide stickers allow consumers to see that the electrical usage will be less than standard appliances, and they will more likely buy the product that helps save energy. Using less electricity helps create less pollution and saves natural resources.

Energy-efficiency labeling is used most frequently on electronic equipment. Energy Star and similar programs are used to show that computers and other electronic equipment use less electricity, even when turned off, than standard equipment. Appliances plugged into an electrical outlet use electricity even when turned off. Some use more than others, such as videocassette recorders (VCRs) that need to power clocks and program timers. A logo is used to identify equipment that goes into a "sleep" mode, or draws less energy from the outlet, even when not in use.

Energy-efficiency ratings are expanding into other equipment used in homes, from refrigerators to air conditioners and windows. New homes are often rated on overall energy efficiency. Improvements to existing homes, such as replacing old windows or insulation, also receive energy-efficiency ratings. Energy used to heat and cool homes depends on how well insulated the house is and whether the windows have gaps through which heat can leak or cold winds can enter.

Lisa A. Wroble

See also: Alternative energy sources; Energy conservation; Fossil fuels.

Energy policy

CATEGORY: Energy

Prior to the 1970's the United States made no effort to develop a national energy policy. Various policies had been adopted earlier regarding individual fuel and energy sectors. The 1973 oil crisis created a new view of energy as a distinct and important policy arena. Successive administrations undertook various initiatives, and Congress enacted several broad laws addressing aspects of energy production and consumption that had not previously been addressed. However, no enduring, coherent national policy resulted.

Energy production and consumption are strongly linked to a nation's environmental quality, economic well-being, and security. Energy use is one of the greatest sources of environmental degradation. Acid rain, along with most other ecosystem and health-related air pollution, originates with power plant and vehicle fossil fuel combustion, as do carbon dioxide emissions, which are the most important contributors to global warming. Energy extraction, transportation, and utilization facilities create land-use and conservation impacts and also use and pollute water. Nuclear power plants generate thousands of tons of long-lived radioactive wastes.

Historically, economic growth has been accompanied by rising energy use. Beginning in the mid-1970's, experience in the industrialized nations demonstrated that the two factors were not necessarily tightly coupled and that using energy more efficiently could significantly slow energy growth rates while still permitting vigorous economic expansion. For less-developed nations, the connection remains firmer until the transition has been made to an industrialized economy.

The national security implications of energy have become increasingly important as more countries have been forced to look beyond their borders for necessary energy supplies. U.S. reliance on imported oil was first underscored by the 1973 oil embargo by the Organization of Petroleum Exporting Countries (OPEC) in response to U.S. support for Israel in the 1973 Arab-Israeli War. It was soon highlighted twice more: by the 1979-1980 "second oil shock" precipitated by the Iranian revolution and unexpected collapse of Iran's substantial oil production, and by the 1990 Iraqi invasion of Kuwait—a major oil producer—and perceived threat to the vast oil fields of Saudi Arabia, which led to the Persian Gulf War.

EARLY HISTORY

Before the 1970's there was no serious effort to craft a national energy policy. The principal reason was that despite transient imbalances, cheap and abundant energy supplies were readily available. Unrelated policy initiatives focused on individual fuels and industries, driven largely by the different characteristics of each and by a desire to keep energy prices low and supplies ample.

Coal was the dominant fuel through the 1940's, having eclipsed fuelwood in the mid-1880's. The coal industry was dispersed and competitive, and there was little impetus for federal regulation. Coal's share of total energy consumption began to decline as oil and gas production rose after World War I.

Oil's share of total energy increased rapidly, and by 1950 petroleum overtook coal as the nation's largest energy source. A complex system of tax subsidies, import quotas, and other mechanisms arose to protect the domestic oil industry. Natural gas use increased sharply after World War II as a consequence of gradual improvements in pipeline technology, wartime pipeline construction subsidies, and a rise in demand during the war. In 1938 the Federal Power Commission was given authority to regulate aspects of interstate commerce in gas; states had exercised some regulatory authority since the late nineteenth century. Federal price regulation continued to be a thorny issue up through the mid-1980's, when most gas prices were finally deregulated.

In the early 1950's the federal government began to strongly push for the development of commercial nuclear power and provided large research, development, and demonstration (RD&D) subsidies in a unique effort to stimulate the industry. No other fuel or technology received such promotional assistance. Nuclear power is also unique in that most state regulation is preempted by federal law. In the 1970's nuclear power stalled in the marketplace for reasons related to costs, safety concerns, lack of public acceptance, and an intractable radioactive waste disposal problem. As of 1998 no firm order for a new plant in the United States had been placed since 1973.

The electric utility industry grew as electricity consumption rose by more than 1,100 percent between 1950 and 1997. This led to an increase in total fuel consumption: When fuels are burned to generate electricity (as opposed to being used directly, such as when natural gas is used for home heating), two of every three units of fuel are unavoidably lost as waste heat during the conversion process. The monopoly market power of electric utilities was recognized early, and both federal and state regulation were imposed. In the late 1990's there was movement in many states to deregulate the utility industry and open up electricity generation to competition.

Prior to the 1970's technologies and initiatives aimed at increasing the efficiency of energy utilization—often popularly referred to as "energy conservation"—and developing renewable energy sources—such as solar, wind, geothermal, and biomass energy—received little attention and no meaningful federal funding.

1973 Oil Embargo

The 1973 oil embargo drove oil prices in the United States up and, combined with other complex factors, created localized shortages. Suddenly energy, never before regarded as a distinct federal policy arena, was thrust to the top rank of the policy agenda in the United States. However, the elements of national policy remained deeply embedded in the unique industries, arrangements, and regulatory regimes that had coevolved with the three fossil fuels (oil, natural gas, and coal), nuclear power, hydroelectricity, and electric utilities. Yet for the first time there was also strong interest in promoting technologies to utilize renewable energy sources and increase the efficiency of energy use through a combination of federal regulation, research funds, tax subsidies, and public education. Federal RD&D funds for these alternative energy sources rose sharply in the 1970's, dropped dramatically during President Ronald Reagan's administration (1980-1988), which opposed support for them, and rose again in the decade that followed. By the mid-1990's, however, funding (in constant dollars) remained far below the peak levels reached around 1980, especially for renewable energy research.

Between 1973 and 1983 several comprehensive, groundbreaking energy policy studies were produced by groups in the private, nonprofit, and government sectors. These served to focus government and public attention on the importance of the issue. Coincidentally, the sudden rise in concern about energy came during the same period that the modern environmental movement arose. Beginning in the late 1960's, a widening recognition of the connection between energy and environmental issues was accompanied by frequent conflicts between energy and environmental policy goals. For example, the 1969 National Environmental Policy Act affected many energy projects and industries, while new coal mining laws and the 1970 Clean Air Act tended to discourage coal combustion.

Beginning in the early 1970's a wide assortment of new energy-related initiatives were adopted, but many—like oil price controls, efforts to jump-start a domestic synthetic fuels industry, and a breeder reactor program—were soon abandoned.

Federal Administrative Structure

Until the 1970's responsibilities for energy matters were scattered among various federal agencies, and there was little coordination from either a regulatory or policy perspective. After the 1973 oil embargo, efforts were made to improve the situation, but there was no encompassing centralization of function until the Department of Energy (DOE) was created in 1977. This reflected the elevation of energy at the national policy level and marked the first time that energy had been afforded cabinet-rank status. Even then, several regulatory commissions remained independent, and some other federal departments retained important responsibilities.

At the state level, public-service commissions exercised long-established jurisdiction over elec-

U.S. Oil Consumption, Production, and Imports 1949-1997

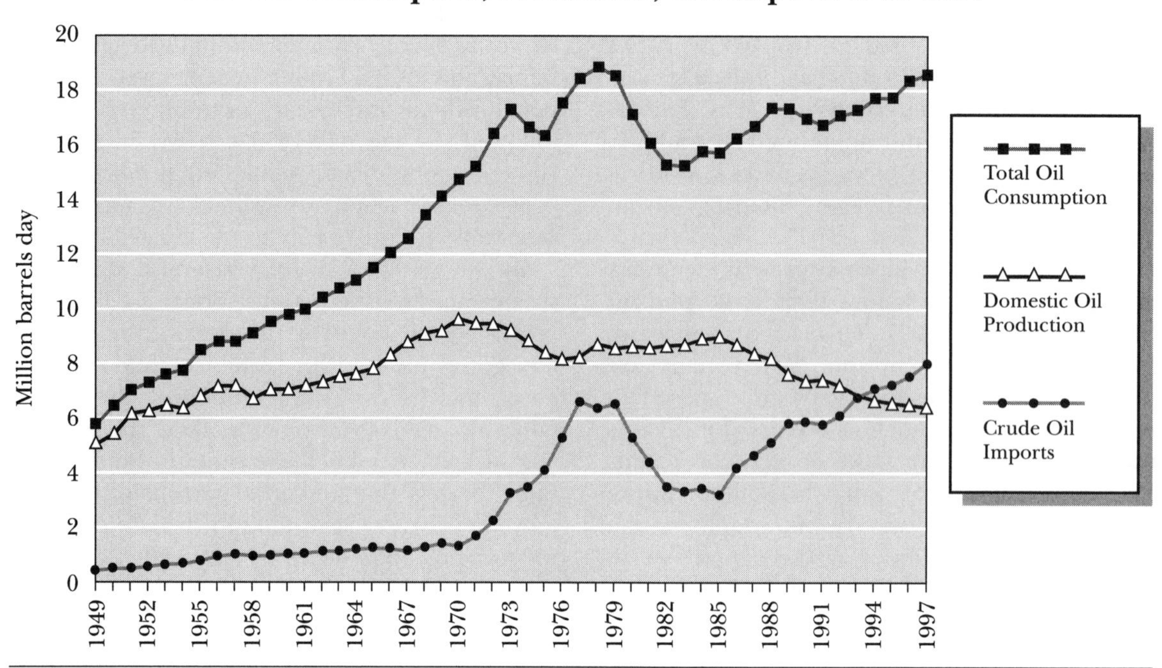

Source: U.S. Department of Energy/Energy Information Administration.

U.S. Energy Consumption, Energy Intensity, and Oil Price 1949-1997

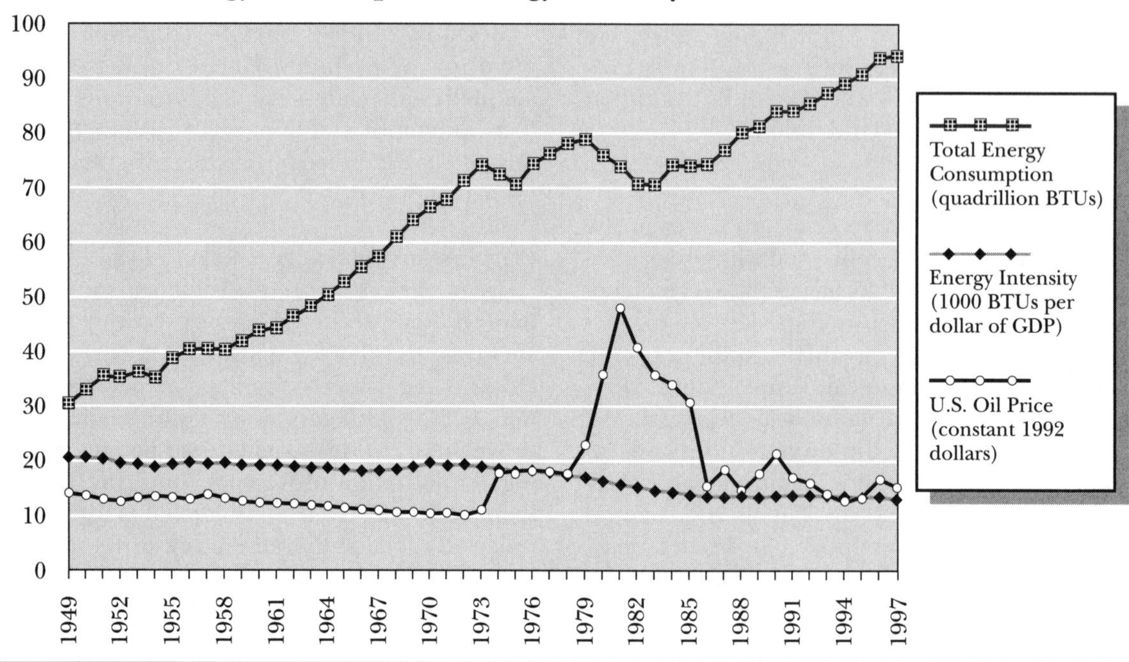

Source: U.S. Department of Energy/Energy Information Administration.

Note: Energy Intensity is a measure of average efficiency of energy use for the nation as a whole, expressed as the amount of energy used per dollar of Gross Domestic Product, annually.

tric utility rates, and other state agencies controlled land-use decisions that could affect power plant construction. Many states established energy agencies, and some began to formulate explicit energy policies and fund energy research.

Following the 1973 oil embargo several administrations drafted national energy plans. The first was President Richard Nixon's Project Independence, which proposed national energy self-sufficiency by the 1980's—a goal soon recognized as politically attractive but unrealistic. President Jimmy Carter's administration (1976-1980) produced the 1977 National Energy Plan, which emphasized short-term strategies to reduce oil imports, reduce total energy demand through increased energy efficiency, raise coal production, and increase the use of renewable energy. Carter's proposals regarding renewable energy sources and increased efficiency marked the first time that the federal government had proposed a serious, tangible commitment to these alternatives.

During the Reagan administration no comprehensive energy policy document was produced, although the administration strongly favored fossil and nuclear fuels and opposed funding for energy efficiency and renewables. President George Bush's administration (1988-1992) put forth a 1991 National Energy Strategy that emphasized oil, gas, and nuclear power production but offered little support for energy efficiency and renewables. The response of the environmental community was sharply critical, but many of the proposals were included in legislation enacted the following year. President Bill Clinton's administration (1992-2000) followed with energy plans in 1994 and 1998, as well as the 1998 Climate Change Technology Initiative. However, an opposition-led Congress was generally unreceptive, the administration did not place high priority on its energy proposals, and through 1998 no major new legislation or initiatives were adopted.

Federal Legislation

Federal legislation sometimes marks national recognition of the importance of a hitherto ignored policy area. Prior to the 1970's, legislation was directed at individual fuels and energy sectors. However, beginning in the 1970's Congress adopted several broad-scope energy laws that, for the first time, addressed disparate energy matters in single pieces of legislation. Some contained new research and regulatory initiatives and addressed issues or technologies that had never before received significant federal attention or support. Among the most important were the Energy Reorganization Act of 1974, which established the short-lived Energy Research and Development Administration (superseded in 1977 by the DOE) and companion legislation that funded a program of energy RD&D that included a renewables and efficiency component. The 1975 Energy Policy and Conservation Act mandated automotive fuel-efficiency standards, authorized a variety of energy-efficiency programs and standards, and established the Strategic Petroleum Reserve as a hedge against future supply disruptions. The 1976 Energy Conservation and Production Act funded state conservation programs and authorized a federal program to weatherize low-income housing.

President Carter signed several pieces of legislation comprising the National Energy Act of 1978. These included the National Energy Conservation Policy Act, which expanded weatherization programs, authorized utility residential conservation programs and an energy grants program for schools and hospitals, and also authorized appliance efficiency standards. The 1978 Public Utilities Regulatory Policies Act (PURPA) paved the way for the tremendous expansion of nonutility, independent electric power producers that occurred in the 1980's and 1990's. The 1978 Powerplant and Industrial Fuel Use Act (PIFUA) barred the use of oil or natural gas fuel in new electric generating plants or major industrial facilities (repealed in 1987).

Although no major energy legislation was passed during the Reagan administration, Congress enacted the 1992 National Energy Policy Act at the end of the following Bush administration. The most wide-ranging legislation adopted since the late 1970's, the law included provisions to help independent power producers, notably by opening access to utilities' transmission systems to nonutility electricity generators. Other important changes included easing licensing re-

U.S. Energy Sources, 1997 (percentages)

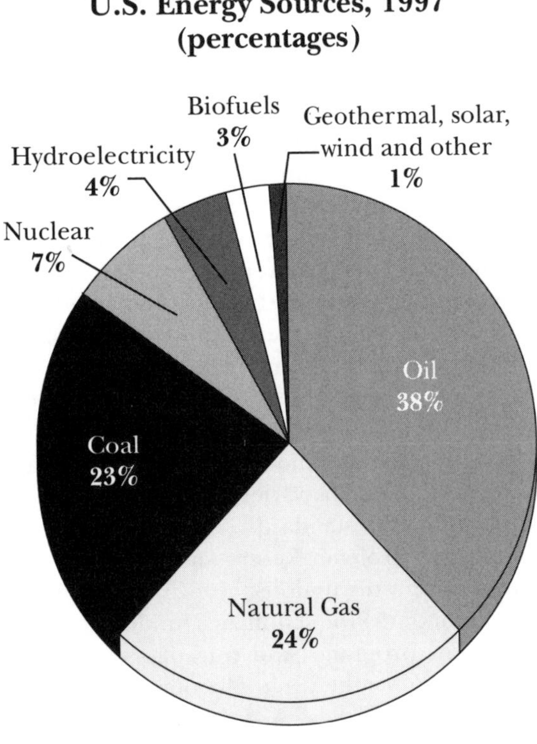

Source: U.S. Department of Energy.

quirements for new nuclear power plants and restricting public access to the process, initiatives intended to foster energy-efficiency improvements and renewable energy resources, and a variety of measures related to alternative fuels, energy-related taxes, coal development, and other matters. Between 1993 and 1998 no major energy legislation was enacted.

U.S. CONSUMPTION TRENDS

As oil prices rose and fell after the events of 1973 and 1979 and other issues began to capture the public's attention, energy began to drop down the list of pressing policy concerns. The 1990 Iraqi invasion of Kuwait and the subsequent Persian Gulf War briefly raised prices, as well as the visibility of energy policy and energy security. In the 1990's, however, friction among OPEC cartel members and increased oil production by non-OPEC countries resulted in unexpectedly abundant oil supplies and low oil prices; by 1998 average real gasoline prices in the United States had fallen to record low levels. In response, public and governmental attention again shifted elsewhere. Yet oil remained the nation's dominant energy source: In 1997 it provided 38 percent of total energy. Transportation accounted for 68 percent of total oil consumption, while oil supplied virtually all the nation's transportation fuel.

U.S. oil production peaked in 1970; since 1993 imports have exceeded production. By 1997 the United States was again heavily dependent on foreign sources of oil, importing more petroleum than ever before and relying on imports for more than one-quarter of its annual energy requirements. Although substantial reserves were discovered and production increased in other areas of the world, in the late 1990's more than 55 percent of known world oil reserves were still concentrated in Middle Eastern OPEC nations. Moreover, the Department of Energy and International Energy Agency projected in the mid-1990's that an anticipated rise in oil demand by the year 2020 (two-thirds of which is expected to come from developing countries) could be accommodated only by an increase in production from the huge reserves of Middle Eastern countries.

Total U.S. energy consumption remained remarkably stable between 1973 and 1990 in response to substantial increases in the efficiency of energy use. A key measure of energy efficiency, energy intensity (the amount of energy used per dollar of gross domestic product), declined by 28 percent during the period but leveled off after 1986, largely in response to falling energy prices. Energy consumption between 1992 and 1997 rose to record levels in each successive year as lower energy prices and vigorous economic growth led industries and consumers to give less consideration to energy efficiency in planning and purchases.

In the late 1990's energy experts remained concerned about the economic, environmental, and national security implications of energy supply and consumption patterns. Environmental concerns were exacerbated by the growing international recognition of the probable impacts of global warming. The ability of most governments to meet carbon emission targets set forth in the international climate agreements adopted in the 1990's hinges on changes in energy technologies and consumption patterns.

Continuing Problems

The United States still faces several fundamental energy policy problems. A heavy reliance on imported oil continues. There has been no real effort to stem rising transportation-sector oil consumption caused by increased total annual miles driven and rising sales of sport utility and similar vehicles that are not subject to passenger car fuel-economy standards. Large quantities of coal are still used to produce electricity. No meaningful policies to curb fossil-fuel use and accompanying greenhouse gas emissions have been implemented; meanwhile, the United States remains the world's largest emitter of such gases. Another problem is the longstanding failure to include the substantial externalized social and environmental costs of energy extraction, utilization, and security in the price of energy. The federal government has also failed to establish stable, consistent policies to pursue opportunities for increased energy efficiency and the promise of renewable energy resources.

Critics argue that such problems are not serious and that an aggressive response is premature. Environmentalists contend that the economic, social, and environmental costs of continuing the status quo will be far more serious than the economic costs of taking steps to mitigate these problems.

Several comprehensive studies by government and nonprofit groups have concluded that an aggressive national commitment to energy efficiency and renewable energy sources could, over time, result in the displacement of significant fractions of fossil and nuclear fuels and a sharp reduction in energy-consumption growth rates, with no severe economic penalty and sizeable environmental benefits. To cite just one example, by the mid-1990's energy-efficiency improvements begun in the early 1970's were saving the nation more than $200 billion annually in energy costs.

Innovative energy policies favored by environmental advocates include imposing meaningful taxes on carbon fuels, including additional gasoline taxes, and abolishing large and long-standing tax and research subsidies for the fossil fuel and nuclear industries. Another strategy would be removing market barriers and providing ample, stable RD&D funds and tax incentives for energy efficiency and renewable energy, including alternative transportation fuels. Environmentalists have also suggested mandating more stringent fuel-economy standards for all vehicles and tighter energy-efficiency standards for appliances, motors, and buildings, and managing the deregulation of the electric utility industry to protect renewable energy and energy-efficiency investments, markets, and potential.

Environmentalists acknowledge that there is strong political opposition to many of these measures, which makes it unlikely that they will be adopted without committed federal leadership and a shift in the national sentiment about the importance of energy.

Energy Policies Around the World

As in the United States, most nations did not develop explicit energy policies until the oil price and supply dislocation that occurred in 1973. As of the late 1990's, there were probably as many different energy policies as there were nations. In every country, energy receives more or less policy attention depending upon availability and cost of domestic and imported fuels, competition from other pressing policy and social concerns, and the degree of industrial development.

European nations have, for many years, pursued markedly different energy strategies. In the 1990's the increasing economic integration of Europe led to efforts to forge a common European Union energy policy. Japan developed the world's third-largest nuclear power program because no significant domestic fossil fuel resources were available to power the tremendous postwar growth of the Japanese economy. However, Japan remains completely dependent on imports for its oil needs. Continued heavy reliance on domestic coal in China, Russia, India, and Germany (along with the United States, the world's second-largest coal consumer) has raised concerns about associated greenhouse gas emissions.

Most developing countries face pressing social and economic problems that eclipse energy and environmental considerations, even though the latter often aggravate the former. For example, the costs of imported energy—usually oil—constitutes a significant drain on foreign exchange for many developing nations. At the same time, scarce capital and technical expertise make it difficult to develop alternative energy resources, technologies, and infrastructure without foreign investment and assistance.

Phillip A. Greenberg

SUGGESTED READINGS: A good overview of international energy policy, its relation to climate change, and the promise of alternative energy sources is Christopher Flavin and Seth Dunn, "Rising Sun, Gathering Winds: Policies to Stabilize the Climate and Strengthen Economies," *Worldwatch Paper* 138 (1997). The September, 1990, issue of *Scientific American*, "Energy For Planet Earth," provides a good introduction to many technologies and policy issues. An insightful political history of fossil fuels and the electric utility industry is *The Politics of Energy* (1992), by David Davis. The Congressional Research Service produces excellent, regularly updated issue briefs on timely energy policy issues, available via the Internet (http://www.cnie.org) or by request from the offices of members of Congress.

SEE ALSO: Corporate average fuel economy standards; Energy conservation; Nuclear regulatory policy; Oil crises and oil embargoes.

Environmental economics

CATEGORY: Resources and resource management

Environmental economics involves the study of the relationship between the economy and the environment. It is concerned with the allocation of costs associated with pollutants, the allocation of natural resources, and efforts to place a value on resources for which there are no markets.

Environmental economics grew out of the environmental issues that impinged upon the social consciousness beginning in the 1960's and 1970's. Visions of a "silent spring," polluted rivers, and smog-filled cities posed questions about whether a free market economy efficiently allocates resources. A fourfold increase in the price of oil, gas lines, and the view of Earth as a small blue sphere against a black void prompted debate about whether sufficient nonrenewable resources exist to sustain economic growth. Deforestation, species extinction, and global warming have raised doubts about whether markets adequately value environmental resources, prompting economists to inquire into how to value resources for which there are no markets.

Traditional economic theory largely ignored the relationship between economics and the environment. The assumption of economic rationality depicts firms as profit maximizers and consumers as pleasure maximizers. The invisible hand of the market conveys the view that voluntary exchange promotes economic harmony. Firms and consumers in pursuit of their self-interest unintentionally promote the interest of all. Nature is reduced to an input into the production process, providing both renewable and nonrenewable resources. Pollution was treated as an aberration, an example of market failure requiring some form of government intervention to restore the harmony of the market.

An alternative view expressed by Herman Daly and John Cobb in their book *For the Common Good: Redirecting the Economy Toward Community, the Environment, and a Sustainable Future* (1989) conceives of the economy and nature as interdependent. Emphasis is placed on the concept of extended rationality; that is, individuals find it in their self-interest to protect the environment and care for future generations. Both the economy and nature are viewed in terms of coevolutionary processes, each affecting the other. This view focuses on creating institutions to channel self-interest in environmentally sensitive ways.

POLLUTION

Despite highly restrictive assumptions, economists most often use the perfectly competitive model to evaluate environmental policy. Under perfect competition, no single agent has the power to influence price, resulting in an equilibrium price that efficiently allocates resources. Efficiency means that the benefit of producing one more unit equals the cost of producing that unit. Social welfare (happiness of the individuals in society) is maximized because each unit produced prior to equilibrium yields more benefit than it costs.

For example, consider a firm engaged in the production of copper by the ton. As the consumption of copper rises, the benefits obtained from an additional ton decline; conversely, as production rises, the costs to the firm of producing an additional ton rise. The intersection of these supply and demand factors results in an equilibrium that sets a price for the copper and also determines the amount of copper that the firm should produce.

Suppose, however, that the firm also produces as a by-product pollution that injures others. In this case, the free-market price would not reflect the external cost imposed on others or on society at large resulting from the pollution. Pollution is a type of externality, a cost involuntarily imposed on one party as a result of the activities of others. If corrections were made to the supply-and-demand equilibrium to account for the external cost, the price of the copper would rise, and production would fall. Note that correcting for external costs does not eliminate pollution; it merely requires that producers and consumers consider the external costs in their decisions.

The existence of externalities implies that the free market misallocates resources: Too much is produced for a price that does not include the external costs. Social welfare is not maximized, because costs to the firm plus external costs of the last units produced exceed the benefit of the last units.

Dealing with Externalities

There are several options available to government in correcting for externalities: standards, taxes and subsidies, property rights, and marketable permits. In the past, environmental laws generally imposed a fixed standard to which all businesses must conform. The simplicity of standards from a policy point of view has made them widely used. However, standards have been criticized for their coercive element, their failure to consider local circumstances, and their apparent arbitrariness.

In the 1930's, Arthur Pigou, one of the first economists to address externalities, recommended that government impose a tax equal to the external cost. Critics, however, cite difficulty in measuring the external costs, measuring the amount of pollution, determining the location to measure the pollution, and so on.

Ronald Coase, in a classic article entitled "The Problem of Social Cost," advocates a free market solution. Coase recommends assigning property rights for the air or water, leading the affected parties to bargain over who pays the costs associated with pollution. If the injured party owns the resource, he or she charges the polluter an amount not less than the damages for using the resource. If the polluter owns the resource, the injured party would bribe the polluter not to pollute. The injured party would not pay an amount exceeding the damages created by pollution, and the polluter would not accept an amount less than the profits foregone. To remain unbiased, Coase advocates the doctrine of ethical neutrality. In the absence of property rights, the polluter is no more responsible for pollution than the injured party. Who pays depends on which method reduces the transaction costs (costs of identifying the party or parties harmed, and transacting the compensation). In most cases, minimizing transaction costs requires the injured party or parties to bribe the polluters not to pollute.

Market permits were adopted by the Clean Air Act amendments of 1990. The government sells permits allowing the purchaser to emit a limited amount of pollution. Advocates argue that permits provide the polluter an incentive not to pollute. If the costs of installing pollution control equipment are less than the permit, the firm reduces its pollution. If the costs exceed the price of the permit, the firm buys the permit. Critics, however, charge that permits imply that government endorses pollution.

Renewable and Nonrenewable Resources

There are two types of natural resources: renewable and nonrenewable. Renewable resources may be replenished, such as a forest.

Nonrenewable resources cannot be replenished.

The conditions for sustaining the environment are easy to identify in theory but difficult to achieve in practice. First, the harvest of natural resources must be less than the growth rates of those resources. Deforestation and depletion of fisheries, for example, indicate that in many instances the harvest rates exceed the growth rate. Second, nonrenewable resources are, by definition, nonrenewable. The World Resources Institute estimates that fossil fuels provide 90 percent of the commercial energy used in the world. Reserves of coal are estimated at five hundred years, and reserves of oil are estimated at less than one hundred years. Ultimately, to sustain economic growth, renewable resources must be substituted for nonrenewable resources. Third, emissions must not exceed nature's ability to absorb wastes. Markets can be used to provide incentives to encourage people to reduce, reuse, and recycle. Innovative businesses have reduced pollution by altering production process and the input mix, thereby also reducing their costs.

Resource depletion was addressed by Garrett Hardin in a classic article titled "The Tragedy of the Commons" (1968). Commons refer to a resource owned by no one and available to everyone. Assuming rationality, individuals exploit the resource as long as the benefit exceeds the cost. The result is that self-interest leads individuals to destroy the resource. Hardin recommended privatizing the resource.

Economists are divided on whether the market alone is sufficient to make the transition from nonrenewable to renewable resources. Conservative economists assert that the market works. As oil production slows, the price of oil rises, providing incentives to entrepreneurs to find alternatives. This assumes that by allowing markets to work, new technologies will be developed, making inputs infinitely substitutable.

Markets, however, rarely allocate resources in ways that preserve the environment. The interest rate, for example, reflects society's preference between allocating resources for present use versus future use. The higher the rate of interest, the more society discounts the use of the resources in the future. If a firm finds that the market interest rate exceeds the rate of increase in the price of a resource, such as oil, then the profit-motivated firm will sell its resource and invest the proceeds.

The market reflects the preferences of those who have the "dollar votes." Nature and habitats, while important, do not vote. The services that nature provides in the forms of recycling wastes, maintaining the climate, providing oxygen, absorbing carbon dioxide, and providing aesthetic pleasure are not reflected in market values. Hence, what is most profitable is not necessarily consistent with environmental preservation.

VALUING NATURAL RESOURCES

Markets allocate resources based on their price or value. How does one place a value on something for which there is no market? The answer better enables policymakers to allocate resources among competing uses, such as whether to use public lands for recreation or oil drilling. Placing a market value on natural resources generally means transforming them into commodities: a forest transformed into lumber for example, or mining gold in Yellowstone. There are exceptions: The Nature Conservancy Fund uses a strategy called debt-for-nature swaps by purchasing land from Third World countries for preservation, thus reducing Third World debt and protecting rain forests. In many cases, however, such as administering public lands, market solutions are unavailable.

In addressing the value of a resource, economists have developed three classifications for value: user value, option value, and existence value. User value refers to the value in using the resource to the individual. User value is reflected in the value of the resource to hikers, recreationists, and skiers. Option value refers to the value of having the option to develop the resource at some future point. Existence value refers to the value of bequeathing the environmental resources to future generations, to habitat, and so on.

To determine the value of a resource, economists have employed a number of approaches. The most widely used are willingness to pay and willingness to accept. Willingness to pay asks individuals how much they would be willing to pay to enjoy environmental benefits. Willingness to accept asks how much individuals would be willing to accept in order to incur some loss. The difficulties with all approaches, however, reveal that there is no objective way to determine the value of a resource.

John P. Watkins

SUGGESTED READINGS: A nice introductory text for the nonexpert is Tom Tietenberg, *Environmental Economics and Policy* (1998). William J. Baumol and Wallace E. Oates offer a traditional approach in *The Theory of Environmental Policy* (1998). For a nontraditional approach, see Herman E. Daly and John B. Cobb, *For the Common Good: Redirecting the Economy Toward Community, the Environment, and a Sustainable Future* (1989). A survey of the subject is offered by Maureen L. Cropper and Wallace E. Oates in "Environmental Economics: A Survey," *Journal of Economic Literature* (1992). For Ronald Coase's classic article on market solutions to environmental problems, see "The Problem of Social Cost," *The Journal of Law and Economics* (October, 1960). For a discussion of the problem of the commons, see Garrett Hardin, "The Tragedy of the Commons," *Science* (December, 1968).

SEE ALSO: Accounting for nature; Debt-for-nature swaps; Pollution permits and permit trading.

Environmental education

CATEGORY: Philosophy and ethics

Environmental education involves structured instruction of environmental topics at all levels of learning. Such programs are designed to inform students about issues associated with natural and built environments and to teach them how to use this knowledge to achieve balanced and sustainable habitats.

Environmental education should not be viewed as a static or monolithic practice but rather as one that encourages and promotes diversity and changes in learning styles and practices. Education faces new challenges to make the increasingly complex field of local, regional, and global environmental issues meaningful and relevant at the individual level.

Environmental educators seek to apply experiential learning strategies—learning by direct experience—to the education mainstream in order to equip students with community participation skills. Integrating "community service learning" into the environmental education curriculum, whereby "hands-on" experiential learning through environmental problem-solving activities becomes the principal student focus, is proving to be a valuable resource for enabling students to become "stakeholders" and participants in community building and environmental responsibility.

Integrating experiential education with classroom learning enables the instructor to bring practice and theory together. Indeed, studies in education methodologies by the National Training Laboratories have demonstrated that students' average retention rates of materials covered ranges from only 5 percent from straight lecture-type instruction to 30 percent from demonstration-type activities to 75 percent from practice by doing and up to 95 percent from teaching others or making immediate use of learning—which is the essence of experiential learning.

The growing trend in the United States for experiential learning at all levels of education is evidenced by the community service learning programs that are appearing on campuses across the country. While this new and responsive teaching-learning process has been quickly adopted by the education and social science fields, it has not yet gained wide acceptance in the academic fields directly related to environmental education. Nationwide service integration data compiled by Campus Compact reveals that less than 5 percent of undergraduate service learning coursework is focused on science, engineering, and mathematics.

Traditionally, environmental science instruction has perhaps inadvertently taken on an adversarial role, training students to become watchdogs for community and regional environmental preservation. The challenge is to find constructive ways to encourage learning in the

"Hands-on" learning experiences are a vital part of environmental education. (Jim West)

environmental field so that the students and teachers can make valuable contributions to their communities. Numerous initiatives for such community-based service learning in the environmental field have been advanced.

The primary goal of service learning is to place students into practical field settings, where the primary motivation is service, followed by developmental learning. Participants perform constructive community service while developing their critical thinking and group problem-solving skills. Internships with public and private agencies or firms involved with environmental activities are a proven education methodology for this type of "learning how to learn" experience. This teaching strategy emphasizes the students' responsibility for self-development. It has proven to be an effective way for undergraduate schools to establish outreach programs and attract underrepresented groups to study in selected areas.

Service learning courses permit faculty to integrate principles across the curriculum, provide leadership training for students, improve written and oral communication skills, and increase awareness of social responsibility. These courses offer a fresh approach to teaching environmental education. As an example, the National Backyard Composting Program, established to help local communities reduce the organic food and yard wastes that now take up 30 percent of the landfill volume, offers excellent initiatives for students to work with local communities in setting up demonstration composting stations for individual residential uses and applications. Students not only learn the intricacies and scientific basis for this transferable technology but also have the opportunity to reinforce this learning by teaching others.

Integrating self-disciplined experiential learning with community service-type activities relating to the environment instills in students a "need to know" that promotes initiatives, creativity, and problem-solving skills that can prove valuable in future employment settings. Students also gain insight about cooperative working arrangements and building consensus in addressing and resolving environmentally sensitive issues or problems.

In formulating a framework for such a learning experience, three basic questions must be addressed to get a clear sense of focus on a community-based environmental service-learning project: What is it we want to learn or do? Why do we want to do it? How will we do it? The emphasis on plural rather than singular underscores the student-teacher partnership that is initially established for effective experiential learning. Indeed, the teacher plays a more important role as a facilitator or coach than as one who merely conveys knowledge to others in a conventional classroom lecture-type setting. This "teacher as a facilitator" model offers exciting promise for instilling into students a sense of appreciation for their own individual learning and how it can make a difference in their world.

Robert B. Seaman

SUGGESTED READINGS: *Rethinking Tradition: Integrating Service with Academic Study on College Campuses* (1995), edited by Tamar Kupiec, is a good overview of service learning approaches.

SEE ALSO: Composting; Environmental ethics; Renewable resources; Social ecology.

Environmental engineering

Category: Resources and resource management

Environmental engineering focuses on using engineering methods and practices to solve problems related to maintaining public health with minimal environmental impact. The discipline encompasses a wide range of projects, such as controlling pollution and providing safe drinking water, waste treatment, and urban and rural drainage.

The environment as a whole is large and complicated. To reduce the level of complexity in analysis, it is helpful to invent relatively simple ways to look at the whole and its constituent parts. This simplification is sometimes called a model for systems analysis. An example of such a model is one in which a researcher considers an environmental system as consisting of distinct elements that interact with one another. Air, water, soil, waste products, and hazardous products may be considered constituent parts of a whole. It is important to understand how each part of the large system behaves. The quality of each system, system interactions, and the impact of human activities on each system are of enormous importance to the well-being of the whole.

Although different in their ultimate goals, environmental science and environmental engineering are intimately related. It is important to understand how the natural environment is constituted, how it sustains itself, and how it is affected by human activities. Based on this understanding, rigorous criteria can be developed to help maintain high standards in the quality of the environment. In turn, such criteria can be used to help establish realistic standards that will guide the development and utilization of new technologies.

Technology has three aspects: The first aspect is helpful to human life and the environment, while the second is detrimental to both of them. The third aspect of technology can help solve many environmental problems, even those that are created by technology itself. In order to realize this last aspect, the talents of scientists, engineers, technicians, social scientists, and health professionals must be combined to gain an understanding of how human activities affect the environment and what can be done to reduce or eliminate the damage that has been done to it.

Human civilization continues to require increasing amounts of fuel for its machines, chemicals for its industrial plants, fertilizers and pesticides for its farms, and paper products for homes and offices. The operation of these machines and the use of these products generate waste products that must be handled safely. The essence of environmental engineering is finding practical ways to help solve environmental problems in an efficient and affordable manner.

Contexts of Environmental Engineering

Environmental engineering operates in three broad contexts: natural systems, engineered systems, and design. Environmental engineering in natural systems is concerned with the chemistry and biology of air, water and soil quality, water and air pollution, limnology (the scientific study of bodies of fresh water), global atmospheric change, the fate and transport of pollutants in the natural environment, hazardous substances, and risk analysis. Environmental engineering in engineered systems concentrates on systems made by humans such as water supply and treatment processes, wastewater treatment processes, processes for air pollution control, groundwater remediation, and solid and hazardous waste management. Environmental engineering design is the application of physical, chemical, and biological operations and processes in natural and engineered systems. It must meet five sets of requirements: legal, public health and sanitary, socioeconomic, aesthetic and sociocultural, and engineering.

The use of water in the home provides a good example of environmental engineering design. Running water in every home is an expected convenience in many countries. After water has been used, however, it must somehow be disposed of. The safe disposal of wastewater is necessary in order to protect the health of individuals, families, and entire communities. This requires the design of treatment, recycling, or disposal systems that will dispose of wastewater in such a way as not to violate any laws, regulations, or ordinances regarding water pollution and sewage disposal (legal requirement). The supply of drinking water or waters that are used for recreation cannot be contaminated by wastewater and must be protected from possible carriers of public health hazards such as animals and insects (public health and sanitary requirement). In addition, treatment systems cannot exceed the cost that the community can bear (socioeconomic requirement) and cannot cause a nuisance such as excessive noise, unpleasant odor, or unsightly appearance (aesthetic and sociocultural requirement). Finally, sewage should be transported, discharged, treated, and disposed of in specified ways. Typically, this is achieved by developing a system that takes into account the four interconnected stages of waste treatment: source of wastewater, wastewater collection, treatment, and disposal or reuse (engineering requirements).

Air Pollution

Carbon monoxide (CO) is a colorless and odorless gas that is produced during most combustion processes. Current technological and agricultural activities are emitting CO in large and increasing quantities. For example, CO is abundantly produced by automobile engines, oil refineries, and other industries, and by the burning of forests, agricultural fields, grasslands, and savannahs around the world. An important issue in environmental science and engineering is what happens to this CO and what can be done to control it.

CO plays a key role in the chemistry of the lower atmosphere, known as the troposphere. Once CO is released into the atmosphere, it can be transported over long distances. Ultimately, it is oxidized to carbon dioxide (CO_2) by the hydroxyl (OH) radical. The OH radical is the key molecule in the breakdown and removal of greenhouse gases, such as methane, which is also important in the chemistry of stratospheric ozone (0_3). However, CO is by far the largest consumer of the OH radical. Thus, as CO emissions increase and OH oxidizes by CO, the amount of OH available to oxidize other gases in the atmosphere decreases. When concentrations of OH decrease, the breakdown and removal of other greenhouse gases also decreases. The decreasing amounts of tropospheric OH may potentially influence stratospheric ozone, the removal of greenhouse gases, and the climate. Thus, researchers use the methods of environmental science to determine whether a problem exists, and, if it does, to determine the extent of its importance and decide what to do about it. Accordingly, the scientific study of air pollution, as well as the design and development of air pollution control technology, allows researchers to identify sources, environmental and health impacts, regulations, and modeling of air pollution; develop processes and alternative strategies for control; and project global climatic implications.

An interdisciplinary team of scientists and engineers at the National Aeronautics and Space Administration (NASA) has designed and built a system that can be placed on satellites to measure air pollution in the atmosphere in collaboration with other systems at a variety of universities in the United States and around the world. The project is called Measurement of Air Pollution from Satellites (MAPS). The MAPS instrument, based on a technique called gas filter radiometry, has been used to determine the global distribution of CO mixing ratios in the free troposphere.

Legislation and Engineering

The increasing awareness of the threat that human activities pose to the shared environment is translated into pressures that are brought to bear on developers of technology, those who sell technology, and those who consume it. This awareness is typically converted into regulations and legislation to which people must adhere. An example of such regulation in the United States is the Pollution Prevention Act of 1990, which made pollution prevention the national environmental policy of the United States. Pollution prevention refers to source reduction. This includes promoting practices that conserve natural resources by reducing or eliminating pollutants through increased efficiency in the use of raw materials, energy, water, and land. These requirements present challenges as well as opportunities.

Concerns for the quality of the environment has added new dimensions to old fields of study in engineering and science. For example, traditional hydraulics engineering now has a subspecialty called environmental hydraulics and environmental water resources. Environmental hydraulics is the study of how problems connected with human activities relate to the quality and motion of fluid in rivers, lakes, estuaries, coastal waters, and air. Water resource agencies have traditionally been concerned with merely ensuring an adequate supply of water. However, the importance of water quality has increased substantially along with environmental concerns. Particular emphasis is now placed upon the studies of pollutant mixing in bodies of water, modeling flow and contaminant transport in surface and subsurface environments, creating methods for control of turbidity in urban runoff, and designing and building outfalls for the discharge of wastewater and thermal effluent. Furthermore, studying these environmental concerns has become increasingly sophisticated, requiring the integration of many new technologies: complex computer models, electronic laboratory and field instrumentation systems, and data acquisition and transmission for real-time control.

Groundwater sources are continuously threatened by excessive pumpage and contamination. Stringent federal regulations on groundwater quality and quantity management have become necessary, and they have created the impetus for technical manpower with strong multidisciplinary backgrounds in areas such as hydrology, hydrogeology, geochemistry, optimization, computational techniques, and software engineering.

Another example is transportation engineering. The movement of people, goods, and services provides for an improved quality of life and strong economic activities. Thus, efficient transportation systems form integral building blocks of a developed society. However, the physical construction of pathways and sites, the operation of transportation facilities and vehicles, and the travel behaviors associated with the vehicles themselves may have negative impact on the environment. Therefore, transportation requires not only efficient design and operation but also the careful linkage among travel behavior, urban life, and environmental quality and policy.

Environmental engineering is a broad and multidisciplinary subject. Generally, academic preparation for work emphasizes basic engineering and scientific principles, as well as the design and application of environmental engineering operations and processes. Thus, at the graduate level, the discipline attracts scientists, engineers, and students from a variety of engineering and science backgrounds.

Josué Njock Libii

SUGGESTED READINGS: Mackenzie L. Davis and David A. Cornwell, *Introduction to Environmental Engineering* (1991), is a good introduction to the subject from a civil engineering point of view. Joseph A. Salvato, *Environmental Engineering and Sanitation* (1992), is an excellent discussion of environmental science and technology with particular emphasis on comprehensive applications of sanitary science and engineering. Bela G. Liptak, editor, *Environmental Engineers' Handbook*, volumes 1, R, and HI, provide extensive discussions and references. Robert A. Corbitt, editor, *Standard Handbook of Environmental Engineering* (1990), contains several useful essays on the subject. *The Journal of Environmental Engineering* is a good reference for technical and current issues in the field.

SEE ALSO: Automobile emissions; Sewage treatment and disposal; Waste management; Water treatment.

Environmental ethics

CATEGORY: Philosophy and ethics

Environmental ethics is the field of inquiry used to evaluate the ethical responsibilities humans have for the natural world, including natural resources. There are many, often conflicting, perspectives on appropriate human responsibilities toward nature and natural resources, including anthropocentrism, individualism, ecocentrism, and ecofeminism; each has strengths and weaknesses.

Anthropocentrism is a human-centered philosophy whose adherents believe that moral values should be limited to humans and should not be extended to other creatures or to nature as a whole. A justification for this perspective is that moral relationships are sets of reciprocal rules followed by humans in their mutual relationships. Nonhumans are excluded from moral relationships because they lack comprehension of these rules. Other anthropocentrists argue that, from an evolutionary perspective, successful species should not work for the net good of another; any species doing so in the past have become extinct.

Some anthropocentrists oppose restrictions on natural resource use to avoid such negative impacts as the loss of jobs or products beneficial to humans. However, other anthropocentrists stress that the natural world is a critical life- support system for humans and advocate effective environmental controls so that it will maintain its full value for present and future generations. This anthropocentric regard for the environment is based on the practical value of the natural world for meeting human needs rather than a belief that the natural world has intrinsic value.

Adherents of individualism believe that humans should extend moral concern to individual animals of certain species. Individualists include advocates of the animal liberation and animal rights movement. Individualists accept that all humans have intrinsic value; they also argue that because some animals share morally relevant qualities valued in humans, these animals should be extended moral concern. Animal liberationists define the capacity for pleasure and pain (sentience) as the morally relevant feature to be considered. Animal rightists value more complex qualities, including desires, consciousness, a sense of the future, intentionality, and memories; they commonly associate these qualities with most mammals. Individualists generally are not concerned with natural resource use unless that resource use involves a direct threat to individuals of a species deserving moral concern, as through hunting or trapping.

Ecocentrism is based on the belief that the natural world has intrinsic value; it includes both the land ethic and deep ecology perspectives. Land ethic advocates believe that moral concern should be extended to the natural world, including natural units such as ecosystems, watersheds, and bioregions. Land ethic advocates emphasize respect (rather than rights) for the natural world. Ecocentrists may justify a land ethic by noting that all living creatures have a common origin and history on the planet and are ecologically connected and interdependent. The notions of common origin and history, as well as interdependence, are viewed as analogous to the human concept of family. Ecocentrists view humans as members of a large family comprising all of nature. Because family relationships entail not only privileges but also responsibilities for the well-being of the other family members and their environment, it follows that humans have responsibility for the natural world.

Impact on land health is an important criterion by which natural resource use is assessed in a land ethic. Characteristics of land health include the occurrence of natural ecological functioning, good soil fertility, absence of erosion, and having all the original species properly represented at a site (biodiversity). From a land ethic perspective, natural resource use should minimize long-term impacts on land health or should even enhance land health.

Deep ecology is often viewed as an ecosophy—an ecological wisdom that calls for a deep questioning of lifestyles and attitudes. Some guidelines that regularly occur among its adherents include living lives that are simple in means but rich in ends, honoring and empathizing with all life forms, and maximizing the diversity of human and nonhuman life.

Ecofeminists believe that many environmental problems are tied to a desire to dominate nature, and this desire is closely linked with the problem of the domination of women and other groups in society. Ecofeminists believe that these problems would decline with a transformation in societal attitudes from dualistic, hierarchical, and patriarchal thinking to emphasizing an enrichment of underlying relationships and placing greater focus on egalitarian, empathetic, and nonviolent attitudes. Ecofeminism emphasizes less intrusive and more gentle use of natural resources.

Many Westerners have reexamined established cultural and religious perspectives for inspiration and insights in developing an environmental ethic. Native American cultures are often viewed as a source of moral insights on human relationship to the environment. While difficult to generalize for the many diverse cultures, several perspectives appear common to many Native American groups: a strong sense of identity with a specific geographic feature such as a river or mountain; the notion that all of the world is enspirited and has being, life, and a self–consciousness; and a strong sense of kinship with the natural world. Such Native American views are commonly associated with reduced environmental impacts and harmonious relationships with the natural world.

Judaism, Christianity, and Islam share common traditions; each contains elements that scholars have drawn upon for insights into environmental responsibility. Some scholars emphasize portions of the biblical book of Genesis where the world is seen as God's creation, which humans should be free to use and enjoy. Subjugation, use, and development are acceptable, but the land also must be appreciated and protected as belonging to God. Others emphasize the special role of humans as caretakers or advocate close relationships to the natural world, as exemplified by Saint Francis of Assisi. Attitudes toward the natural world and natural resource use may vary widely among the various groups of Jews, Christians, and Muslims. Some Eastern philosophies, such as Taoism and Buddhism, contain insights for environmental ethics. Both encourage a caring behavior toward nature.

Richard G. Botzler

SUGGESTED READINGS: Information on environmental ethics can be found in the quarterly journal *Environmental Ethics,* edited by Eugene Hargrove. Two good general introductions to the major ideas in environmental ethics, including many of the classic articles, are contained in *Environmental Ethics: Divergence and Convergence* (1998), edited by Richard Botzler and Susan Armstrong, *The Environmental Ethics and Policy Book: Philosophy, Ecology, Economics* (1997), edited by Donald VanDeVeer and Christine Pierce, and *Environmental Ethics: Reading in Theory and Application* (1997), edited by Louis Pojman.

SEE ALSO: Animal rights movement; Deep ecology; Ecofeminism.

Environmental health

CATEGORY: Human health and the environment

Environmental health—one of the major disciplines in the public health field—applies scientific study to environmental agents that have a detrimental effect on the health and well-being of human populations.

Environmental health involves protecting the general public from contaminated air, water, and food, and from arthropods and other vectors that carry pathogenic organisms; ensuring safe handling and disposal of nonhazardous and hazardous wastes; and reducing risks from contaminated surroundings. Since World War II the field of environmental health has been broadened to include noise pollution, radiological health and safety, the environmental impact of large construction projects, and the impact of environmental disasters on large populations.

In the United States the responsibilities for ensuring the environmental health and safety of the population are shared by the National Institute of Environmental Health and Safety (NIEHS), the U.S. Public Health Service (US-PHS) and its Centers for Disease Control (CDC), the Environmental Protection Agency (EPA), the Nuclear Regulatory Commission (NRC), the U.S. Department of Agriculture (USDA), the Food and Drug Administration (FDA), the Federal Emergency Management Agency (FEMA), and numerous other government agencies. Each state has mechanisms for environmental health education, enforcement, and oversight. Substantial responsibilities fall on local public health inspectors, sanitarians, coroners, animal control officers, and a host of elected and appointed officials at the city and county level.

Within the United States corporate enthusiasm has grown for environmental audits. These audits validate corporate compliance with federal, state, and local environmental laws and regulations, and help clearly define and publicize policies and procedures within the corporation. An audit allows the corporation to recognize environmental risks, bring them under control, and adjust resources and personnel needed to complete environmental work.

Water Treatment

Most cities in industrialized countries have municipal water treatment plants, which try to ensure that drinking water supplies are free from pathogenic organisms and harmful substances. In addition, many communities fluoridate drinking water to prevent tooth decay. During the 1990's a number of large U.S. cities experienced epidemics resulting from inadequate water treatment. Some of the most persistent problems occurred in Milwaukee, Wisconsin, where agricultural runoff resulted in heavy loading of Cryptosporidium, a parasitic protozoan that was not destroyed during water treatment.

Several cities with aging water distribution systems suffered epidemics related to contamination that occurred after the water left the treatment plant but before it reached users. Water leaving treatment plants has usually been disinfected using a chlorination or ozonation process, leaving a residual that prevents growth of pathogens. However, when water pressure is low at the fringes of the distribution system, the residual disinfectant may be insufficient to prevent growth of pathogenic microorganisms. Some of these microorganisms may cause serious infections in people with damaged immune systems, especially acquired immunodeficiency syndrome (AIDS) and chemotherapy patients; these people must either use bottled water or boil their drinking water.

The goals of sewage treatment are the elimination of pathogenic organisms and the reduction of the amount of organic materials in the wastes discharged into the environment. Because of the potential for the cultural eutrophication of lakes by nutrients in treated wastewater, most communities discharge treated wastewater into rivers or oceans. Sewage normally contains a number of heavy metals—including mercury, lead, copper, and iron—which become concentrated in sewage sludge. Sludge must be disposed of properly; most is dewatered and landfilled.

Arthropod and Animal Control Programs

Insects and other arthropods may serve as vectors in the transmission of disease; environmental health programs are crucial in controlling these threats. Effective vaccines are available for some diseases, such as yellow fever, and mass vaccination programs are recommended for high-risk populations. In areas of the world where malaria, yellow fever, dengue fever, filariasis, viral encephalitis, and other mosquito-transmitted diseases pose public health threats, environmental health officials seek to eliminate mosquito breeding grounds and control mosquitoes near population centers using pesticides.

Prairie dogs, rats, mice, rabbits, and deer harbor a number of infectious diseases that may be transmitted to humans via arthropods; the most important bacterial pathogens include plague, tularemia, and Lyme disease. Rickettsia, small microorganisms similar to bacteria, cause a number of infectious diseases, including Q fever and Rocky Mountain spotted fever. More than one thousand cases of Rocky Mountain spotted fever occur in the United States each year, mostly on the East Coast; a vaccine is available.

Wild and feral animals harbor microorganisms that can produce disease in domestic animals and humans. Veterinary public health efforts seek to limit the spread of zoonosis (epidemics of disease among animals), and environmental health efforts are directed at preventing the spread of disease from animals to humans. Bovine tuberculosis (TB) in deer and buffalo, which may be transmitted to cattle and then to humans, is a national problem. Cattle herds are routinely tested for TB in the United States. Cattle ranchers near Yellowstone National Park complain that they must shoot buffalo that stray onto their property to prevent their cattle herds from becoming infected. Each year people who hunt deer, moose, and elk are advised to have their kills inspected for TB lest they become infected by handling and consuming contaminated meat.

The prevention of accidental injury and death is an important environmental health concern. Roughly 50 percent of all accidental deaths in the United States involve automobiles. (Jim West)

Periodic epidemics of rabies among raccoons are a continuing problem. People often will try to help an obviously sick animal, not realizing the risk to themselves. Vaccination of all pet dogs and cats is an important step in controlling the spread of rabies. Animals that bite humans are usually euthanized and examined for evidence of rabies; if the test is positive, or for some reason is not performed, rabies vaccine should be administered to the bite victim.

Rodents, especially rats and mice, carry diseases such as *Salmonella typhimurium.* Environmental health programs designed to control rodent populations in urban areas help prevent epidemics. Recognition of hantavirus epidemics in the American Southwest has led to increased efforts to control wild rodent populations.

FOOD SAFETY

The FDA is responsible for approving food additives; substances that are poisonous or carcinogenic may not be added to foods. The pesticides used on fresh produce must also be approved. The CDC is responsible for identifying the strains and origins of microorganisms that cause nationwide food-borne epidemics.

Food poisoning can be broken down into two categories: noninfective and infective. Noninfective food poisoning is caused by contamination with a toxic substance such as an insecticide or a bacterial toxin such botulism. Infective food poisoning is caused by a pathogenic organism, most commonly a bacterium, virus, or parasite, that infects people who consume contaminated food. Such epidemics have involved a variety of products, ranging from alfalfa sprouts and breakfast cereal to fast-food hamburgers and ice cream. In early 1996 cyclospora-contaminated strawberries sickened many people in the eastern United States. In late 1996 and early 1997, frozen strawberries grown in Mexico and prepared in California were contaminated with the hepatitis A virus, causing a food-borne epidemic in several states; more than one hundred school children in Michigan contracted the disease.

In the United States a number of epidemics

are caused by food and beverages contaminated with a pathogenic strain of *Escherichia coli* known as O157:H7, resulting in a CDC estimate of 20,000 to 40,000 cases each year; of these, 250 to 500 end in death. The best-known epidemics have involved hamburger; such outbreaks are referred to as "hamburger disease" in Canada. Although efforts have been made by various government agencies and the media to educate the public about the necessity for thoroughly cooking hamburger, epidemics repeatedly forced recalls during the 1990's, involving thousands of tons of hamburger. The National Institutes of Health (NIH) is working to develop a vaccine against *Escherichia coli* O157:H7.

Meat safety issues during late 1990's centered on the communicability of bovine spongiform encephalopathy (BSE), or mad cow disease, a prion neurodegenerative disease that entered the European food chain as British cattle were given a commercial feed made from animal remains contaminated with scrapie. A number of human deaths from BSE, classed as new-variant Creutzfeldt-Jakob disease (nvCJD), caused Great Britain to ban the sale of beef brains, marrow, and spinal cord beef products. In 1989 the United States banned the import of British beef because of the BSE outbreak, and Canada banned British beef in 1990. To prevent a similar outbreak of BSE in the United States, use of ruminant remains in animal feed was banned.

Radiation, Noise, and Accidents

Radioisotopes are widely used in medical treatment, by industries, and by governments throughout the world. Wastes generated by the mining and purification of radionuclides are a worldwide problem. Within the United States, many tons of radioactive waste in tanks at various government laboratories require constant monitoring. Within the United States the greatest exposure to radioactive substances is from radon (Rn), an inert gas that is produced from the decay of radium in soil, rocks, and building materials.

Following a number of accidents at nuclear reactors, including Three Mile Island in Pennsylvania in 1979 and Chernobyl in the Soviet Union in 1986, most people began to recognize that living near a nuclear plant has some risk. The NRC monitors nuclear power plants to ensure that they are functioning correctly and that all safety equipment is in place to prevent unplanned releases of radioactive materials. Prior to licensing a nuclear power plant, an evacuation plan for the community must be in place.

Many environments in industrialized societies experience sound levels that approach the threshold of pain. It is difficult and expensive to control noise pollution, whether background levels resulting from community sources or intrusive noise from aircraft and car alarms. Many communities try to minimize noise complaints by limiting construction to daylight hours and banning airplane traffic between certain times. In 1982 there were about 1,100 noise-control programs at various governmental levels in the United States. By 1990 fewer than 20 programs were left; the rest had fallen victim to budget cuts. One key requirement for noise abatement is the existence of a reasonable and enforceable antinoise statute.

Preventing accidental injury and death and protecting the environment through implementation of safety programs have long been mainstays of environmental health. Accidents are a leading cause of death for Americans between the ages of one and thirty-seven, with automobile accidents accounting for roughly one-half of all accidental deaths. Efforts to reduce injury and death in automobile accidents include mandatory driver education in most states, mandatory seat belt use in many states, air bags for front seat passengers, and side bars or side air bags to protect against broadside impacts. Gunshot wounds are also a significant source of injury and death among people aged five to twenty-four. Efforts to reduce this problem have centered on making guns more difficult to buy; however, studies indicate that many young people have access to guns in the home.

Continuing public education on accident prevention is essential in industrial societies, where many individuals have access to hazardous substances and large-scale transport of potentially hazardous substances is routine. This responsibility usually rests with police and fire departments on the local level.

Land Contamination and Air Pollution

Appropriate land use has emerged as a major American concern. Epidemiological studies in Europe have demonstrated that living close to landfills heightens the risk of neural tube birth defects. The U.S. National Environmental Protection Act of 1969 requires an environmental impact statement (EIS) on federal actions that affect the human environment. Among the legislation that helps prevent land contamination in the United States are the Toxic Substances Control Act (1976); the Resource Conservation and Recovery Act (1976); the Comprehensive Environmental Response, Compensation, and Liability Act (1980), also referred to as CERCLA or Superfund; the Hazardous and Solid Waste Amendments (1984); the Superfund Amendments and Reauthorization Act (1986); and the Pollution Prevention Act (1990). However, abandoned hazardous waste sites continue to be discovered next to schools and playgrounds, while heavy metals contaminate dust and soil with which the children come into contact. Remedial actions to mitigate health threats from contaminated land require extensive funding to be effective.

The Clean Air Act amendments of 1970 and 1990 have resulted in steady improvements in air quality in most urban areas in the United States. The EPA monitors and reports on selected air pollutants (carbon monoxide, nitrogen dioxide, ozone, lead, sulfur dioxide, and minute respirable particles) nationwide. The prevalence of asthma in inner cities has continued to grow at an alarming rate and is a major public health concern.

Anita Baker-Blocker

Suggested readings: Short reviews of many of the topics relevant to environmental health are available in the *American Medical Association Encyclopedia of Medicine* (1989), edited by Charles B. Clayman. *Mad Cows and Mother's Milk: The Perils of Poor Risk Communication* (1997), by Douglas Powell and William Leiss, examines BSE and *Escherichia coli* food poisoning. Richard A. Lovett, "Training a Molecular Gun on Killer *E. Coli*," in *Science* 282 (November 20, 1998), provides a succinct look at the problems of food contamination and hopes for a vaccine. *Environmental Auditing: Fundamentals and Techniques* (1985), by J. Ladd Greeno, Gilbert S. Hedstrom, and Maryanne DiBerto, is a classic work on corporate responses to environmental problems. *Contaminated Land: Problems and Solutions* (1993), edited by Tom Cairney, offers an international view of the problem of land pollution.

See also: Air pollution; Birth defects, environmental; Environmental illnesses; Water treatment.

Environmental illnesses

Category: Human health and the environment

Environmental illnesses are ailments caused by exposure to chemical agents, radiation, physical hazards, and nature's reactions to invasions by humankind. Included within this category of illnesses are occupational diseases.

Environmental and occupational illnesses include noninfectious and infectious diseases caused by environmental exposures, in addition to injuries caused by physical hazards considered beyond the immediate control of the individual. Nonoccupational environmental diseases identified by Healthy People 2000 include asthma, heatstroke, hypothermia, heavy metal poisoning, pesticide poisoning, carbon monoxide poisoning, acute chemical poisoning, and methemoglobinemia.

Physicians in ancient Egypt noted environmental conditions that negatively impacted health, and some historians believe that lead poisoning was a strong contributor to the fall of the Roman Empire in 476 c.e. Awareness of environmental illnesses intensified during the Industrial Revolution with the realization that some diseases were strongly associated with specific occupational settings. Some early examples include silicosis, a lung disease contracted by large numbers of industrial workers, miners, and potters who were exposed to silica dust, and a delayed form of bone disease in laborers working within manufacturing plants that contained phosphorus.

Most industrial countries had implemented early forms of environmental protection laws by the 1920's, but the increased use of caustic chemicals and radioactive materials made research involving ecology (scientific study of how living organisms are affected by their environment) increasingly complex. The ecology of infection involves interactions among the climate (as shown by the seasonal increases in influenza and pneumonia); contaminated air, water, and food; and nature itself, with many serious diseases such as tuberculosis, cholera, malaria, and typhoid fever significantly decreasing in incidence upon implementation of appropriate changes within the environment.

Environmental illnesses can affect every organ and system of the body in both mild and severe forms, with diagnosis made more difficult when specific exposures cannot be identified or when symptoms of the illness are delayed. The onset of some disease symptoms occur immediately, but many occur long after exposure; some forms of cancer, for example, have latency periods exceeding thirty years. Epidemiological studies of exposed populations are complicated by the fact that clinical features are often nonspecific. Furthermore, many illnesses can be enhanced by both the environment and personal habits such as smoking and medication abuse.

Chemical Agents

Environmental hazards that influence health and disease processes include natural stressors such as heat, cold, altitude, relative humidity, and wind speed. Unnatural environmental illnesses are created by humans rather than nature and are generally caused by one of three factors: chemical agents, radiation, or human-made physical hazards. Exposure routes include direct or indirect contact with toxins and contaminated air, water, and food. Risk is greatly increased when multiple toxic agents act together, as illustrated by the increased risk of lung cancer in asbestos workers who also smoke or inhale secondhand smoke. Toxic waste dumps pose considerable environmental risks since exposure to multiple hazardous chemicals can occur simultaneously. Thousands of hazardous chemicals have been introduced into the environment with advances in industry; common inorganic examples include dichloro-diphenyl-trichloroethane (DDT), vinyl chloride, and polychlorinated biphenyls (PCBs), while common organic examples include asbestos, mercury, lead, and arsenic.

The pesticide DDT is the most widely referenced example of the danger of introducing synthetic compounds into the environment before long-term effects have been researched. Used for years following World War II, DDT nearly eliminated malaria worldwide. However, it is not easily biodegradable and persists in the environment for years. DDT was banned from nearly all developed countries following its detection in essentially every living organism tested. Many other agricultural pesticides are designed to deter or eliminate weeds, insects, fungi, or rodents that pose a threat to crops. When these toxins drift with the wind or are absorbed into the crops they are designed to protect, illnesses such as cancer and birth defects can result, with their extent related to dosage and exposure duration. Many chemicals and chemical combinations have the potential to produce delayed forms of cancer. For example, exposure to asbestos may lead to lung cancer and mesothelioma, vinyl chloride may cause liver cancer, and benzene may cause leukemia. The expression of diseases that come from chemical agents and radiation depends upon agent entry into the body, metabolic processes within the body, routes by which the body attempts to excrete the substance, and medical treatments.

Airborne pollutants have a much greater influence on the body during physical exertion than during rest because the increased rate and depth of breathing exposes more particulate matter to the delicate tissues of the lungs. Physical work requires a transition from nose breathing to mouth breathing, thus bypassing the body's natural air purifying system in the nasal hairs and mucous membranes, which are generally very effective at removing pollutants at low ventilation rates. Tiny, industrial-generated particulates are more dangerous than larger particles because they are not trapped in the upper respiratory tract, they attach solidly to alveoli in the lungs, and they cannot be effectively exhaled.

Ozone is extremely toxic to humans, causing lung irritation, chest pain, broncho-spasm, headaches, and nausea. Long periods of breathing ozone combined with hydrocarbons, aerosols, and sulfur and nitrogen dioxide may be a contributing factor to allergies, asthma, emphysema, bronchitis, and lung cancer. Temperature inversions in cities located at high altitudes and in basins surrounded by mountains that block winds and trap pollutants produce a greenhouse effect. Temperature inversions cause a reversal of the normal atmospheric temperature gradient that heats the harmful chemicals, thus enhancing their effects upon the body. For example, the strong eastern winds blowing toward Denver, Colorado, trap a brown cloud of pollutants against the Rocky Mountains, requiring the daily broadcasting of air-quality reports for senior citizens and cardiorespiratory patients. Such broadcasts frequently suggest that these populations stay indoors.

Radiation, Physical Hazards, and Nature

Ever since the 1945 atomic bomb attack on Hiroshima, Japan, scientists have become increasingly concerned about the health effects of radioactive pollution. Even small-scale testing of nuclear weapons directly affects the environment, a realization that led the United States, Great Britain, and the Soviet Union to sign the Nuclear Test Ban Treaty in 1963. Both ionizing and nonionizing radiation can cause acute and chronic health problems such as chromosome damage, with workers continually exposed to radioactive metals and X rays being most susceptible. The 1986 Chernobyl nuclear plant malfunction in the Soviet Union, which was the worst peacetime nuclear disaster in history, caused cancer, birth defects, and skin disease among those exposed to radiation. The disposal of nuclear wastes also poses health concerns since many radioactive substances have a half-life of more than ten thousand years.

The predominant source of physical hazards that cause environmental illnesses are human-made environments that increase the incidence of traumatic injuries and create noise pollution. Accidents occurring in unsafe work surroundings account for a large proportion of prevent-

Toxic waste dumps have the potential to cause severe and widespread environmental illness because they can expose people to many different chemicals at the same time. (Reuters/Petr Josek/Archive Photos)

able injuries. Noise in the workplace can cause hearing loss, the most prevalent occupational impairment, which can progress to permanent deafness. Health problems related to noise pollution are increasing among musicians and their audiences, as well as in urban environments where the constant din of traffic and construction contribute to illnesses such as headaches, depression, and insomnia.

Nature can also take revenge against ecological imbalances caused by humankind in the form of diseases such as hantavirus, rabies, giardia, poison ivy, Rocky Mountain spotted fever, and Lyme disease. Hantavirus does not cause obvious illness in its host, but its effects are transmitted to humans when they inhale dust or mist containing dried traces of the urine or feces of infected mice. Being a distant cousin of the fearsome ebola virus, hantavirus outbreaks must be handled much like an outbreak of hepatitis. Rabies is transmitted to humans by bites or scratches containing the saliva of rabid animals such as dogs or bats. The disease attacks the nervous system. Giardiasis is a nonbacterial intestinal illness caused by a parasite found in untreated or improperly treated surface water taken from streams and lakes. Symptoms of infection include diarrhea, nausea, reduced appetite, abdominal cramps, bloated stomach, and fatigue.

Poison ivy, poison oak, and poison sumac have considerable value as wild plants but contain an oily resin called urushiol, which causes a rash, severe itching, and blisters. Rocky Mountain spotted fever is an infection caused by a dog tick, resulting in fever, headache, rash, and nausea or vomiting. As infection progresses, the original red spots may change in appearance to look more like bruises or bloody patches under the skin. Lyme disease was classified following a mysterious juvenile arthritis outbreak and has since accounted for more than 90 percent of vector-borne illnesses in North America. Spread exclusively through bites from infected ticks, its early stages are marked by fatigue, malaise, chills, fever, headaches, muscle and joint pain, swollen lymph nodes, and skin rashes. Later stages may include arthritis, nervous system abnormalities, and heart conduction disturbances.

Agencies and Legislation

Agencies that have federal authority to investigate environmental issues related to disease include the Department of Labor, under which fall the Environmental Protection Agency (EPA) and the Occupational Safety and Health Administration (OSHA), and the Department of Health and Human Services, under which fall the Food and Drug Administration (FDA), the National Institutes of Health (NIH), the Centers for Disease Control (CDC), and the Health Resources and Services Administration (HRSA). The National Institute of Occupational Safety and Health (NIOSH) conducts ongoing research to identify hazards and develop safety standards, with many companies now employing industrial health advisors.

International coordination regarding environmental and occupational health concerns is provided by the World Health Organization (WHO), founded in 1942 as an agency of the United Nations. The WHO is extremely active in developing countries as industrialization, poverty, and population growth continue to increase. Its broad scope of activities includes controlling widespread disease such as malaria and tuberculosis, establishing purified water supplies and sanitation systems, and providing health education and health planning assistance.

Legislation that regulates workplace practices and sources of pollution that could lead to environmental illnesses escalated during the 1960's. Federal laws that remain the most relevant include the Occupational Safety and Health Act of 1970, the Environmental Pesticide Control Act of 1972, the Toxic Substances Control Act of 1976, the Resource Conservation and Recovery Act of 1976, and the Comprehensive Environmental Response, Compensation, and Liability Act (known as CERCLA or Superfund) of 1980. In 1985 several "right-to-know" laws went into effect, which required manufacturing plant managers to make health and safety information regarding toxic materials available to employees.

Daniel G. Graetzer

SUGGESTED READINGS: For excellent reviews of environmentally induced pollutants and their impact on health and disease, see J. Stephen Kroll-Smith and H. Hugh Floyd, *Bodies in Protest: Environmental Illness and the Struggle over Medical Knowledge* (1997); Stephen Barrett and Ronald E. Gots, *Chemical Sensitivity: The Truth About Environmental Illness* (1998); and Kenneth Wark, Cecil F. Warner, and Wayne T. Davis, *Air Pollution: Its Origin and Control* (1998). Normal J. Vig and Michael E. Kraft, *Environmental Policy in the 1990's* (1990), and John Wargo, *Our Children's Toxic Legacy: How Science and Law Fail to Protect Us from Pesticides* (1998), provide information about legislative actions and professional organizations that combat environmental illnesses. Stephen Edelson and Jan Statman, *Living with Environmental Illness: A Practical Guide to Multiple Chemical Sensitivity* (1998), and Ann Louise Gittleman, *How to Stay Young and Healthy in a Toxic World* (1998), provide perspectives on environmental issues related to health care.

SEE ALSO: Air pollution; Asbestos; Dichlorodiphenyl-trichloroethane (DDT); Environmental health; Lead poisoning; Particulate matter; Smog; Water pollution.

Environmental impact statements and assessments

CATEGORY: Land and land use

Environmental impact statements and assessments involve the evaluation of the environmental impacts of proposed or existing human activity—including construction, resource extraction, and land-management policy implementation—and the reporting of those impacts in a formal, written document for public review.

An entity planning a development can choose to do an environmental impact assessment (EIA), but formal EIAs are a mandated response to specific legislation. Under some legislation, an initial scoping process is done to determine whether a more lengthy process, the EIA, is required. The results of a legally mandated or regulatory EIA are documented in an environmental impact statement (EIS). EIAs are becoming more popular because of increased pressures to improve resource management and conservation. The United States National Environmental Policy Act of 1969 (NEPA) helped usher in the era of environmental assessment for government decision making. By 1998 more than one hundred countries had established EIA processes. Ideally, administration of an EIA program promotes government and corporate accountability for environmental alterations and institutionalizes systematic, science-based policy analysis.

The environmental movement, spurred by Rachel Carson's book *Silent Spring* (1962), influenced U.S. legislators to reconsider the lack of a national policy for the environment. NEPA, signed into law on January 1, 1970, heralded a new role for citizens to participate in reviews of government decisions. The EIA process produces draft environmental impact statements (DEIS) and final environmental impact statements (FEIS) for public comment. However, once the FEIS has been accepted, NEPA has been satisfied, but this in itself does not constitute approval or denial of a proposed project. By focusing on the process rather than the end result (as most environmental permit programs do), the EIA reflects a compromise between environmental and political interests. The goal is to ensure that a suitable EIS or "finding of no significant impact" (FONSI) is prepared. A well-planned project should be able to withstand the public scrutiny. Other laws and permitting processes may be required before a proponent can actually go ahead with a proposed development or action, but these processes can build upon or use the data gathered in the EIA.

In the United States, NEPA marked a change for federal agencies because it added environmental accountability to every agency's mission, along with a specific method to carry out environmental reviews. NEPA not only provided a common thread among agencies, but also comprehensively linked various categories or media in which environmental impacts occur. The EIA process examines impacts ranging from archaeological resource depletion to air pollution. Social impacts resulting from proposed actions are as much a part of the EIA as water resources issues, noise, solid waste disposal, and other common types of environmental impacts. Aesthetics has also proved an important if initially nebulous category of impact, although a significant body of literature has arisen to treat the need to quantify impacts normally considered subjective. Socioeconomic values of environmental resources (for life support, amenities, and raw materials) are also used in evaluating trade-offs among alternatives. Formal EIAs in most countries generally include consideration of similar wide arrays of environmental impacts.

The broad categories of impact are intended to reflect the interconnectedness of environmental settings and allow the interplay of social, economic, political, and environmental issues. The breadth also extends to the type of projects that require an environmental assessment. In the United States, any action by a federal agency; involving a federal license, permit, or funding; or taking place on federal property is subject to the EIA process. Approximately two dozen states have their own equivalent assessment requirements. Nations that use the EIA process are more readily able to participate in global trade, qualify for funding, and meet the increasing international demand for and appreciation of environmental quality.

An EIA assesses more than the proposed action; it also looks at the impacts of legitimate alternatives to the action as well as the impacts of not doing the project. The EIA process includes comparing the costs and benefits of the alternatives and the various impacts. The EIS documents the impacts, costs, and benefits for the project and its alternatives. Under NEPA-type legislation, a DEIS is circulated, and public comments are solicited either in writing or at hearings. The FEIS is issued after consideration of public and other agency input. The courts provide a forum for class-action suits and other assessment-related disputes. The majority of challenges have been based on allegations of either a failure to prepare an EIS or a failure to fully consider the proper alternatives. From a public policy perspective as well as that of peer-reviewed science, it is the public nature of the EIS that determines the success of the EIA process.

The EIA process ideally is part of the planning process rather than an afterthought for projects that have already commenced. Yet the EIA must be conducted late enough in the planning stages to have a sufficient description of the project to assess its impacts. One response is the strategic environmental assessment (SEA). Forecasting impacts, especially for SEAs, cannot simply be done by direct observation. In conducting assessments, a team of professionals will use physical models, mathematical models, qualitative models, checklists, and expert opinions.

Starting with a project description, an assessment proceeds to identification of associated or expected direct and indirect impacts. Next, the existing environmental conditions are described. Then relevant laws and regulations are examined for standards and applicability. Specific environmental impacts are predicted, and their significance is evaluated. The final step is the incorporation of results into the project to reduce or mitigate the impacts; this includes monitoring, reporting, and responding to post-construction impacts. EIA and EIS notification appears in legal sections of major newspapers, at government and corporate Internet sites, and at agency offices. EISs are generally available from the preparers upon request or can be viewed by concerned citizens at various public locations.

R. M. Sanford and H. B. Stroud

SUGGESTED READINGS: A thorough overview of the environmental impact process from the perspective of the assessment team member is provided in Larry W. Cantor's *Environmental Impact Assessment* (1996), which emphasizes the United States. J. Glasson, R. Therivel, and A. Chadwick's *Introduction to Environmental Impact Assessment* (1994) emphasizes European environmental assessments. The periodicals *Environmental Impact Assessment* and *Impact Assessment Journal* provide technical, academic, and professional literature on impact assessment in the United States and other countries.

SEE ALSO: Environmental policy and lobbying; National Environmental Policy Act.